PRINCIPLES OF COGNITION, LANGUAGE AND ACTION

PRINCIPLES OF COGNITION, LANGUAGE AND ACTION

Essays on the Foundations of a Science of Psychology

by

N. Praetorius
University of Copenhagen, Denmark

SPRINGER SCIENCE+BUSINESS MEDIA, B.V.

A C.I.P. Catalogue record for this book is available from the Library of Congress.

ISBN 978-0-7923-6230-2 ISBN 978-94-011-4036-2 (eBook)
DOI 10.1007/978-94-011-4036-2

Cover Design:
'Le Cabinet du Naturaliste' by Pierre Roy.
© Tate Gallery, London 1999

Printed on acid-free paper

To the memory of my beloved son David,
who taught me what it is to be a person.

Contents

Foreword

This book addresses a growing concern as to why Psychology, now more than a hundred years after becoming an independent research area, does not yet meet the basic requirements of a scientific discipline on a par with other sciences such as physics and biology. These requirements include: agreement on definition and delimitation of the range of features and properties of the phenomena or subject matter to be investigated; secondly, the development of concepts and methods which unambiguously specify the phenomena and systematic investigation of their features and properties. A third equally important requirement, implicit in the first two, is exclusion from enquiry of all other mattes with which the discipline is not concerned. To these requirements must then be added the development of basic assumptions about the nature of what is under investigation, and of principles to account for its properties and to serve as a guide as to what are relevant questions to ask and theories to develop about them.

One of the major obstacles for psychology in establishing itself as an independent scientific discipline has been that of delimiting its subject area from those of other sciences, in particular physics and biology. This is understandable, since human beings have bodies made up of physical matter, which, like all other matter, obeys the laws of physics, and to this extent may be described and accounted for in terms of physics. Likewise, human beings are living organisms like other living organisms, built up of cells and organs, which, in order to stay alive and procreate, organize their interaction with the surrounding environment in ways which serve biological ends.

Whether for these or for other reasons, it has apparently been difficult to reach agreement on the *psychological* properties that uniquely characterises human beings as opposed to physical systems and biological organisms, and thus to agree on what properties and features human being have apart from or in addition to those of physical systems and biological organisms. Admittedly, most psychologists to-day can agree that human beings are persons who may act, cognize, think and feel, and who develop languages in which together they may reflect on and communicate about their actions and what they cognize, think and feel. Likewise, most psychologists would agree that such psychological phenomena fall within or even define the area of psychological inquiry.

But, when it comes to *assumptions* about the *nature* of these psychological phenomena and properties of persons, or the concepts, models and principles by which they may be adequately described, there is little common ground among psychologists. Even worse, discussion about issues as fundamental as these, which has proved to be essential for the development and progress of the natural sciences, is almost totally lacking within psychology. Lacking also is debate as to whether the wide variety of (often incompatible) models and conceptual systems, currently used to describe psychological phenomena, are indeed adequate, just as is reflection about the tenability of the assumptions on which current conceptual schemes and models are based.

It is the intention of this book to alert psychologists to the importance of such debate and reflection, and in particular to serious flaws in the assumptions of traditional and more recent philosophical positions and theories about cognition, perception, language and action, by which current psychological research and theorizing are influenced.

To this end a thorough, critical discussion will be carried out, on the one hand, of the assumptions of the *Naturalist* positions (Physicalism, Scientism, Eliminative Materialism, Computational Functionalism, Biologism, etc.) which influence research and theorizing within traditional, so called "experimental" psychology on perception, cognition, language and action, and on the other, of the radically different assumptions of the *Constructivist* positions (including Structualism and various of its Deconstructivist, Relativist, Anti-realist and Irrealist successors) which influence the so called "social constructionist movement" and its research and theorizing.

Just like their Materialist and Idealist forerunners, the assumptions of Naturalism and Constructivism represent attempts to solve the problem about the relation between Mind and Matter, or Mind and Body - and with it the related problem about the relation between Mind and Reality, or between cognition and description of reality and the reality being cognized and described. Within Naturalism (as within Materialism) attempts have been made to solve these problems by proposing that mental states and processes in some way or other may be reduced to or deduced from physical, biological or physiological states or processes of the body and its interaction with physical reality. Conversely, within Constructivism (as within Idealism) attempts have been made proposing that the world of matter is nothing but a product of conceptual and linguistic schemes which somehow or other develop in our minds, and thus is reducible to something mental. In effect, solutions to these problems have been attempted by Naturalism and Constructivism by denying the independent existence of Mind and Matter respectively.

In the book I argue that the reductionist solutions to the Mind-Matter problems proposed by Naturalism and Constructivism imply assumptions which are untenable, and that their solutions to the Mind-Matter problems have consequences which are absurd – and therefore cannot possibly qualify as assumptions on which the development of a scientific psychology may be founded. During these arguments it will be shown, furthermore, that the Mind-Matter problems entail conflicting assumptions which prevent them from being consistently stated, let alone solved – be it by the natural sciences or psychology or by philosophical investigations. Rather, they are problems which, to use Ryle's words, have to be "dissipated". More importantly, however, I shall attempt to show how the arguments refuting the assumptions and solutions of the Mind-Matter problems, in themselves point to alternative assumptions about the relation between Mind and Reality and Mind and Body, which are not only tenable, but which must be presupposed and taken for granted by both psychology and philosophy.

Arguments in refutation of the assumptions behind the Mind-Matter problems and their proposed solutions are not particularly complicated, nor are the arguments for the alternative assumptions by which they must be replaced. The Mind-Matter problems arose as a consequence of the Cartesian division of reality into two separate parts or "realms": the realm of material reality with its physical things and their "objective", quantifiable properties; and the realm of the mind to which belong the "subjective" non-quantifiable properties of things, as well as the feelings, thoughts and beliefs of people about reality and things in reality. The problems incurred by this division were, first, how a reconciliation could come about between the "material" and the "mental", i.e. between our bodies and minds; and, secondly, the problem with which Materialism and Idealism initially struggled, namely, if the perceptions and beliefs of objects residing in our minds to which we have immediate access, are the only phenomena the existence of which we may be certain, then what guarantee do we have that those beliefs and perceptions are *true*, and that the objects perceived are *material*?

The main line of argument against these problems of Cartesian dualism is that they imply conflicting assumptions. On the one hand, the possibility of a polar opposition between Mind and Matter is assumed, which necessarily implies that we may indeed talk consistently about both Mind and Matter;

and, on the other, it is assumed that Mind and Matter are independently determinable "realms" in the sense that each may be talked about and characterized independently of referring to the other. That these assumptions are conflicting becomes obvious when we consider the impossibility of talking consistently about *material reality*, and what exists in material reality, independently of or without referring to our cognition and description of it and, furthermore, without assuming that material reality exists as something about which we may have knowledge and put forward propositions which are true. Consider, conversely, the impossibility of assuming that we can talk consistently about our *knowledge* and *perception* of reality and things in reality, without referring to reality and these things; i.e. the impossibility of talking about *what* we cognize or see, without at the same time describing *that* which we cognize or see.

Consequently, I argue that the very possibility of a division between Mind and Matter presupposes the assumption of an interdependency or *necessary relation* between Mind and Matter, i.e. between our cognition and description of material reality and the material reality which exists independently of our cognition and description of it. Hence, for the sake of epistemological consistency, we shall have to assume that we do have knowledge about and a language in which we may put forward propositions about reality which are true. Indeed, I argue, this must be taken for granted as a matter of principle, and be the point of departure for all further investigations into both reality and our knowledge and description of it.

This principle, which I call the principle of the *general correctness of language and knowledge*, is a principle which implies that to be a person and a language user is to have knowledge of and a language in which one may put forward propositions which are true - or false - about reality and the situations in which one finds oneself in reality. However, it is a principle the validity of which cannot be proved. It can only be shown that if we do try to prove it - or, worse, try to doubt or deny it - we will involve ourselves in circularities, contradictions or absurdities. However, this suffices. It is sufficient to show that attempts to prove the principle would have to presuppose the principle, and that, conversely, attempts to doubt or deny the principle would amount to assuming that we may use language to doubt or deny that we can say anything about anything, which is true.

By applying the same *reductio ad absurdum* arguments it can be shown that a logical relation exists between the notions we use to characterize our cognition, description and action, and that which our cognition, description and action concern, i.e. notions such as 'knowledge', 'true', 'objective', 'propositions', 'act', 'intention', 'reference', and 'reality', 'things', or 'facts of reality'. This relationship is to be understood as logical in the sense that none of these notions have well defined meanings independently of reference to well defined meanings of the others, nor may any of them be reduced to or deduced from any of the others. For this reason alone, it will be equally impossible to attempt to reduce that to which any of these notions refer, to that which any of the others refer, and thus to reduce Mind to Matter - and vice versa.

To this point may be added that *intentionality* and *truth* are logical properties of knowledge and linguistic propositions, but not of physical, biological or physiological states or processes - by any definition of physics, biology and physiology. It may be true or false that physical and biological and physiological states and processes exist, but such states and processes cannot themselves be true or false, nor be *about* anything in the sense that knowledge and descriptions may be. Neither is there any way in which these logical properties of knowledge and propositions - intentionality and truth - may be reduced to or explained in any of the terms we use to account for physical, physiological or biological states or processes. Indeed, *reasons of principle* exist why these logical properties of knowledge and propositions cannot be reduced to nor explained in terms of processes and states which are more fundamental than intentionality and truth. Among these reasons is the logical impossibility of accounting for such more fundamental processes and states without describing them, and thus without implying the existence of intentionality and truth. To assume otherwise would be just as absurd as assuming that logic and its principles, on which *par excellence* the language of science relies, could be reduced to or explained in terms of something more fundamental or elementary without using logic.

From these arguments it follow that, on the one hand, the very possibility of an ontological distinction between Mind and Matter precludes *epistemological Mind-Reality dualism*, i.e. the assumption that Mind and Matter are two independently determinable entities or realms. And it follows, on the other hand, that the epistemological conditions for talking in well defined ways about and distinguishing between Mind and Matter, at the same time necessitates the assumption of *ontological Mind-Body dualism*, which precludes reductionism. That is, precludes psychological states and properties of Mind, which uniquely distinguishes Mind from Matter, from being reduced to, derived from, or explained in terms of Matter - and vice versa.

Central to these arguments and assumptions is that epistemological and ontological issues and concepts are inter-related. Thus, any consistent ontological determination and distinction between Mind and Matter involves *epistemological commitments*, i.e. presupposes the assumption of a necessary relation between Mind and Reality (or between our cognition and description of reality, and the reality cognized and described). Conversely, any epistemological consistent account of either Mind or Matter presupposes the assumption of a logical relation between concepts we use to characterize our cognition and description of reality, and the reality which this cognition and description concern.

So far I have outlined examples of the arguments, assumptions and principles about the cognition and language of persons to be presented in this book. However, similar analyses of the action and co-action of persons whereby they may acquire knowledge of and together develop languages to describe things in reality, reveal still other assumptions and principles which have to be taken for granted. To these belongs the assumption about the *inter-subjectivity* of the notion of truth. This is the assumption that what is true or false for a person about the things he or she cognizes and describes, would also be true for others, could they cognize and describe what the person cognizes and describes. Hence, it is not because persons together may agree and make conventions about what is true and false about the things they cognize and describe that they may have or come to acquire a notion of truth. On the contrary, it is because persons have a notion of truth of which it is presupposed that what is true for oneself will also be true for others that they can come to agree and begin to make conventions about the truth of anything. Put differently, the notion of truth both implies and presupposes a notion of 'others' or 'other persons'.

Another principle concerning the identity of things states that the notion of 'the same thing', and hence of 'same' and of 'thing', necessarily implies that identifying a thing as the *same*, is to identify it as that which in different situations, e.g. of observation, description and cognition, may be differently observed, cognized and described. It is not because a thing retains the same, unchanged "essence" or "substance" in different situations in which it may be correctly cognized and described in different ways that we have a notion of 'the same thing'. Rather, a thing may be identified as the same thing precisely in virtue of being the thing which in different situations may be correctly cognized and described in different ways.

I contend that both the assumptions concerning the inter-subjectivity between persons and the notions of the identity of things are assumptions, which have status as principles, just as for the same reasons do the assumptions of the general correctness of knowledge and language. That is, we cannot talk consistently of either inter-subjectivity between persons or of the identity of things without presupposing these assumptions. Neither can these assumptions be proved nor disproved without being conceded.

If the arguments and assumptions in this book are tenable, and the principles derived from them are valid, the consequences are far-reaching. A number of questions traditionally asked and answers attempted in classical as well as more recent theories of Mind, Language, Reference, Identity and Truth will have to be reconsidered - and so will suggestions that solutions to these questions may be either obtained or verified by empirical psychological investigations. On the other hand, the assumptions and principles proposed here open up a whole new range of questions about key-issues concerning cognition, language and action which *can* be consistently stated, just as they suggests ways in which these issues may be dealt with in much more straightforward and consistent ways - be it philosophically or psychologically, theoretically or empirically. If so, I think we shall be able to see the beginnings of a consistent foundation of assumptions and principles that psychologists will have to take for granted and agree, and on which a science of psychology may be based. These are assumptions and principles which may serve as a guide as to what questions to ask and hope to answer by psychological investigations, and as to what concepts and theories about psychological phenomena will stand up.

Acknowledgements

The seven years of drafting, rewriting and putting together the final version of this book, would have been unbearable without the loving support from my family and their understanding of the importance to me of being able to carry this project through. It is with great pleasure that I take this opportunity to thank them and to let them know just how much it meant to me. Thanks are also due to my colleagues at the University of Copenhagen, Jesper Hermann, Niels Engelsted and Arne Poulsen, who were good enough to read through and comment on parts of earlier drafts. However, most of all I want to thank Simo Køppe for his unfailing encouragement and interest during all those years. Not only did he read all the earlier drafts of this book, but he also alerted me to important points – and mistakes – which otherwise I would have overlooked. But this is not all; at a time when he knew that I needed time to think and write, he volunteered to take over some of my teaching commitments. The final draft was read in its entirety by Allan Costall whose informed criticism and suggestion I gratefully acknowledge.

Special thanks are also due to my husband, Keith Duncan, for his painstaking and time-consuming attempts to teach me how to put my thoughts and arguments into understandable and readable English. I am deeply indebted to Pusjka Helene Cohn for her conscientious assistance in all the details associated with the preparation of the book. Last but not least, I thank the Danish Research Council for the Humanities who awarded a grant in support of publishing this book.

Copenhagen, October, 1999

Nini Praetorius

PART I

THE DEVELOPMENT OF A SCIENCE OF PSYCHOLOGY

Chapter 1

Introduction to assumptions and arguments

In his "mountaintop view" of the state of psychology Bruner (1990) laments the fact that - "the science of Mind as James once called it" - has become fragmented as never before in its history. Psychology to day is a "patchquilt", an "aggregate of parts which has lost its *center*", and thus the cohesion needed to assure the internal exchange that might justify a division of labour between its parts. Each part has its own organizational identity, its own parish of authorities, its own journals and its own theoretical apparatus, and still worse, its own less and less exportable rhetoric by which it seals itself off from other parts of psychology - as well as from the intellectual community at large. The consequences are that "outsiders" increasingly ignore their journals which, at any rate, "principally contain intellectually unsituated neat little studies, each a response to a handful of like little studies".

However, Bruner does not content himself with lamenting this unfortunate state of psychology, but joins the growing number of his colleagues who react to the prevailing empiricist ethos of "neat little studies" and their simplifications; colleagues who like he himself are willing, once again, to ask the "great psychological questions" about the nature of man and his mind - how man perceives and understands himself and his fellow beings, and the meaning he makes of his and their actions in the situations in which he and they finds themselves. In this pursuit, says Bruner, it is necessary that we venture beyond the conventional aims of positivist science with its ideals or *reduction*, *causal explanations* and *prediction*, and stop thinking of man in terms of yesterday's physics - not to speak of today's information processing computers. Information is *indifferent* to meaning, and the concept of meaning cannot be *replaced* by or *reduced* to "computability". If what we are looking for are ways of understanding man as a "meaning-making" *agent* of his own action and cognition, says Bruner, we necessarily have to deal with questions about the ways in which the history, culture and language he shares with others shape this meaning and understanding. To this end we need other ideals - and all the inspiration we can get from other disciplines within the humanities, the social sciences and anthropology of how to deal with these questions with a subtlety and rigour that will yield rich and generative answers. For,

> "to reduce meaning or culture to a material base, to say that they "depend", say on the left hemisphere, is to trivialize both in the service of misplaced concreteness. To insist upon explanation in terms of "causes" simply bars us from trying to understand how human beings interpret their world and how *we* interpret *their* act of interpretation. And if we take the object of psychology (as of any intellectual enterprise) to be the achievement of understanding, why is it necessary under all conditions for us to understand *in advance* of the phenomena to be observed - which is all that prediction is. Are not plausible interpretations preferable to causal explanations, particularly when the achievement of a causal explanation forces us to artificialize what we are studying to a point almost beyond recognition as representative of human life?" (Bruner, ibid. p. viii)

As so many of Bruner's colleagues, who share his concern about the current state of psychology, I can only agree with his diagnosis and the need for reformulation of ideals for scientific psychological inquiries.[1] But although the results of the growing number of studies carried out in the spirit suggested by Bruner are encouraging,[2] I fear that it takes more than quality research to free academic psychology from the inadequate ideals by which it is currently dominated - just as in physics it took more than just good research, but a Maxwell, a Planck, an Einstein, a Niels Bohr, a Heisenberg, a Schroedinger, a Dirac and other highly dedicated theoreticians to revolutionize the concepts and assumptions of classical physics, and to make possible the development of modern physics. Likewise, it will take more than just examples of good research to provide psychology and its various disciplines with a *center*, i.e. a basis of generally accepted fundamental concepts, assumptions and principles by which appropriately to define and delimit the *particular* subject matter and phenomena which fall within psychology, and which may guide its practitioners as to what are relevant questions to ask and theories to develop about the particular phenomena they are studying.

[1] Similar concerns over the fragmentation, lack of structure and "chaotic" status of the various disciplines of psychology, as well as the uncertainty as to what is the subject matter of the science of psychology and how its limits are defined, have been expressed by among others Giorgio (1985), Koch & Leary (1985), Miller (1985), Pylyshyn (1987), and Smith, Harré and Van Langenhove (1995).

[2] In this respect the psychological research on the child's development of cognition and language over the last 10 to 15 years deserves special mention.

What is needed in order to establish a similar center for psychology and its disciplines is that, once again, we start seriously to discuss the assumptions and problems in the foundation of psychology, which has been lacking since the days of Wundt, James, Piaget, Vygotsky and more recently Roger Brown - i.e. start critically to discuss issues of a *fundamental theoretical* nature which, from the very beginning, have formed a crucial part of the development of the Natural Sciences and Biology both in establishing these areas of research as *sciences*, and their astonishing progress. First and foremost, however, to change the ideals adopted by the current psychological establishment, we need to uncover the false assumptions about the cognition, action, and language of persons underlying these ideals, and to make it obvious why, in Bruner's word, these assumptions "force us to artificialize what we are studying to a point almost beyond recognition as representative of human life". This is a task, however, which necessitates a closer scrutiny - and refutation - of assumptions of *philosophical* positions by which academic psychology is currently influenced. This will be the primary task and aim of this book.

To this end I shall first present and discuss the consequences of the assumptions of *Naturalist* philosophical positions, which dominate current theories and research within so called *experimental psychology* about the perception, cognition, and action of persons, as well as the consequences of the radically different assumptions of *Constructivist* philosophical positions, which dominate the current so called "Social Constructionist Movement" and its research and theorizing about the same issues. Both philosophical positions - which appear in a host of different versions dressed up in as many linguistic guises - originate in the attempts to solve the Mind-Matter problem, which dates back to Galileo and first formulated as a philosophical problem by Descartes. It is the contention of this book, firstly, that these attempts to solve the Mind-Matter problem are based on assumptions which are untenable and lead to consequences which are self-defeating and absurd - just as did their classical Materialist and Idealist forerunners. And it is the contention, secondly, that the adoption by mainstream academic psychology of the assumptions of Naturalism and Constructivism as a basis for its research and theorizing, the problems and consequences of these positions cease to be of merely 'philosophical' interest, but have become problems and consequences inherent in the foundation of academic psychology.

Now, it is by no means the first time it has been argued that the assumptions of Naturalism and Constructivism about the relation between Mind and Matter are untenable, and that their solutions of the problems to which these assumptions give rise, have self-defeating and absurd consequences. Some philosophers have long since realized that these problems *cannot* be solved,[3] just as it has been argued that the "Cartesian Mind-Matter myth" and the problems it involves needs to be "dissipated" rather than solved (Ryle, 1949). What is new in this book is that it shows that *reasons of principle* exist why the Mind-Matter problem and the host of derived epistemological and ontological problems to which it gives rise, entail assumptions about the relations between Mind and Matter, *which prevent these problems from being consistently stated*, and, therefore, from being solved - be it by the natural sciences or by empirical psychological research or philosophical investigations.

In this chapter, I shall first give an account of how the Mind-Matter problem and some of its related epistemological Mind-Reality problems have been formulated, and present some of the assumptions inherent in their solutions suggested by Constructivism and Naturalism. In the course of this presentation I shall give an outline of the main arguments of this book against these and related assumptions; assumptions which are mainly of an epistemological kind concerning the conditions for having and acquiring knowledge and a language in which correct or true propositions may be put forward about material, physical reality as well as about ourselves, our thoughts, feelings, emotions - and whatever else goes on in our minds. More importantly, though, I shall argue that following a careful, critical analysis of the epistemological assumptions underlying prevailing philosophical positions on the Mind-Matter problems, it will become obvious that alternative, *tenable* assumptions do indeed exist, which not only have to be taken for granted by epistemology, but which must necessarily be fundamental for the development of consistent and coherent theories and research in psychology about the cognition, language and action of persons.

It is my contention that only if psychology begins to free itself from the influence of traditional and contemporary philosophical positions and their solutions to unsolvable epistemological problems, will there be any hope for psychology to develop into a systematic, cumulative science in its own right. That is, a science with its own basis of commonly agreed concepts, which are relevant and appropriate to account for phenomena and problems of a psychological nature. In this respect and to this end, there may be lessons to be learned for psychology from how other sciences have proceeded - and succeeded. In Chapter 2, I shall argue why. There I shall set out a set of basic concepts and principles for a science of psychology which follow from the alternative assumptions argued in the present chapter.

[3] For more recent arguments to this effect, see e.g. McGinn (1989, 1991), Nagel (1986), and contributions in Warner and Szubka (1994), by, among others, Bealer, Myro, Searle, and Kim.

1.1 Basic assumptions of Naturalism and Constructivism

The Mind-Matter problem implies the assumption, widely accepted since Descartes, that reality may be divided into two *independently determinable* parts or "realms". On the one hand there is the realm of material reality with its physical things and their quantifiable properties, the behaviour and structure of which may be exhaustively accounted for in the (mathematical) terms of physics. And, on the other, there is the realm of the mental to which belong the "subjective" non-quantifiable properties of things, as well as the feelings, thoughts and beliefs of people about reality and things in reality.

With this polar division of reality into a "physical" and a "mental" part arose a number of puzzling and intractable epistemological problems which have kept philosophers busy - and secured their profession - for almost 400 years. There was first the problems of how a *reconciliation* could come about between the "material" and the "mental" - if indeed there could - and with it the problem, partly epistemological and partly ontological, that if the *ideas* and *perceptions* of objects residing in our minds to which we have immediate access, are the only phenomena the existence of which we can be *certain*, then what guarantee do we have that those ideas are *true*, or that the objects perceived are *material*? And there was the problem, secondly, that if any one of us only has direct access to the thoughts, feelings and cognition our own minds, then how can we be certain that the thoughts, feelings and cognition of other persons are the same as those we ourselves entertain. Indeed, how can we be certain that the world itself, including other people and their minds, exists as other than ideas in our own private minds?

Present-day Naturalism and Constructivism share the assumption with classical Materialism and Idealism that the Mind-Matter problem and its derived problems about the relation between cognition and reality and language and reality, may actually be solved - and that the way to do so is by denying the independent existence of either one or the other "realm" - in effect, by reducing the one to the other. The basic assumptions of Constructivism (including Structuralism and its Deconstructivist, Relativist, Anti-realist and Irrealist successors) is that what we take to exist in material reality is entirely dependent on, varies with and, thus, is a mere *product* of the conceptions, categorisations and descriptions we happen to produce or "invent" about reality and things in the course of our action in different situations - scientific and everyday situations alike. Hence, those things do not exist independently of such situations, categorisations and descriptions, but only *in virtue* of our descriptions of them and the conceptions about them invented by our minds.

The basic assumption of *Naturalism* (including Physicalism, Scientism, Eliminative Materialism, Essentialism, Identity Theory, Biologism and

Functionalism) is that only that exists *objectively* in reality which may be accounted for in terms of the Natural Sciences. Furthermore, it is the intuition of the steadily growing number of (especially) Anglo-Saxon philosophers who subscribe to Naturalism that all objectively existing phenomena in the final analysis must be of a *physical* nature, and thus that Physics is an adequate explanatory basis for all objectively existing phenomena. About Physics it is assume that

> "[it] is a unified body of scientific theories [...] which together provide a true and exhaustive account of all physical phenomena (i.e. all phenomena describable in physical terms). They are unified in the sense that they are cumulative: the theory governing any physical phenomenon is explained by theories governing phenomena out of which that phenomenon is composed and by the way it is composed out of them. The same is true of the latter phenomena, and so on down to the fundamental particles or fields governed by a few simple laws, more or less conceived of in present-day theoretical physics." (Lewis, 1971, p. 169).

The arguments for the *explanatory adequacy* of physics for all sciences, and not merely for the kind of phenomena which traditionally have been studied within the Natural Sciences, go in simple terms like this. Because everything physical has physical causes and effects, then everything having physical causes and effects must (itself) be something physical - and thus must be truly and exhaustively accountable in physical terms. Since mental phenomena - such as pain or beliefs about reality - have physical causes (i.e. they are caused by physical states and operation of our brains) and effects (pains as well as beliefs about reality may make us act in particular ways which have physical effects), then mental phenomena such as pains and beliefs must be physical phenomena *completely* reducible to and accountable in physical terms. Indeed, so it is claimed, they must *in principle* be accounted for and explained (away) in terms of "fundamental particles or fields governed by a few simple laws, more or less conceived of in present-day theoretical physics". (Lewis, op.cit.). According to this intuition it does not make sense to talk about the objective existence of such things as beliefs, thoughts and other mental phenomena - *except* as phenomena which may be reducible to and be conceived of and accounted for as purely physical phenomena. (Theories espousing these views are discussed in Chapters 9 and 10).

However, far from solving the Mind-Matter problem and the range of related epistemological problems arising from the Cartesian division of reality into a physical and a mental part, each with its own independently determinable existence, the "reduction" of the physical to mental phenomena by Constructivism, and vice versa by Naturalism, have only made these problems even more intractable - and added a few more to the original list.

What has not always been realized, however should become clear during the discussions of this book: Despite the apparently completely opposed assumptions of Constructivism and Naturalism, most of the problems and consequences to which they lead are exactly the same - the problem of *solipsism* already mentioned is a case in point. I shall first outline some of the most obvious problems and consequences of the assumptions of Constructivism - and thereafter of those of Naturalism.

1.2 Implications of Constructivist and Naturalist assumptions

Since, according to *Constructivism,* what we take to exist in reality is but a product or "fabrication" of the categories of our minds and of the terms and notions of our language, there can be no *criteria* in the "outside" world with which to determine the truth or correctness of our cognition and linguistic description of it. Natural languages, it is assumed, are fairly autonomous systems of arbitrary signs and terms, the correct use of which is a matter of what its users may come to agree and make conventions about. Hence, the notion of *truth,* i.e. what is true about reality and things in reality, is itself purely a matter of conventions and agreement among language users. And since the conventions for cognition and descriptions of reality and things in reality varies from culture to culture and are different for people living under different historical and socio-economic conditions, the notion of truth also varies with those conditions and cultures (Gergen, 1985, Burr, 1995). Accordingly, 'truth' is just as *relative* as are all other terms and notions of our language and cognition, and just as *arbitrary* as are the conventions for how those terms and notions should be used correctly to pick out things in reality (cf. Chapters 5 and 19).

An obvious problem with the view that the notion of truth is *conventional* is that it leaves completely unexplained and unexplainable how persons and language users can *begin* to develop conventions and agreement about the correct application of linguistic terms and notions *without* presupposing that things in the world, to which these notions and terms may be correctly applied, exist as things about which something is the case, or true, and something else is not the case, or false. And *without* assuming, furthermore, that persons and language users together may correctly identify and cognize those things and thus what is the case or true, and not the case or false about them. And without assuming, therefore, that they already have notions of true and false *prior* to making conventions about the correct applications of terms and the use of the notion of *truth.* If the problems of neglecting these presupposition do not appear immediately obvious, one only has to try to make sense of this summary of assumptions of Constructivism: To be a person and a language user is to share with others concepts of things in the world, and a language which may be used to make conventions about the correct use of terms and notions to pick out those things. However, this does not imply that in the world of which they themselves are part, identifiable and determinable things actually exist about which something is the case or true - and something else is not - and to which these notions and terms may be correctly applied. Nor does it imply that persons and language users may together determine these things in the world (including themselves, other persons, their bodies, arms and legs, and so forth) as things which exist *independently* of such cognitive notions or linguistic terms.

A related and certainly no less serious problem of Constructivism arises from the assumption that the notions of 'true' and 'correct' are purely *relative* notions - thereby rendering accounts of *anything* merely relatively true and correct. An obvious problem with this assumption is that it is inherently self-defeating, since it would have to apply to accounts made by Constructivism about its own fundamental assumptions that they could also only be relatively true. It may, of course, be suggested that this "relativity" of the notion of truth does not apply when used by Constructivists to talk about the variability of our cognition and use of language, nor when used to talk about the cultural and historical conditions which determine this variability. However, so to suggest would be to suggest that Constructivists, when talking about and determining such variability and conditions, use a concept of truth which is different from the concepts of truth applying to the different forms of cognition and description that different cultural and historical conditions give rise to, i.e. a concept of truth which is itself "immune" to cultural and historical conditions. But so to suggest would obviously be no less self-defeating.

As will be argued (in Chapter 5, 8 and 14) there are compelling reasons for the impossibility of the cognitive and linguistic *relativism* proposed by Constructivism. The very notion of cultural, social and historical differences of our cognition and description of reality, i.e. the fact that reality and things in reality may be cognized and described from different points of view, and with regard to different purposes and intentions, and using different means of observation and descriptive systems, etc., *precludes* the assumption that the things described and cognized do not exist independently of, but only as products of our language and mind. For, if things in reality did only so exist, there could be no comparison between historical, social or cultural differences in describing and cognizing them. Indeed, without the assumption that the *same* things or phenomena may be described *correctly*, though differently in different situations (or, say, in different cultures and under different historical conditions), we would have no notions of the *same thing*, nor therefore of comparative studies of differences in our cognition and descriptions of them.

It is not difficult to show that the consequence of the Constructivist assumption about the relativity and variability of the notion of truth renders every concept of language and cognition totally arbitrary, thereby making it impossible to talk meaningfully about anything. It is an assumption to the effect that although we can use language to talk correctly about both language and cognition, as well as the conditions for cognition and use of language, we cannot use language to communicate or talk correctly or truly about that which these conditions concern, *i.e. ourselves and the reality of which we are part*. Indeed, since nothing exists *independently* of our cognition and language, but only *in virtue* of the conceptions and notions of our minds and language, then, according to Constructivism, it simply makes no sense to say that the categories of our minds and the notions of our language may be used correctly - or incorrectly - to pick out anything which is *different* from and which may be *distinguished* from language and those terms, concepts and categories. In which case the key concepts of language and cognition such as 'reference', 'intentionality', 'aboutness', 'true' and 'false', as well as of 'reality' and 'things', become meaningless concepts - and with them all other concepts of cognition and language.

The assumptions of *Naturalism* and its attempt to solve the Mind-Matter problem do not offer any better prospects, but on the contrary, have consequences which are no less absurd and self-defeating than those of Constructivism. Let me start with what may be considered the basic assumption of Naturalism, namely that the only things existing *objectively* in reality are those which may be accounted for in the scientific terms of Physics (whether in its current most advanced forms, i.e. quantum mechanics - or in its future versions). Consequently, only those descriptions of reality and things in reality are *true* descriptions of what exists, which are based on observations within the scientific context of physics. It follows automatically from this assumption that descriptions and observations of reality and things in reality obtained in all other and non-scientific situations - for example our everyday observations and descriptions of things in everyday situations - must be *incorrect* or *false*, and also that *what* is described in such other descriptions have no claim to *objective* existence. According to this view, knowledge and descriptions of things in our familiar world (such as tables and chairs, billiard balls and rocks), i.e. reality as it appears to our "scientifically unaided" senses, is a product of the human mind; indeed it is a reality, a world of things, which does not exist independently of the "subjective, species-specific points of view of the human mind", and the like (cf. Chapter 8).

The main problem with this view is that it renders both our conception of what *exists* in reality and our notion of *truth* and *objectivity* dependent on *particular* situations, i.e. scientific physical situations, in which things in reality are observed with particular opportunities of observation, and using particular descriptive systems. However, by so doing, this view ignores the dependency of scientific descriptions of physical phenomena on non-scientific descriptions of the things, to which those scientifically described phenomena are claimed to apply. The problem entailed becomes obvious if we can agree *both*

1. that the point of departure for any scientific investigation must be ordinary everyday description and determinations of the situations in which we find ourselves in reality and carry out investigations - scientific or otherwise - on things which exist in those situations

and

2. that any scientific description and explanation of some particular phenomena or property of things rely on and are logically related to other descriptions (non-scientific and scientific alike) of the thing to which the phenomena and explanations of its properties apply.

This is to be understood as follows: To the extent that it makes sense to say that the general laws and explanations of e.g. classical mechanics are about the behaviour of *material objects* of our familiar world, such as tables and chairs, billiard balls and rocks, human as well as other physical "bodies" (what else?) then the laws and explanations of classical mechanics necessarily rely on ordinary everyday descriptions and determinations of such "material objects" and "bodies", existing in particular places in space at particular times, etc. To give another example, indefatigably stressed by Niels Bohr throughout his works and now accepted by every physicist: the conditions for carrying out, observing and describing experiments in *quantum physics* (including the descriptions of the research design, the working of measuring instruments etc.) rely on descriptions which are expressed in terms of the laws and explanations of *classical physics*.

This dependency or logical relation between scientific descriptions - as well as between scientific and non-scientific description of reality - has important consequences. *First*, it means that no particular description put forward within any particular situation, using any particular descriptive system - for example a quantum physical description of the behaviour and phenomena of elementary particles of which things in reality are composed - can be *sufficient* or *exhaustive* of the things to which the description applies.

But nor can the descriptions on which such quantum physical descriptions rely be reduced "away" or replaced by quantum physical description. For, that would amount to reducing away the descriptions of the very conditions for carrying out experiments in quantum physics. In other word: no description of anything - scientific or non-scientific - can be *identical* with what it describes. To claim the reductionist assumption of Naturalism, however, is precisely to claim such an identity. It is to claim that that of which we can give *both* an ordinary everyday description[4] *and* a physical description[5] is the same as that which can *only* be correctly described in terms of physics. However, so to claim would render our notions of 'same objects', and thus of 'same' and 'objects', completely ill-defined.

That scientific theory building is *cumulative* (cf. the quotation from Lewis, op.cit.) means precisely that the descriptions at one level of the system rely on and, thus, are logically related to descriptions at other levels. Therefore, no description at any one level within the system can be *sufficient* or *exhaustive*, nor can descriptions at any of the levels of the system be reduced to descriptions at any other level of the system. The chain of dependency between scientific and non-scientific descriptions - starting with everyday descriptions of things in our familiar world, and finishing with the quantum physical description of the structure of elementary particles of those things - has the consequence, *secondly*, that it would make no sense to claim that the description of phenomena and properties of things at any particular level is a *more* true or *more* objective - or *true* or *objective* - as opposed to the descriptions of phenomena and properties of things at other levels of the system. Nor would it make sense to claim the only *objective* existence in reality of phenomena or properties of things observed and described at any particular level of the system, and therefore to claim that only such descriptions are *true* description of reality. In particular, it would not make sense to say that everyday descriptions of things in our familiar world on which scientific descriptions rely, are *not* true description, nor descriptions of what exists objectively in reality - but are rather the products of our "subjective" (as opposed to "objective") points of view, and thus, in effect, are things which reside in our minds. For so to say would amount to saying both (1) that it is possible to develop true description of reality and its physical properties on the basis of descriptions of things which are not themselves true, and (2) that the general explanations and descriptions of physics about the properties and structure of physical phenomena apply to and concern things and phenomena, which do not themselves exist objectively in reality, but rather in the "realm" of our minds - and so on ad absurdum.

[4] In terms of objects such as tables and chairs, cannon balls and rocks.

[5] In terms of e.g. the molecular structure or properties of particles of which these objects are built up.

The impossibility of denying the dependency between descriptions in the above chain of descriptions, and of denying the truth and objectivity of any of the description in this chain, save for those belonging within the most advanced disciplines of particle physics, is perhaps most simply illustrated in the following example: Such a denial would imply that a physicist, once he had discovered the laws and structure of particle physical phenomena, would not be able to give correct or true description of the laboratory in which he carries out his experiments, the measuring instruments, their position in space relative to one another, or, for that matter, of any other part of the equipment of an experiment in particle physics. In short, would imply that he would no longer be able to give correct descriptions or accounts of the conditions for carrying out experiments in particle physics. What Naturalism completely fails to see is that by making our notions of *true* knowledge and description of reality, as well as our notions of what exists *objectively* in reality, dependent on the particular possibilities of observation and description of scientific situations of physics, the notions of 'true' and 'objective' are made ill-defined in all other situations, and that, *as a consequence*, are made ill-defined in the scientific situations of physics as well.

The epistemological assumptions of Naturalism so far outlined are clear enough - and so (I argue much more extensively in Chapter 8) are their unfortunate consequences. However, the further assumption, i.e. that all phenomena which have physical causes and effects, including mental phenomena in virtue of having such causes and effects, are *in principle* truthfully and exhaustively accountable as purely physical phenomena, is *not* at all clear. Now, the notions that something is *in principle* explainable in terms of some general scientific laws, or would follow *in principle* from some particular scientifically determinable conditions or presuppositions, are well known notions in physics - as well as in other areas of inquiry on matters which we can talk logically about. Furthermore, the conditions for the use of the notion of what *is* or *can* be shown to be in principle the case within these areas may have precise definitions, hence allowing those notions to have well-defined meanings. This is so when, for matters of concern to e.g. classical mechanics, we say that, given particular initial conditions, it is in principle possible to give exhaustive, causal accounts of the movements in space and time of physical objects. And it is so when, say, in Einstein's Special Theory of Relativity, particular phenomena are said to follows in principle from the basic presuppositions and conceptions about physical phenomena of concern to this theory. Indeed, it may even be strictly defined how the truth of what is in "principle possible", or "follows in principle", may be subjected to *empirical* confirmation (whether with available technological means or some which may be developed in the future).

However, despite the widespread agreement to the contrary - even among strong opponents of Naturalism - what is meant by the claim that mental phenomena are *in principle* truly and exhaustively accountable in physical terms is not at all clear; nor is it clear how this claim could be subjected to empirical testing - let alone clear in the sense of the uses of the notions 'in principle' or 'empirical' in the cases just mentioned. But neither has anybody succeeded in making it the least bit clear what is meant by the claim that mental phenomena *are* physical phenomena, let alone come up with any precise formulation of the conditions under which this claim may be amenable to empirical confirmation - whether at present or in some future development of physics.

Put in concrete terms, what is not at all clear is what it means to say that, e.g. the feeling of pain in my thumb, or my seeing the object over there as a letter box, is in principle completely accountable in terms of, say, quantum physics, and thus that some particular quantum physical description is a description of the pain felt in my thumb, and that some other quantum physical descriptions is a description of my perception of a particular object as a letter box. Nor is it at all clear how it could be determined whether either of these particular quantum physical descriptions are *correct* or true descriptions of my experience of pain in my thumb or my perception of a letter box respectively.

Well, what *may* in principle be possible, is to give an account of, say, the quantum physical structure of the particles of which a letterbox is made up. But that the object in question is a *letter box*, i.e. an object in which I can put *letters*, which will be sent to the persons carrying the names and addresses written on them, cannot possibly be exhaustively accounted for in terms of quantum physics. For a complete quantum physical description to be a correct description of a letter box would, as far as I can see, require *either* that notions such as 'letter boxes', 'letters', 'persons', 'addresses', and so forth, were part of the conceptual apparatus of quantum physics - which they are not. (Nor are there any reasons to believe that such notions will be part of the vocabulary of any future development of quantum physics). *Or*, would require that it were possible to determine when and whether a particular quantum physical description is a *correct* description of a particular thing, e.g. a letter box, as opposed to some other thing, e.g. a table or a chair. However, since such determinations could not be made independently of "folk" notions about such things as "letter boxes", "tables" or "chairs", we would still be in need of concepts of, as well as ways of determining things and phenomena in terms of letter boxes, tables or chairs - and with them all other notions implied in the meaning of such notions in our everyday language.

It may of course also be in principle possible that, one day, we may be able to give a complete quantum physical descriptions of the physical and physiological working of the brain *when* we perceive an object as a letter box - or when we feel pain in our thumbs. However, mere *correlations* between such *physical states* in our brain and the concurrent *mental state* of having knowledge or beliefs about objects in the world, or experiences of pain in the thumb, do not amount to *explanations* of how such mental states or experiences arise out of purely physical phenomena or states (for extensive arguments of this point see e.g. Nagel (1986) and McGinn (1991)). In particular, it does not explain how mental states, such as belief or knowledge of object may be *about*, let alone be *true* and *false* about what exists in reality. Nor do such correlations explain how language and description put forward in language, arising out of some other physical states of the brain, may be used to describe the *same* objects, let alone explain how language may be used to *refer* to and be *true* about those things and beliefs. That is, such correlations do not explain how *referentiality* (or "aboutness") and *truth*, and thus the crucial *logical* properties of beliefs and linguistic propositions, arise out of physical phenomena and states *which do not have those properties*.

But neither does it make the least bit of sense to say that physics in its currently most advanced form, i.e. quantum physics, could in principle exhaustively and truly account for the referentiality and truth of beliefs, knowledge or linguistic propositions. That would require *either* that referentiality and truth were part of the descriptive and explanatory vocabulary being used to account for the properties and behaviour of quantum physical phenomena - and thus that quantum physical phenomena are the sort of things which may refer and be truth functional - which they are not. *Or*, it would require that mental phenomena, such as beliefs, knowledge and propositions put forward by persons about things in the world - including, of course, beliefs, knowledge and propositions about quantum physical phenomena - could be reduced to, and thus be expressed in terms of physical phenomena, which do *not* have the logical properties of referentiality or of being true or false. In effect, it would require that, just as quantum physical phenomena, our beliefs, knowledge and propositions about quantum physical phenomena thus reduced and expressed did not refer to anything, nor would they be true or false about anything.

Well, if we assume the latter, and thus that the notions of 'truth' and 'reference' which language share with logic, could be reduced to quantum physical phenomena, we might as well assume, indeed we would be assuming that logic, on which relies *par excellence* the language of the sciences, could be reduced to and explained in terms of something more elementary or fundamental which did not imply the existence of logic - and hence that such reduction and explanation could be accounted for without using logic. (This point and its consequences in general, as in particular for different versions of functionalism and computationalism, will be discussed in Chapters 7 and 11.)

The points made so far about the assumptions of Constructivism and Naturalism may be summarized in this way: What *Constructivism* has failed to see is that the possibility of talking consistently about our *cognition* and *description* of ourselves, our minds, and the situations in which we find ourselves in reality - culturally, historically or otherwise - hinges on the possibility of giving consistent, objective and true account of *reality*, ourselves and these situations, and thus *hinges on the presupposition that reality, we ourselves and such situations exist objectively and as thing about which we may have knowledge*. What *Naturalism* has failed to see, conversely, is that the possibility of talking consistently about *reality* and the properties of things existing objectively in reality hinges on the possibility of talking consistently about our *cognition* and thus *hinges on the various correct or true ways in which we may cognize and describe reality and these properties in different situations*. The two positions have in common that they both fail to see that, generally, we cannot talk consistently about *reality* and things in reality without or independently of referring to our cognition and descriptions of them, nor without presupposing that reality and these things exist as things of which we may have knowledge and put forward propositions which are true or correct - in both everyday and scientific situations. And they fail to see that neither can we, conversely, talk consistently about our *cognition and description* of reality and things in reality without or independently of referring to reality and those things, nor independently of presupposing that reality and these things exist objectively and independently of this cognition and description - in both everyday and scientific situations.

And this brings us right back to the argument of why it is logically impossible to state the Mind-Matter problem consistently. The assumptions behind this problem is precisely the possibility of a *polar* opposition between a material "realm" of physical objects and a mental "realm" of our cognition and description concerning these objects, the contents of each of which may be determined and talked about *without* or *independently* of referring to the contents of the other. Hence, what makes the assumptions behind the Mind-Matter problem logically inconsistent, and thus the formulation of this problem logically inconsistent, is that it ignores the above interdependence or *necessary* relation which exists between our cognition and description of reality and reality (itself) *without which neither our notions of 'reality', nor of 'cognition', 'description of reality', 'true' and 'false' may have well defined meanings*. This point and its consequences for theories of language and cognition will be thoroughly discussed throughout the chapters of Part II.

But first, in Part I, Chapters 3 and 4, I shall illustrate how the logically inconsistent assumptions of the Mind-Matter dichotomy underlie and lead to inconsistencies and problems in theories of perception within psychology. It has seemed obvious to me to start my analysis of the problems and inconsistencies of the Mind-Mater dichotomy here, for, in my view, in no other areas of research within psychology and philosophy does it become so immediately clear why the assumptions behind this dichotomy are untenable and lead to inconsistencies. And in no other area of research and theorizing will an analysis of the problems incurred by these inconsistencies make it so immediately clear why the assumption of an *interdependency* or a *necessary relation* between Mind and Matter, is the only tenable alternative on which to base theories and research within psychology and philosophy. Furthermore, all the fundamental implications and principles of this alternative assumption, to be developed in later chapters of the book, appear in *embryo*, so to speak, during the analysis in these first chapters on perception. In the section which follows, I shall give a short summery of the main results of this analysis.

1.3 Consequences for Psychology of Perception

Current theories of perception in psychology may be seen as a continuations of a tradition started by Helmholtz, Hering, Wundt, Koffka and Koehler in that they have in common with this tradition the assumption that the epistemological problem of how it comes about that we acquire knowledge about and perceive the world as it is - or as we do - is a problem which may be accounted for and explained *empirically* by perception psychology. The central tenet of this tradition is that the process of perception may be described as a causal process - starting with stimuli "impinging" on our sense organs, which may be independently described in purely physical, mathematical (or geometrical) terms - and terminating with something quite different, i.e. visual experience of things in our familiar world. Granted a complete account may be given of how the "visual system" detects and its internal, functional "machinery" processes the information inherent in the stimuli, then a complete (or general) causal explanation may be given of how our visual perception of the world and things in reality comes about - thus overcoming the Mind-Matter problem and the polar opposition between mind and matter it entails.

As examples of such attempt to provide general or complete accounts of how visual perception of the world come about, I have chosen two – on the face of it – very different theories of perception. The one is Gibson's theory of "direct perception", which aims to explain how features and structures of the world and its objects are presented in the stimuli, i.e. in the structure of the light emitted from objects in the world, to be detected and picked up directly by the perceiver, and thus providing the perceiver with perception of these features of the world and its objects. The other is Marr's computational model of vision which aim to explain how the visual system, by a series of computations of structures in the light emitted from the world and its objects, produces representations of structures of features of the world and its objects, and thus providing the perceiver with perception of these features of the world and its objects.[6] Needless to say, any *causal* account of how perception of the world comes about requires that both the *cause*, i.e. structures of light in the stimuli or computed representations of them, and the *effect*, i.e. perception of world and its objects, be *independently* determined and described. In other words, such causal account requires that it be possible to determine and describe the structures of the stimuli (or computed representations of them) *representing* the structures and feature of the world, independently of or without referring to how we actually perceive the world and its features and structures.

However, the analysis in Chapter 3 of both Gibson‘s and Marr‘s theories clearly shows that these requirements *are* not, indeed, *cannot* be met – and thus that their attempt to provide general and complete causal theories of perception inevitably fails – i.e. due to the inherent circularity of their explanations. Moreover, it shows that the much debated differences between Gibson's and Marr's theories notwithstanding, the reasons for the problems to which they both give rise – among them the problem of circularity - are *in principle* the same.[7] Thus, in the case of Gibson's theory of perception, the analysis shows that it has to be *presupposed* that the visual system, in order to detect and pick up information in the stimuli emitted from the environment *as* information about things in the world, is already equipped with a considerable amount of "knowledge" about how we actually perceive features and properties of things in our familiar world - and, therefore, his theory of perception has to presuppose what it was supposed to explain. In

[6] My reason for choosing Gibson's and Marr's theories, first presented some twenty years ago, is that they are still by far the most thoroughly developed attempts to provide *general* or *complete* theories and explanations of visual perception. Furthermore, they are well known to most psychological readers - and still widely discussed. However, I hope it will be clear from my analysis of Gibson's and Marr's theories that the fundamental assumptions behind more recent computational and connectionist theories of perception, are not different from those underlying Gibson's and Marr's, but just "more of the same" - and hence I hope it will be clear that the general results of my analysis apply equally well to these other more recent attempts to explain how perception of the world comes about.

[7] It may be objected that Gibson did not only commit himself to the views of traditional *realism*, but also contributed a non-dualistic, non-causal *ecological* account of perception, which does not entail the problems discussed in what follows, among them the problem of circularity. In Chapter 2 I shall thoroughly discuss - and refute - this objection.

the case of Marr's computational model of vision, the analysis shows that in order for the model to determine the true nature of the information and solve the ambiguity problems of the representations computed at the various levels of processing, it has to be provided with rules and a data base of "knowledge" about how we actually perceive features and properties of things in the familiar world - and thus, as in the case of Gibson's theory - will have to presuppose what it was supposed to explain.

Equally importantly, the analysis shows yet another reason why the basic assumptions underlying Gibson's and Marr's causal theories of perception, i.e. the possibility of determining the *cause* of perception (i.e. stimuli in the world), independently of the *effect* (i.e. the perception of the world), does not hold. This has to do with the fact that a technical description of the stimuli (or computed versions of it) thought to underlie and cause our perception, necessarily rely on an ordinary everyday description and perception of the world and objects from which the stimuli are emitted. Thus, in an experimental test of the explanations and predictions of either theory, the technical description of the stimulus situation (or computed versions of it) may be conceived of as a way of representing the reality and objects of which the experimental scenes are made up (i.e. rooms with walls and pictures on the walls, cf. Gibson, or various shaped blocks put on a surface, cf. Marr). However, such a technical description of the stimuli cannot be given independently of an ordinary everyday description and perception of these experimental scenes. Indeed, it would not be possible to apply the technical description to these scenes, nor to determine whether, or to what extent, or for what purpose, the technical description is an *adequate* or *relevant* description of – precisely – these scenes. We may contend, therefore, that the technical description of the stimulus situation – or of computed representations of it – rests on an *abstraction*, i.e. in the sense that it depend on an ordinary everyday description and perception of the world.

However, this dependency of a technical description of the stimulus situation on an ordinary everyday description of the world, is a condition which applies *in general* for the implementation and test of *any* account or theory of the processes of perception. If so, the assumption of a Cartesian partition of "reality" and "perception of reality" into *independently determinable* parts, on which rests not only Gibson's and Marr's theories, but all psychological theories aiming to explain causally how it comes about that we perceive the world as it is – or the way we do – simply does not hold. And because it does not hold, such theories will inevitably be circular.

The implications of the above presuppositions and conditions for implementing and testing theories and accounts concerning the processes of perception, may be expressed in general terms thus: It seems that we cannot talk consistently about *reality* and things in reality independently of or without referring to our perception of it, nor independently of or without presupposing that reality and things in reality exist as things we may perceive and talk correctly about. But neither can we talk consistently about our *perception of reality* independently of or without referring to reality and things in reality, i.e. we cannot talk about and describe *what* we perceive (our perception of reality) without or independently of talking about and describing *that* which we perceive (i.e. reality and these things). Because of this inter-dependency between reality and our perception and description of it, without which neither the notions of 'reality' and 'things in reality', nor of 'perception of things in reality' would have well defined meanings, no *general* questions can be asked about the relation between reality and our perception and description of it. On the contrary, it seems that we shall have to assume that a *necessary relation* exists between reality and our perception of it, which precludes any explanations and question of a *general* kind of how it comes about that we perceive reality as it is. If we do attempt so to explain - or even worse, attempt to doubt or deny that we perceive reality as it is - we would involve ourselves either in circular explanations or in absurdities. For both such explanations and doubts, in order to be well defined, would have to rely on the presupposition that we are indeed capable of correctly perceiving and describing reality and things in reality.[8]

Now, the implications of the above presuppositions and conditions for implementing and empirically testing accounts and theories of the processes of perception, may well be that the assumption of a Cartesian Mind-Matter dichotomy between "reality" and "perception of reality" does not hold, and as a consequence, the epistemological problem of perception to which it gives rise, cannot be consistently stated, nor therefore be solved or overcome empirically by perception psychology. However, it may be objected (and so it has) that this does not mean that *causal functional theories* about the processes taking place in the "machinery" of the perceptual system, and the stimulus conditions which cause them, are not possible. It is only, so the objection goes, if we "mix up" or confuse philosophical *epistemological* issues about the nature and status of perception with *psychological* issues about the causal functioning of the perceptual system that the problems (among them the problem of circularity) arise. However, if e.g. Gibson's and Marr's theories are freed from their undeniable epistemological pretension and instead interpreted as *functional theories*, so the objection continues, the

[8] The presupposition of a necessary relation between perception and reality does *not* imply that we cannot distinguish between perceiving things, and the things being perceived. Nor does it mean that things in reality do not exist *independently* of our perception. On the contrary - as should become clear in the discussion in Chapter 4 - what it does mean is that I cannot describe my perception e.g. of the apple on the table in front of me independently of or without referring to and describing the apple on the table in front of me.

critique I raise, and problems I point out in Gibson's and Marr's theories are not longer valid. To this I can certainly agree - provided it is understood that it would be just as problematic to maintain that our perception of reality may be accounted for adequately, let alone exhaustively, in terms of functional theories describing the causal processes between stimuli in the world and some functional features in the "machinery" of the perceptual system.

Apart from the reasons already given, this has to do with some quite elementary conditions for carrying out and describing experiments in perception psychology. An analysis of these conditions, carried out in Chapter 4, shows that investigations of any concrete problem of perception necessarily rests on and presupposes *background knowledge* as well as descriptive and perceptual capabilities on the part of the experimental subjects, which go well beyond the perceptual phenomena being studied, but which form a substantial part of subjects' perception during the experiment. To this presupposed background knowledge and these capabilities belong not only the subjects' ability *correctly* to describe what they perceive during the experiment, but also their ability correctly to perceive, identify and describe the *circumstances* in which the experiment takes place - including the position in the room of themselves, the experimental set up, or the part of it to which they have access. To this must be added the presupposition that the subjects understand the experimenter's instruction about what - according to the purpose of the experiment - are considered relevant descriptions of the perceptual phenomena to which they are supposed to attend. In this respect an experimental situation resembles an ordinary everyday situation in which two persons have established a basis of correct descriptions of the things they perceive in the situation in which they find themselves, just as the subjects' descriptions of what they perceive in the experimental situation resemble, indeed *are part of* their everyday language and descriptions of things and events occurring in everyday situations. Therefore, the descriptions by subjects of what they perceive – in experimental situations as in everyday situations - rely on the presupposition that *in ordinary everyday situations under normal conditions of observation, the subjects may correctly perceive and observe reality and things in reality, and that, as users of the same language as the experimenter's, they know correct descriptions of what they perceive and observe.*[9]

Because of these presuppositions for carrying out psychological experiments on perception, it would be misleading to claim that the perception

[9] Let it be clear that this does not mean that errors of perception - or mis-perceptions of things - cannot or do not occur in everyday perceptual situations. However, we cannot talk consistently about perceptual errors or mis-perceptions, let alone determine the nature of such errors or mis-perceptions, without presupposing that we know of, or may come to know of, correct perceptions of the thing being mis-perceived.

reported by the subjects during the experiments could be accounted for adequately, let alone exhaustively in terms of causal processes between some or other "local" stimuli of the experimental set up and some or other local functions of their perceptual system.[10] Indeed, neither the perceptual phenomena reported by the subjects, nor the processes of such perceptual functions may be *determined* and *accounted* for adequately, *independently* of referring to the subjects' *general* knowledge and perception of the experimental situations; nor without referring to the purposes and conditions of observing and describing the things presented to them in those situations.

The fact that part of the conditions and presuppositions for carrying out experiments in perception psychology is that in ordinary everyday situations under normal conditions of observation, we may perceive and describe reality and things in reality correctly, means that perception psychology cannot explain, nor give proof of the correctness or veridicality of our everyday perception. On the contrary, *it must be taken for granted*, and be the point of departure for the investigations of any *concrete* problem of perception within psychology - including investigations of the processes in the "machinery" and functions of the perceptual system - albeit rarely if ever made explicit in attempts to account for perceptual phenomena in terms of such processes and functions.

The conditions so far outlined for the subjects' and investigator's descriptions respectively, mean that the subjects' description of their perception during the experiment relies and depends on their general knowledge and correct description of reality and the features of things existing in reality, just as the experimenter's technical description of the stimulus situation relies and depends on his general everyday perception of it. This is just another way of saying what was said earlier in general terms about the *interdependence* or *necessary* relation between reality and our perception and description of it – and hence about the impossibility of assuming a Cartesian division of reality and our perception of it into independently determinable parts or "realms".

For some readers the main result of the analyses in the chapters on perception may well be that we shall have to accept that *any* psychological theory of perception, as well as any investigation of the processes of perception, necessarily build on and, thus, inevitably *are* "mixed up" with assumptions and presuppositions of a philosophical, epistemological nature. However, given the assumptions are correct and made explicit, they may provide invaluable insights into the kinds of questions and problems that consistently

[10] In Chapters 10, 12 and 15 of Part II, I shall show that the same applies *mutatis mutandis* to research and theories of the "machinery" and functions of the cognitive and linguistic "systems", thought to be implemented in our brains and underlying our cognition, language and use of language.

and meaningfully *can* be addressed by perception psychology - and which cannot. I shall have more to say about this in Chapter 4, but let me already say this much here: In view of the necessary conditions and presuppositions for investigations on perception so far outlined, accounts of the relation between the perceptual phenomena being studied and the stimulus conditions under which they occur, cannot be of a causal, but, as a matter of principle, can only be of a *correlational* kind. However, given our observations and theories about these correlations are correct, the outcome of such research may be the development of technologies and means to solve problems of a *practical* nature - just as is the case in other sciences. Of more general importance, though, once we have understood the conditions and presuppositions of an epistemological kind for carrying out investigations on perception, there is hope that perception psychologists may come to agree on a set of basic concepts, assumptions and principles for describing phenomena and problems concerning perception of a genuine *psychological* nature, and to develop theories which may appropriately and consistently account for such phenomena and problems. If so, we may begin to have a situation within psychology of perception resembling the situation in the natural sciences, who do have a common basis of concepts, assumptions and principles about the nature and properties of the phenomena falling within their subject area. But also, as will be the topic of the chapters of Part II, a situation in which we shall come to realise *that* and *why* the basic concepts, models and theories to describe psychological capabilities and phenomena are necessarily fundamentally different from the concepts, models and theories of the natural sciences.

Chapters in Part II of this book on theories of language, mind and cognition will show that the fundamental conditions, requirements and presuppositions summarized above, apply *mutatis mutandis* for the research and theorizing within these other areas. In these chapters I shall discuss their implications and consequences for a variety of recent so called causal, computational theories and models of cognition, and similar models and theories of the acquisition and use of language.[11] A thorough analysis of these theories and models will show that the problems and inconsistencies from which they suffer, are quite similar, if not identical to the problems encountered in perception psychology. This is not surprising, since these current theories and models of cognition and language are based on the same fallacious assumptions of a Mind-Matter dichotomy, as are current theories and models of perception. In the next chapter, I shall outline some further formal reasons for the problems and inconsistencies inherent in the assumption of this classical dichotomy, as well as some of the formal consequences

[11] Among them *The Syntactic Theory of Mind* by Stich, (1983) and Fodor's *Causal Theory of Content* (1987, 1994).

and implications for theories of perception, language and cognition of the alternative assumption of a *necessary relation* between Mind and Matter presented in this book.

Chapter 2

Alternative assumptions and principles

2.1 Criteria for a research area to attain the status of a science

It is well known that the psychological establishment has no taste for debates about problems of a fundamental theoretical nature of the kind, which have proven so essential for the founding and progress of the natural sciences. With a few exceptions, there has been no debate among psychologists and little reflection as to whether the models and conceptual schemes used to account for psychological phenomena and properties are indeed adequate, or as to whether the assumptions about the nature of these phenomena and properties, on which the models and schemes are based, are indeed tenable. Just as unfortunate, there has been little understanding of the necessity of the to-ing and fro-ing between theoretical considerations and interpretation of data from empirical work, which has been so crucial for the progress made within the natural sciences.[1] It is also well know that academic psychology, in order to be accepted as a science on a par with the natural sciences, has attempted to take over the models, assumption and principles from those sciences to describe and explain phenomena of a psychological nature - thereby violating what since Galileo has been considered as criteria for an area of study to attain the status of a science. To understand what is needed for psychology to be a cumulative empirical science on a par with the natural sciences and biology, and yet a science which fundamentally

[1] For an excellent account of the necessity of theoretical discussions of basic assumptions and principles for the interpretations of result from experiments in physics, I refer the reader to Abraham Pais' book on the development of theories in Quantum Physics (Pais, 1991).

differs from those other sciences, we shall have to turn to those criteria, and to how they originated.

So, let us turn to Galileo, the founding father of what is now considered *empirical* science, and to the approach he developed for studying physical phenomena, thereby establishing physics as the first of the empirical sciences. His interests were to determine the "physical causation" of the behaviour of "physical matter" (by which he meant the movements in space and time of material objects), and for the systematic study of these phenomena he proposed three methods, namely, quantifying of observations, deduction from hypotheses, and experimental testing of hypotheses. However, as Galileo pointed out, observation of "physical matter", will provide a lot of information which is irrelevant for the physical causation of its behaviour. Properties like taste, smell, colour, for example, do not affect the behaviour of material objects, and he suggested that observations of the properties of things be divided into two groups: the qualitative and *physically irrelevant* on the one hand,[2] and on the other, the quantitative or *physically relevant* observations. Equally importantly, he had the insight that not *all* quantitative observations of physical objects and their properties are relevant in every context or for any purpose. The only way to discover what are the *relevant* observations, and thus the ones playing a part in the causation of the particular kind of "movement" of a physical object, is by careful *reflection* and *analysis* of observations; one has, as Galileo puts it, to separate the observed compounds into relevant quantitative dimensions, which thereafter may be isolated and handled abstractly by means of mathematics. By using this strategy of *reduction* he found that quantitative determination of e.g. uniform rectilinear motion of an object required the measurements of two dimensions only, namely distance and time - and thus that its motion may be unambiguously expressed as a function of distance and time measurements.

This strategy of reduction came to determine the criteria for an area of study to qualify as a science. These are, first, that it be possible to define what constitute the features and properties of the object (or phenomena or subject matter) being investigated. Secondly, that the concepts developed to describe and determine the *particular part* or *aspects* of the object or subject matter being studied, be so precise as unambiguously to delimit these parts and aspects from other parts or aspects of the object with which the area of study is not concerned. Thirdly, that the area has developed *appropriate* and *well defined* methods whereby those particular parts and aspects may be systematically observed, and fourthly, that the purpose of the observations themselves be well-defined. These criteria must be fulfilled by any scientific discipline which claims that its inquiries, as well as the results emerging

[2] This group also includes volition, intentions, feelings and the like which, in Galileo's view were "names" for phenomena residing in the "human mind".

from them, make a difference vis à vis common sense and superstition. However, *how* and *in what way* a discipline defines its area of research and distinguishes it from those of others studying radically different phenomena, is just as essential to secure this difference.

That a particular area of knowledge and inquiry may attain the status of scientific research, then, depends on the possibility of indicating precisely which - limited - part or aspects of reality the particular science concerns. We may say not only what it is about, but equally importantly, what it is *not* about. And we can indicate with what part of language and what concepts a science is particular concerned - and with what part or concepts it is not, i.e. isolate them from other parts and concepts of language. However, the criteria which provide a science with a well defined and limited subject area at the same time provide the descriptions, models and explanations of the science with a *limited field of application* - thereby limiting the kind of phenomena it may explain and account for. This is to be understood in the sense that e.g. descriptions, concepts and general explanations within Classical Mechanics of "movement" in time and space of macroscopic material objects are not applicable to the description of "movement" of electrons and other particles within Quantum Physics - and vice versa.

However, this does *not* mean that what physics is saying about physical matter is not applicable to the physical properties of phenomena being studied within other areas of research. To the extent that e.g. human beings, birds, plants, or even spoken or written language have physical properties, those properties are of course amenable to physical description. And to the extent that all matter is built up of atoms and molecules, what is said about atoms and molecules of material things being studied within physics applies to everything material, including material components of humans, birds and plants. Likewise, to the extent that living cells, genes and hormones in birds or plants are similar to living cells, genes and hormones in the human body, what is said within biology about these matters - and by the Natural Sciences about their physical properties - applies equally well for human beings.

What it *does* mean, however, is that what e.g. Classical Mechanics says about the movements of material objects, or what modern physics says about atoms, molecules or elementary particles, *only* applies to those features and properties of humans, birds, plants etc. which they have in common with all other material objects. By definition, physics as a well defined, delimited research area, has nothing to say about all other properties which humans, birds or plants may have *apart* from and in *addition* to their physical properties. Likewise, what modern biology says about cells, hormones, genes and other properties of organisms only applies to those properties of human beings, which they have in common with other living organisms. By definition, biology as a well-defined, delimited research area, has nothing to say about all the other properties which human beings may have *apart* from or *in addition* to other living organism. Hence, neither biology, nor physics have anything to say about the *psychological* properties and phenomena of human beings, such as their use of language to put forward propositions about reality, which may be true or false, nor about their ability as self-designating *agents* to carry out intentional *movements* or actions - properties and abilities which have been *discarded as irrelevant* to and not part of the matters studied within physics and biology. Nor may the concepts used to characterise and determine psychological properties and phenomena of human beings be *reduced* to or *deduced* from the concepts used to describe and determine phenomena studied in biology and physics, i.e. *concepts which have acquired well defined meanings precisely because it has been possible to isolate them from, among others, the concepts and terms used to describe psychological properties and phenomena of persons*.

It may be said that two kinds of *reduction*, in the sense of *abstraction*, are involved in establishing an area of empirical study as an area of scientific research. First, an *external* reduction is involved in *delimiting* the area from others areas of scientific research, and the discarding as irrelevant phenomena and properties of things belonging within other areas of research. And secondly, an *internal* reduction is involved in the "careful reflection and analysis of observations" of the phenomena and properties of things, which may be relevant for the purpose of the studies of those things, and by which concepts and dimensions these properties and phenomena may be adequately and unambiguously determined and explained. To deal with the problems entailed in both kinds of reduction, and to engage in the reflection and abstraction it involves, is to deal with and to be engaged in what was referred to above as *fundamental theoretical issues* of a science. However, as pointed out by Galileo, this business of clarifying assumptions, principles and concepts applying to the phenomena and properties of things being studied by a science, is not only necessary in the early phase of establishing the science; it is a never ending process of clarifications *and* revisions, necessitated, among others, by results accumulated from empirical tests of hypotheses and deduction from observations as the science progresses.

Despite the necessarily limited field of applicability of any research area, academic psychology has attempted right from the start to take over the concepts, methods and models from those of the natural sciences and biology to account for and explain psychological phenomena. Apart from the wish to be considered as a science on a par with these sciences, there are probably several other reasons for this. First, although it may not seem immediately obvious how to account for the cognition, perception or action of persons in terms of the laws and descriptions of the natural sciences and biology, it remains a fact that persons are *also* physical things and physiological and biological organisms, part of the functioning of which *can* be accounted for in terms of the descriptions and laws of physics and biology. Secondly, experience from other areas of research seems to prove that the method of reducing things to their elementary properties has lead to far more accurate accounts of the structure and behaviour of things - in our familiar world as well as in the not so familiar part of reality described by physics and biology. And since the physical, physiological and biological functioning of the organism of persons is a *necessary condition* for persons to function as cognizing and acting agents, having languages and concepts of true and false, and all the rest, it is not difficult to understand why it has seemed obvious that these psychological properties of persons in the final analysis may be accounted for satisfactorily and exhaustively in terms of the physical, physiological and biological functioning of their bodies.

To these reasons must be added, thirdly, that if such reduction of psychological phenomena and events to purely physical, physiological or biological phenomena and events were possible, the advantage would be considerable. We would then be in a position in which orderly, causal physical, physiological or teleological biological explanations and accounts could be given of the seemingly unpredictable and unorderly cognition, thinking, reasoning, memory, action and communication of persons - *that is*, if only such thinking, action, and the rest, could be reduced to ever more precise accounts of the causally predictable physical reactions and the teleologically appropriate biological responses of the organs of the human body and brain.

However, the lessons to be learned from Galileo is not that *any* reduction of the properties of phenomena or events being studied will do, nor that *any* points of view from which they are observed may be arbitrarily discarded as irrelevant. More specifically, let it be assumed that what uniquely characterizes *persons* is that they have knowledge of and a language in which they may talk correctly about themselves and the things existing in the situations in which they find themselves - which we cannot deny without denying the possibility of discussing this or any other issues. And let it be assumed that what characterizes persons is that they may *act* and *co-act* with other persons, and - as opposed to causally determined reactions of e.g. physical and physiological systems - to act implies having *possibilities of action*. And let it be assumed, furthermore, that to carry out acts implies having made *choices* between different possibilities of action - and thus that actions of persons rely on the *intentions* of self-designating *agents* having knowledge about their possibilities of action with things in the situations in which they find themselves; indeed, let it be assumed that part of our knowledge about things in the world entails knowing what one may or may not do with them in different situations, and what happens to the things – and to ourselves - as a consequence of the acts we carry out with or on them. And let it be assumed, therefore, that to carry out *intentional* act implies having made a choice in order to bring about particular changes in our (future) possibilities of action with things, thereby providing new possibilities of observing and acquiring knowledge about them. Lastly, let us agree that persons may reflect on this knowledge, as well as on the role played by their needs, emotions and purposes for the choices of action they make, and that in a language they share with other persons they may discuss the *implications* of this knowledge and these acts; and agree, furthermore, that this not only provides persons with the possibility, but is a *presupposition* for persons to develop new situations with new possibilities of action, observation and description - none of which we can deny without denying the possibilities of persons to develop sciences. Well, given these assumptions it would *not* do for a science of psychology to discard any of these properties and features of our cognition, description and action, and thus *any of the psychological features and properties which uniquely constitute human beings as persons*. But nor would it do, let alone be possible, to reduce any of these features and properties of the language, cognition and action of persons to purely physical, physiological or biological processes or functions of the human bodies and brains, *to which those features do not apply*. If in the endeavour of psychology better to understand and more objectively, scientifically account for the action and co-action of persons such reduction be attempted, it would amount to discounting or doing away with the very phenomena of which such better understanding and objective accounts are sought.

Now, it may well be true that the processes and functions of the bodies and brain of persons are *necessary conditions* for persons to acquire and develop knowledge about reality and the situations in which they find themselves, just as such processes are necessary conditions for acquiring a language in which they may describe and put forward propositions about this knowledge and these situations. However, neither this knowledge or language, let alone the logical properties of referentiality and truth of propositions put forward in language, can be explained in terms of, be reduced to or derived from physical, physiological or biological functions and processes; i.e. from something more basic or elementary, *which does not imply the existence of knowledge, language, referentiality and truth*. That *reasons of principle* exist for the impossibility of such explanation and reduction will be further argued in Chapter 8. But let me already say this much here:

Even if it were possible that one day exhaustive accounts could be given, say, of our ordinary everyday cognition and perception of reality in terms of some particular physical, physiological causal processes or biological teleological functions, this would only leaves us with the problem of explaining how the cognition and accounts of *these* processes and functions are caused, i.e. by yet some *other* and *different* causal physical or physiological processes or biologically teleological functions. And, to repeat a point already made, proponents of the view that cognition, language and their logical properties of referentiality and truth in some future developments of physics and biology could be explained in terms of or be reduced to more basic or fundamental physical processes and biological functions, *which does not imply the existence of cognition, language and these logical properties*, would have to concede that *accounts* of such more basic or fundamental physical processes and functions could only be given in a language with descriptions or expressions which *were* referential and truth functional - and thus would have to be accounts *which relied on the existence of referentiality and the notion of 'truth'*. In other words, so to concede, is to concede that, as persons and language users we may well acquire knowledge and develop scientific theories and descriptions about the physical, physiological and biological processes and functions of our organism and brain, which are necessary conditions for our cognition and language. However, neither this knowledge or description, nor the logical properties of referentiality and 'truth' of cognition and language, can be explained in terms of or be reduced to any of the elementary processes and functions of which persons may have knowledge and talk correctly about (cf. especially Chapter 12 for further arguments to this effect).

It could be objected that mental processes such as consciousness and cognition do not arise out of nowhere, and that, somehow, the physical processes of the brain must possess the "potentiality" for mental processes to occur - and with them the fundamental properties of intentionality and truth of cognition. What I am arguing, however, is that how physical processes of the brain come to give rise to such mental processes and their fundamental properties, cannot be explained in terms of nor be derived from, nor reduced to these physical processes. With regard to this impossibility of explaining mental processes and their fundamental properties in terms of something more basic, the situation for psychology would seem comparable to that of physics. For almost half a century, physicists have had to accepted that the fundamental properties of matter, space and time (or of space-time) applying to our physical universe, actually appeared out of nowhere – in the sense: arouse out of a "big bang" of a "singularity" (i.e. an infinitely compressed "point" of energy-matter), which defies characterization in terms of space and time (or of space-time). Neither these properties, nor the laws or structure governing the quantum physical states and processes occurring after the "big bang" existed prior to this event – indeed nothing determinate can be said about the physical states or structures prior to 10^{-43} seconds after the "big bang". Thus, although space, time and matter of our physical universe arose out of a "big bang" of a singularity, it does not make sense - not even as a figure of speech – to talk about these properties as inherent potentialities of the singularity, even less to attempt to explain them in terms of whatever properties the singularity might have possessed.

The same applies to mental processes and their psychological features and fundamental properties, i.e. in the sense that none of these features and properties exist as properties and features of the physical processes of our brain, nor do any of the notions of such features and properties have any meaning at the level of description of such physical processes. Hence, it would not make sense – not even as a figure of speech – to say that the brain processes out of which these psychological features and fundamental properties of mental processes arise, inhere the potentialities for their existence. But nor would it make sense to attempt to reduce these mental processes, features and properties to physical processes or states of the brain, even less to explain them in terms of or derive them from such physical processes and states. In this respect and for this reason psychology will have to accept that the fundamental features and properties of psychological events do indeed arise out of nowhere – just as in physics this is equally accepted about the fundamental features, properties and laws applying to our physical universe. Any attempts to explain fundamental features and processes of a psychological kind in terms of more elementary physical processes and states of our brain would amount to attempts to explain away these psychological features and properties – and, by the same token, to remove the logical possibility of accounting for the physical processes and states thought to underlie them.

All this may be expressed in a straightforward and intuitively understandable way. As persons and language users we may come to know about and correctly describe the physical processes and states of our brain and *therefore* we cannot be reduced to what we know and may correctly say about such processes and states of our brain.

Let me summarize the points so far made: Granted that none of the above mentioned psychological capabilities and features of persons (cf. p. 31) can be reduced to, or be derived from, the physical, physiological and biological functioning of their bodies or brain, and granted that no science about anything could be developed without persons having these psychological capabilities and their logical properties, then I think we shall also have to concede that no consistent science of *Psychology* about persons could be developed, which ignored or discarded these capabilities and properties of persons. Or, to put it the other way round: No consistent psychological science about persons can be developed, which does not accept and presuppose that these psychological capabilities and their logical properties must be *fundamental* and *constitutive* for persons. If so, we can begin to see an outline of what may be a generally accepted basis of concepts and notions to describe and delimit the subject area and phenomena which fall within a scientific psychology, as well as a set of necessary and fundamental epistemological assumptions about these phenomena on which this science must build.

2.2 Basic assumptions for a science of Psychology

Now, further analysis of the constitutive psychological capabilities and properties of persons (cf. p. 31) reveals that it is impossible to talk sensibly about any of them without or independently of referring to the others. Thus, it is not possible to talk about persons carrying out deliberate acts on thing in reality, without referring to persons having knowledge of the reality and things on which they may carry out acts. Nor, conversely, is it possible to talk about this knowledge of persons, without referring to persons who may carry out deliberate acts on things in a reality in which they find themselves, thereby providing themselves opportunities of observing and thus of acquiring knowledge of them. But neither is it possible to talk about language, and the use of language by persons to communicate about things in reality of which they have knowledge and may act with, without referring to reality and things existing in reality, nor therefore without presupposing that these things exist as things with which they may act, have knowledge of, and may use language to talk correctly about. Indeed, none of the notions used to characterize our cognition, description and actions with things in reality, have well defined meanings independently of reference to well defined notions of reality and these things. If so, a logical relation exists between notions such as 'persons', 'knowledge', 'action', 'possibilities of action', 'intentions', 'language', 'propositions', 'true and false', 'reality' and 'things'. It is not difficult to show that this is the case, nor to argue, therefore, that the existence of a logical relation between such notions must be among the basic assumptions of a science of Psychology.

The necessity of this assumption and what follows from it may become immediately clear if we consider the impossibility of talking about persons (for example, myself and the readers of this book), without referring to concrete situations in physical, material reality in which they find themselves, and of which they are part. Nor would it be possible to talk about persons, who can act with material things and co-act with other persons in physical and social reality, without assuming that both the things with which they may act, and other persons with whom they may co-act, exist in reality. Conversely, we cannot say without talking nonsense that although it is correct that persons exist, who may act with material things and co-act with other persons, this does not imply that material objects or other persons exist in reality with whom they may act or co-act; nor say that we cannot be *certain* that material things or other persons exist with whom they may act or co-act. But neither would it make sense to say that persons are "the sorts of things" which may find themselves at particular places and times in reality relative to other things and person with whom they may act, without referring to persons having material, physical bodies. To paraphrase Zinkernagel (1988), it would be nonsense to say about a person that he went to a party, but he forgot to bring his body.

And if, as already argued, it is the case that the action of persons (in contrast to the causally determined reaction or responses to things and events of physical or biological systems), logically implies that persons have possibilities of action, and also that to carry out an act implies having made a *choice* between different possibilities of action; and if, furthermore, to act with or on things may be understood logically as that of changing ones future possibilities of action with things or other persons with the purpose to bring about some particular future possibilities of action, then we cannot talk about the action of persons without referring to the *intentions* that persons may have for making those particular choices of action, nor without referring to the *knowledge* of persons about the things and other persons existing in the situations in which they find themselves and carry out action; persons, moreover who, at least to some extent, are able to *anticipate* the future consequences of their action.

Hence, we cannot have any well defined meanings of any of the concepts we use to characterize persons and their acts without or independently of referring to well defined meanings of a whole series of other concepts and notions, among them notions such as 'possibilities of actions', 'choices between possibilities of action', 'intentions', 'knowledge', 'reality', 'concrete situations', 'material things' and 'bodies'. That is, independently of referring to things and properties, which concern *both* the action of persons *and* the reality in which they act and of which, therefore, they have knowledge.

So, although we have to assume that reality and things in reality exist just as independently of the cognition, descriptions and actions of persons, as the bodies of persons exist independently of the things with which they may act, we cannot talk meaningfully about reality and these things, i.e. what they *are* and are *not*, what is the case about them and what not, or what acts we may carry out with them and what not, etc., *without referring to persons, who have knowledge about reality and those things*, nor, therefore, *without assuming that reality exists as something which persons may have knowledge about*. If so, it has to be assumed that a *necessary relation* exists between the knowledge of persons about reality, and the reality they have knowledge about.[3] (For extended arguments for this point and consequences thereof I refer the reader to Chapter 7.)

A similar analysis of the fact that persons have developed natural languages in which they may put forward propositions about things in reality will show a similar necessary relation between language and that which language is about, as well as a logical relation between concepts used to characterize language and that which expressions and assertions are about. Thus, I think we shall have to agree, first, that it must be a defining property of a natural language that it may be used to put forward propositions *about* something, and a defining property of propositions that they may be *true* or *false*. We cannot sensibly say: "Here we have a language, a natural language, which may be used to put forward propositions, but propositions in this language are not *about* anything, i.e. nothing exists to which these propositions may *refer*". Nor can we sensibly say that in this language propositions may be put forward which are *true* or *false*, however, nothing exists about which they may be true or false. But neither can we talk about propositions put forward in language by its users about reality, themselves, and other things existing in the concrete situations in which they find themselves, independently of presupposing that language users have *knowledge* about reality, themselves and those things and situations. We cannot say without talking nonsense, "here we have a language in which language users may put forward propositions which may be true or false about reality, themselves, other things etc., but they do not have - or we cannot be certain that they have - any knowledge about the things they may talk correctly about". That is, we cannot talk about language and true or false assertions put forward in language about anything independently of talking about *knowledge*.

It could of course be objected that it is a defining property of a natural language that it may be used to talk about language and the conditions for talking correctly about that which we speak. However, this does not warrant the assumption that we can use language to talk correctly about ourselves, other people or things existing in the situations in which we find ourselves in reality - including the situations in which we use language to talk about language. Now, so to say would clearly be absurd. For one thing, how could it possibly be maintained and assumed that we can use language to talk correctly about language and the conditions for using language correctly, without assuming that we can use language to talk correctly about that which

[3] This assumption does not imply that our cognition is infallible or that we do not make mistakes or may be wrong. We know from experience that we may be wrong and that we may make mistaken – indeed we are even able to determine how we are wrong or mistaken. However, as Wittgenstein has pointed out in "On Certainty" (1969), we cannot talk consistently or in a well defined way about mistakes or faults of knowledge, nor determine faults or mistakes of knowledge, without presupposing that we have some true knowledge of that about which we may be wrong or mistaken.

these conditions concern, *in casu* reality, ourselves and things in reality? For another, no language exists without language users, nor without concepts of such matters as space and time by which they may locate and refer to themselves, other persons and things existing in material reality of which they themselves are part. (Arguments for this and the following points about language and its use are presented in Chapter 5.)

The assumption of the necessary relation between language and what language is about does not imply that any proposition put forward in language about reality, ourselves and other things, is true or correct - on the contrary, it may be incorrect or false. However, it is an assumption to the effect that we cannot determine the correctness or truth of any *concrete* proposition without presupposing that, *generally*, language may be used to put forward propositions, which may be true or false about that to which they refer. Indeed, it is because of this presupposed *general* correctness of language that the truth of any concrete proposition or assertions may be determined. So, although we may give reasons for what I shall call *the assumption of the general correctness of language*, we cannot prove this assumption. We cannot do so any more than we can prove the correctness of any particular propositions put forward in language *without* referring to other propositions or assertions which are presupposed to be correct - and thus without referring to a language in which true propositions may be made. In other words, any attempts at proving the general correctness of language would be circular in that it would have to presuppose what it were supposed to prove. Conversely, attempts to doubt or disprove this assumption would be to doubt or disprove the very condition for putting forward the doubt or disproof. Hence, we would contradict ourselves and end in absurdities, i.e. the absurdity of assuming that we can use language to doubt or disprove that we can say anything about anything, which is true.

The set of assumptions about the necessary relation between, on the one hand, our knowledge, action and language and, on the other, the reality of which we have knowledge, may act in, and put forward propositions which may be true or false, are *compelling* in the sense that *they have to be presupposed and taken for granted* and be the point of departure for any further investigation of our cognition, language and action, *as well* as that which our cognition, language and action concern. Since these assumptions are about the necessary relation which exists between knowledge, language and reality, they cannot be characterized as assumptions which concern knowledge and language *in contradistinction* to reality, nor conversely, as assumptions which concern reality in contradistinction to knowledge and language. Neither does *knowledge* about these assumptions - and thus knowledge about the conditions for language and knowledge - concern cognition and language in contradistinction to reality or that which we may cognize and use language to put forward propositions about.

The implications of the assumptions thus far presented may be summed up in this way: Any theory of knowledge as well as any general theory of language and action (if indeed such theories were possible), would have to be just as much theories about *reality*, since we cannot talk consistently about knowledge, language and action without or independently of referring to that which our knowledge, language and action concern, *in casu* reality and things in reality.

The important consequence of this is that we do not *need* a theory of how statements put forward in language about reality correspond to or match with reality, nor a theory of how knowledge about things or facts in reality correspond to or match those things or facts. It *cannot* be justified or explained, but on the contrary, it has to be taken for granted and be presupposed for any of our notions of 'knowledge', 'descriptions', 'reality' and 'facts of reality' to be well defined. Nor can we *question* or *doubt* this correspondence without questioning or doubting that we have knowledge about and may use language to talk correctly about reality and things in reality – and thus in effect, without questioning or doubting the very condition for formulating the questioning or doubt.

Neither can we *justify* or *explain* this correspondence, i.e. justify or explain that the language we speak may be used to put forward true – and false – statements about things or facts in reality, and thus that we have a notion of 'truth'. We cannot do so, for example, by investigating the things or facts of reality of which true and false statements may be put forward. For, any such investigations would require that we could identify and determine the things to be investigated, and thus knew what they *are* or are *not*, i.e. what is the case – or true – and what is not the case – or false – about them, and *therefore*, that we already had a notion of 'truth' *prior* to those investigations.[4] But nor can we justify or explain that we have notions of true and false by referring to *conventions* made by language users as to what may count as true or false about things - since we cannot begin to apply those conventions and assertions to anything, without presupposing that we can identify and determine the things to which they are supposed to apply - and thus that we necessarily have notions of true and false *prior* to the formulation of any such convention.

What we *can* say, however, is that to acquire a language and to be a user of language in which one may talk about reality and things in reality, and also communicate to others what one knows about reality and these things, presupposes that one has a notion of 'truth' which is such that what is true

[4] Cf. the attempts by logical positivism thus to make "truth" a function of what we can talk correctly about. I shall have more to say about this in Chapter 5.

about one's cognition of reality would also be true for others - could they know what oneself knows and cognize what one cognizes. Thus, it is not because persons together may come to agree and make conventions about what is true and false about the things they cognize, experience and describe that they have or come to acquire a notion of truth. On the contrary, it is because persons have a notion of truth of which it is presupposed that what is true for oneself will also be true for others that they can come to agree and begin to make conventions about the truth of anything. Hence, the notion of 'truth' both implies and presupposes a notion of 'others', or 'other persons'. In this sense, and in this sense *only*, the notion of 'truth' is fundamentally social. (The arguments for this point and consequences of it are further developed in Chapter 20, Part IV).

2.3 Principles of Cognition, Language and Action

The assumptions of the inter-relatedness between language and reality, and between knowledge and reality, are *necessary* in the same sense and for the same reasons as are the principles of formal logic. Thus, it can be shown that attempts to prove or doubt or disprove these assumptions cannot be made without conceding them. For this reason I have formulated them in a principle, which I call (1) the *principle of the general correctness of language and knowledge*. It is a principle which implies that to be a person and a language user is to have knowledge of and a language in which one may put forward propositions which are true - and false - about reality, oneself and the situations in which one finds oneself in reality.

Furthermore, the assumption of a logical relation between the concepts we use to characterize, on the on hand, knowledge, language and action, and on the other, the reality and things that our knowledge, language and action concern, I formulate in a principle which I call (2) The *principle of the logical relation between concepts of language, knowledge, action and reality*. It is a principle which implies that we cannot have well defined meanings of the concepts we use to characterize the cognition, use of language and action of persons and the reality that this cognition, language and action concerns – independently of or without referring to well defined meanings of the others. And it is a principle which implies, therefore, that none of these notions may be deduced from or reduced to any of the others, nor may that to which any of them refer be reduced to that to which any of the others refer. (For extended arguments to this effect, I refer the reader to Chapter 12.)

The fact that persons may act intentionally and, in so doing, may create new situations with new possibilities of action and observation, and thus may come to know more about the *same* things in different situations, means that no description of a thing put forward in any particular situation can be identical with the thing. Indeed, to say that some particular descriptions of things and their constituents, put forward under some particular conditions of observations and description, concern things which in other situations may be correctly described in different ways, inevitably implies that such particular descriptions rely on other descriptions of the same things. This dependency of descriptions of things on other descriptions of the same things, I have formulated in a principle of identity which says (3) we *cannot say anything about anything without being able to say more about the same.* It is a principle which states that the notion of 'the same thing', and hence of 'same' and of 'thing', necessarily implies that identifying a thing as the *same,* is to identify it as that which in different situations, e.g. of observations, description and cognition, may be differently observed, cognized and described. Thus, it is not because a thing retains the same, unchanged 'essence' or 'substance' in different situations in which it may be correctly cognized and described in different ways that we have a notion of 'the same thing'. Rather, a thing may be identified as the same thing precisely in virtue of being the thing, which in different situations may be correctly cognized and described in different ways. (Arguments for this principle and its consequences are presented in Chapter 16.)

Lastly, the fact that to acquire a language and to be a user of language presupposes a notion of truth which is such that what is true for oneself is also true for others, I formulate in (4) *The principle of the logical relation between the notion of truth and the notion of 'others', or 'other persons'.*

Let me finish this introduction to arguments, assumptions and principles developed in this book, by contending that for the same reasons that none of the psychological capabilities and phenomena discussed above can be accounted for, nor have well defined meanings independently of referring to the others, it would seem highly unlikely that any of the physical, physiological or biological processes and mechanisms of the brain underlying these capabilities and phenomena could be accounted for without referring to the others. Nevertheless, according to the newly revised version of the so called Cognitive Revolution, human cognition, language and action may be accounted for and explained in terms of "specialized neural automata" of the brain which, according to this version, works as a multi-modular computational system. Among the apparently endless numbers of modules or "cognitive instincts" of the brain (Tooby and Cosmides, 1996, p. vii) we find a module which "invents" mental properties such as colours, which we "project onto our percepts of physically colorless objects". And we find specialized systems for grammar induction, for dead reckoning, for construing objects - each of which "has its own agenda and imposes its own exotic organisation on different fragments of the world". We also find a "language-acquisition" device whose primitives include elements such as "noun phrase" and "verb phrase"; and we find a rigid object mechanism that "construes" the world in terms of "solid objects, relative location and mutual exclusivity within volume boundaries" - just as we may find yet other modules containing "social-exchange algorithms" that define a social world of agents, benefits, requirements, contingency and cheating, and a "theory of mind module" that "speaks of agents, beliefs, and desires". Now, in order to solve the so called "characteristic domain of problems" with which this multi-modular brain system may be faced, its modules, so it is claimed, are designed to "interpret the world" in its own "*pre-existing terms and framework*", and to operate "*primarily and solely*" with its own specialized "*lexicons*", that is, sets of procedures, formats and representational primitives "closely tailored to the demands of its targeted family of problems" (ibid. p. xi - xiv). - Just name a problem you think may fall within this family, and you may rest assured that there will be a pre-programmed module in your brain designed to solve it.

Well, for a start, it would seem that difficulties are bound to arise should attempts be made to account for the working of, say, the specialized language module containing concepts of 'nouns' independently of the working, say, of the "rigid object" module, which "construes" solid objects, and thus without or independently of notions of 'objects' to which these nouns may referred; likewise it would seem difficult to account for the concepts of 'verbs' contained in the same language module independently of some other module containing a "lexicon" of notions of what objects may do - or what we can do to them - to which these verbs may refer. And it would not only seem *difficult*, but downright *impossible* to explain how human beings may act and reflect on their possibilities of action with things and co-act with other persons in material and social reality, and in so doing create not only different cultures and socio-economic systems, but also new situations in which new and more knowledge may be acquired about themselves, other persons and things in the world - in everyday as well as in scientific situations - *if* it were really the case that the functioning of our minds relied on a brain with universal, pre-programmed specialized automata, each having a fixed "lexicon" of procedures, formats and representational primitives "closely tailored to the demands of some *particular targeted family of problems*". Indeed, it would be impossible to explain or account for how human beings *differ* so fundamentally from mere biological organism and mechanical systems (e.g. computers), which do *not* find themselves in concrete situations in reality, and which are *not* self-designating, intentional agents to which the principles of language, cognition, action, identity and truth apply. So, unless Tooby and Cosmides and other like minded scientists assume about their own brain that it is equipped with a very special and immensely complex "module", which is *not* pre-programmed and fixed, they could not possibly deliver the following "manifesto" about the present and future state of psychology:

> "The realization that the human mind is densely multimodular has propelled modern psychology into a new theoretical landscape that is strikingly different from the standard empiricist approaches of the past, [by which they mean the first phase of the Cognitive Revolution and its model of the brain as a digital computer, my addition]. In consequence, the outlines of the psychological science of the coming century are getting clearer. In this new phase of the cognitive revolution, discovering and mapping the various functionally specialized modules of the human brain will be the primary activities. Even more fundamentally, psychologists are starting to put considerable effort into making their theories and findings consistent with the rest of the natural sciences, including developmental biology, biochemistry, physics, genetics, ecology, and evolutionary biology: Psychology is finally becoming a genuine natural science. [...] As the operation of the genetic code is tracked through molecular biology and cell biology to developmental neurobiology, the processes that organize the developing nervous system are becoming increasingly intelligible. These developmental programs were "designed" by selection to build a physical structure that realizes certain functional informational relationships. Discovering what these relationships are is the province of still other fields, such as evolutionary biology and cognitive psychology." (ibid., p. xiv - xv.)

Apart from the fact that what Tooby and Cosmides claim to be both known, clear and increasingly intelligible, is not at all known, clear or intelligible,[5] the impossibility of this "reductive naturalization" of psychology and the challenge with which this impossibility leaves psychology, is no better summed up than in these words by Nagel:

> "The limits of the classical methods of objective physical science are not surprising, since those methods were developed to deal with a definite, though universal, type of subject matter. If we are to take the next great step, to a truly theoretical understanding of the mental, we must proceed by regarding this limitation as a challenge to develop a new form of understanding appropriate to a subject whose exclusion from physical science was essential to its progress. [...] But until we discover a way to stand theoretically astride the boundary between objective spatiotemporal physical reality and the subjective content of experience, we cannot claim to be in possession of the basic intellectual tools

[5] The reader may think that I have chosen an extreme example in the passage just quoted, but the view it represents is by no means unique (see e.g. the collection edited by Broadbent, 1993).

needed for a comprehensive understanding of conscious life." (Nagel, 1994, p. 67 - 68).

By attempting to determine some of the principles and logical conditions for the cognition, language and action of persons, it is the intention of this book to provide the epistemological foundation on which such intellectual tools may be developed. This attempt will go hand in hand with discussions and refutations of the Naturalist and Constructivist position and the ever-more unintelligible and nonsensical theories and "models of mind" developed by its adherents among philosophers, who are queuing up to offer "philosophical credibility" to the fallacious and inadequate assumptions on which are based so much of the research and theorizing of current academic psychology.

Chapter 3

Problems of explanations and theories of visual perception

3.1 Introduction

For millennia it has been one of the main problems within philosophy to find satisfactory answers to the questions of whether our knowledge about reality is true or in accordance with reality "itself" - and if so, how it comes about that we acquire this knowledge. Ever since the development of sensory physiology from which perception psychology evolved during the last half of the 19th century, psychology has considered these questions a matter to be explained by empirical investigations. In current psychological research on visual perception this task is often formulated as the problem of explaining how the "visual system" detects and processes the information emitted from the physical world.

It will be argued here that attempts to explain how it comes about that we perceive the world as it is, or the way we do, invariably rest on the Cartesian assumption of a polar opposition between an external, "physical" reality and an internal, "mental" (or experienced) reality, which are *independently determinable*. According to this assumption it must somehow be possible to give an account of how stimuli, emitted from the physical world and "impinging" on our sensory systems, by some causal process give rise to perception of the external physical world. Furthermore, granted a complete account may be given of how the visual system detects and its "machinery" processes the information inherent in the stimuli, then a complete (or general) causal explanation may be given of how our visual perception of the world and things in the world come about - thus overcoming the Cartesian Mind-Matter division and "the problem of perception" to which it gives rise.

Examples chosen for a discussion of this basic assumption are J. J. Gibson's and D. Marr's attempts, within this dualistic framework, to develop "complete" or general theories of perception. In the course of this discussion I shall argue that the assumption of a polar opposition between perception and reality inherent in such attempts is untenable. Thus, this assumption wrongly implies that it is possible to talk consistently about *reality* and things existing in reality, without referring to our perception of reality and these thing, and also that it is possible to talk consistently about our *perception* of reality and things in reality, without or independently of referring to the things we perceive. Against this assumption I shall argue that we cannot give any accounts of or describe *what* we see (i.e. of our perception of things), independently of or without at the same describing it as *that* which we see (i.e. the things), nor talk correctly about reality and the things which exist in reality, without or independently of presupposing that reality and these things exist as something we may perceive, and about which we have knowledge and may talk correctly about.

Hence, in contrast to the assumption of the traditional Mind-Matter dualism, I shall argue that an inter-dependency or *necessary relation* exists between reality and our perception of reality, which makes impossible any *general* non-circular causal theories of perception. How and why this is so, should become apparent in the analysis (in Chapter 3) of Gibson's theory of perception and Marr's model of vision. Equally importantly, I shall argue (in Chapter 4) why, for similar reasons, perception cannot be explained adequately (let alone exhaustively) in terms of causal functional theories about the processes in the "machinery" of the visual system and their causal link to stimuli in the world.

First, a word about my choice of Gibson's and Marr's theories of perception to demonstrate problems and inconsistencies to which the Cartesian Mind-Matter dichotomy gives rise. It could well be argued that their theories, first presented some twenty ears ago, are rather dated, and that I should have chosen more recent examples. My reason for choosing Gibson's and Marr's theories, however, is that they are still by far the most thoroughly developed attempts to provide *general* or *complete* theories and explanations of visual perception. Secondly, both theories are well known to most psychological readers – and still widely discussed. Thirdly, and not least, I hope it will be clear from my analysis of Gibson's and Marr's theories that the fundamental assumptions behind more recent psychological theories of perception – including computational *and* connectionist models developed within robotics – are not different from those underlying Gibson's and Marr's theories, but just "more of the same" – and hence I hope it will be clear that the general results of my analysis apply equally well to these other more recent attempts to explain how perception of the world comes about.

It is necessary to stress at the outset that the argument for the existence of an inter-dependency or *necessary relation* between reality and our perception of it, is *not* that we cannot distinguish *perceiving an object* from *the object* being perceived. Nor does this argument imply that reality does not exist independently of being perceived. We can certainly distinguish between the act of visually experiencing an object "out there" and the object in question - just as we can distinguish between remembering, imagining, thinking of an object and the object itself, or distinguish between a linguistic description of an object and the object being described. Indeed, there are obvious reasons (to which I shall come back in what follows) why we have to presuppose that we can make such distinctions, and obvious that things we perceive exist just as independently of our perception of them as our linguistic descriptions of the same things exist independently and as something different from linguistic descriptions of them.

It is worth stressing, furthermore, that my discussion of Gibson's and Marr's works primarily concerns their attempts to develop *general* and causal functional theories of vision, and the assumptions that such attempts entail. It is not to dispute the results or data of their research. Rather, I shall try to show that the procedures whereby these results are obtained preclude the *epistemological* interpretation suggested by Gibson and Marr. Thus, I shall argue, that the very implementation and experimental tests of their theories rest on conditions and epistemological presuppositions which are in direct conflict with their interpretations. These condition and epistemological presuppositions, which apply *in general* for implementing and testing account and theories of the processes of perception, I shall present in Chapter 4 in connection with an analysis of the requirement for carrying out investigations and experiments on visual perception. In this chapter, I shall also try to point out the general epistemological consequences and implications of these conditions and presuppositions, and show that they apply equally well for research and theorizing within cognition and language. Lastly, I shall attempt to give an outline of the kind of theories we *can* develop, and questions we *can* ask within perceptions psychology - and hope to answer by empirical research

But first, a short presentation of the origin of the causal dualistic approach to perception and some of the problems it has given and still gives rise to.[1]

[1] The main points and arguments against dualistic, causal theories of perception presented in this and the next chapter was first published in English in Praetorius, 1978, and in a paper read at The Boston Colloquium for the Philosophy of Science, May 1981 (Praetorius, 1982).

3.2 Mind-Body dualism

Mind-body dualism was Descartes' solution of the problems which arose at a time when it seemed that a mathematical, quantitative description of the world was sufficient to describe the behaviour of objects in material reality. It seemed that knowledge of an object's geometrical form, motion, mass, and solidity was all that was necessary to describe its behaviour relative to other objects. It was natural to suppose, therefore, that these properties were the only ones that objects "really" possessed, and that the other characteristics of the objects, e.g. their colour, smell, warmth, coldness, taste, etc., were only apparent - and in some way or other dependent on ourselves. It had to be admitted, however, that in a number of respects our bodies are also entities and mechanisms resembling lifeless, material objects - but if we ourselves are a kind of mechanism, how then could any event not describable in physical terms take place in us?

Descartes "solved" these problems by modernising the Christian concept of the Soul, making it a "realm" for all the faculties, qualities, and phenomena for which there was no longer room in the mechanistic world-view. The soul became the domain to which experiences, sensing, feeling, free-will and thought belonged (Descartes (1637/1967).

Locke adopted essentially the same view of the soul, and his belief in the quantitative, mechanistic nature of material reality was strengthened by Newton's impressive theory. He laid the foundation, however, of a revision of Cartesian and Newtonian epistemology, according to which our knowledge of the physical world is ultimately based on innate insight into its geometrical structure and physical laws, supplemented by experience gained through experiments. Locke criticised the assumption that knowledge of the world could be based on innate insights or knowledge about the world: The only knowledge which the human Mind or consciousness obtains, he said, must be knowledge derived directly or indirectly from sense-experience. Moreover, at birth the Mind is a *tabula rasa*, on which only the senses can write. Our knowledge of the external world must therefore be built up by means of processing sense-impressions. No other source of knowledge about reality but sense-experience of it is to be found (Locke, 1690/1961).

In Descartes' and Locke's views we have the fundamental features of the theories of perception and cognition which still dominate psychology. We still assume that perception and cognition of the world around us can be accounted for in terms of causal processes which can be roughly described as starting in a physically describable external world - a world that can be *correctly* described as it is *only* in terms of physical concepts and descriptions - and which terminate in something entirely different, i.e. a visual perception of the world, a phenomenon of consciousness, a "percept" or representation residing in our minds. However, the physical world can be described in a multitude of ways, even in strictly scientific physical terms, so one of the main problems for a psychology of perception and cognition in describing the various steps in the process between the physical world and our "percepts" has been to find the *adequate* description of the stimulus. That is, a description of the stimulus which our sense organs are capable of picking up and which "carries" information about reality. A related problem has been the problem of specifying how such sensory information is processed in our Central Nervous System (CNS), or in some fashion by the corresponding different "layers" of cognitive states.

The assumption that a causal account of perception may lead to an answer to the question of how it is that we perceive the world as it is, or the way we do, i.e. given that the various steps in the process have been adequately and exhaustively described, has not been without problems. Admittedly, we have a much more detailed picture of the intermediate stages of the processes than Locke had, but we are still struggling with the problems which for Descartes' and Locke's critics were the stumbling block, namely how physico-chemical processes in the organism, no matter of what type, can possibly give rise to the experiences of the everyday world of objects which occur in our consciousness.

Since the definition of stimuli by Müller in 1834 as any "force" capable of exiting the sensory systems it has been a problem, moreover, how stimuli can possibly refer to the physical world from which they are emitted. Thus, the description or definition of stimulus in purely physical terms as a quantity of physical energy (or "patterns" of such energy) seems to imply that stimulus has no significance by which to specify the world from which it originates. Packages or patterns of energy bear little resemblance to objects, space, persons, language or symbols - but nevertheless appear to be the only entities capable of exiting our receptors.

By stressing the necessity of a formal distinction between the "proximal stimulus" and the "distal stimulus", widely accepted by perception psychologists to day, Koffka (1935) contributed, perhaps more that any one else, to emphasising the paradox that the behaviour and perception of persons are obviously determined by distal stimuli, though our senses can in fact only be affected by proximal stimuli. The paradox is rendered even more intractable by the fact that the proximal stimulus for a given object is continuously changing as a result of, for example, the continuously changing position of the observer, so that one must assume that countless different stimulus patterns can cause or give rise to the same perception. Perception psychologists have fought an almost hopeless battle to construct theories explaining how different stimuli can cause the same percept, i.e. theories which may explain the phenomena of constancy.

3.3 Gibson's theory of perception.

It is exactly on this point, J. J. Gibson suggested a solution in his theory of perception. His optimistic view was that the invariance in our perception of reality somehow must "correspond" to the existence of *invariant components* in the confused multiplicity of proximal stimuli. That such an assumption had not previously suggested itself to perception psychologists, according to Gibson (1959), was chiefly due to the tendency to consider and classify stimuli exclusively on the basis of the sense organs and the type of energy which can affect the individual receptors, but never on the basis of the *relation between patterns* of physical stimuli and their source in the external world.

To Gibson the physical basis or stimulus for our visual perception is rectilinear propagation of light in the *Optic Array*, which is emitted from the surfaces of objects in the world and projected onto the retina in accordance with the laws of geometrical perspective. (Gibson, 1950, 1959, 1966). In accordance with the view that it is the conditions and relations governing normal everyday perception that perception psychology must concern if we are to gain any knowledge of how and why we perceive the world and objects as we do, it must, furthermore, be the geometrical projection on the retina of the objects *and their background* which must be the starting point of our analysis of the stimulation - and not the objects in isolation from their background, as was previously customary.

According to this intention, Gibson's procedure for determining the information available in the stimulus is as follows (Gibson, 1950, 1966): Here is a room with four walls, a table, a chair, three pictures on the walls, a carpet on the floor etc. An analysis of the optico-geometrical projection of this room, or of a number of such projections corresponding to the retinal image, now reveals that in the individual projection, as well as in a series of projections corresponding to the movement of the objects or the observer, there are many types of "gradients" and "high order variables", that is patterns of light the variations or changes of which can be described as constant or invariant. This suggests, according to Gibson, that in the optic array emitted from the surfaces of the objects and their background, invariant structures exist which correspond to *structures of the world*.[2] Most of the traditional distance cues, as well as the other cues of visual perception, says

[2] Different surfaces reflect light differently, for example a smooth surface such as a polished table top reflects light more uniformly than e.g. the texture of the floor covering (wool carpet, cork tiles etc.), which reflects light unevenly and in a characteristic way. If one looks across the floor to the edge of the room there will be systematic changes in the pattern of reflected light along the line of sight from the eye to the wall - yielding what Gibson calls a texture gradient.

Gibson, can be considered as invariances or constant rates of changes in the structures of the optic array. In this way the optic array provides invariant information which *specifies* the shape, sizes and layout of objects in space in an unambiguous way. That is, information in terms of invariances and constant changes is there in the optic array to be picked up by the perceiver or his perceptual system, which specify the properties and changes of properties of objects.

According to Gibson's theory of perception, based on the analysis of invariant structures to be found in the optic array, correlates with the structures of the world exist which are *sufficient* to render our perception of the world and objects possible. Moreover, the information in the optic array can be "picked up" directly by the perceiver - or rather his sensory system (Gibson, 1979). To Gibson the answer to the classical philosophical question of why we see the world and objects as we do, and with the properties it has, is that correlates of structures in the world are to be found in the stimulus. The stimulus specifies the world - and as perception is directly related to the stimulus, the world is specified in our perception. Gibson also describes this relation between the world, the stimulus and perception by saying that stimulus is a *function* of the world, and that perception is a *function* of the stimulus. (Gibson, 1959, p. 459.) Thus, the objectivity and veridicality of perception is guaranteed by these very relations - for when stimulus is a function of the world, and perception is a function of the stimulus, then perception will be unambiguously related to the world. In Gibson's own words:

> "The objectivity of our experience is not a paradox of philosophy, but a fact of stimulation. We do not have to learn that things are external, solid, stable, rigid, and spaced about the environment, for these qualities may be traced to retinal images or to reciprocal visual-postural processes. "(Gibson 1950, p. 186.)

But does Gibson's model of perception which can be described by the symbols $W \rightarrow S \rightarrow E \rightarrow P$ (W = world, S = stimulus, E = excitation, P = perception) really explain how perception of the world comes about? That is, is it really the case that by giving an unambiguous description of the relation between structures in the world and structures of light in the stimulus, a *sufficient* stimulus description has been provided which for the perceiver unambiguously *specifies* the world with objects and their properties? How, for example, can we be certain that a description of an optico-geometrical projection with all its invariance is an *adequate* stimulus for the organism or its sensory system? It seems to me that by re-describing the world in terms of a stimulus pattern of high order variables and gradients in the light emitted from the world and projected on to the retina, Gibson has only succeeded in moving the original question one step nearer, so to speak, to the organism and its sensory system. Thus, the question of how it is that we can perceive the objects of reality as they are, has similarly been transformed into the question: how is it that the sensory system can pick up the information, gradients, high order variables etc. in the optic array?

Now, Gibson answers this question by a postulate, namely that the sensory system in fact has the required capability to do so. But what compelling reasons does he have to stop what is beginning to look like a regress of questions at this stage? If he does not have any, he might as well have answered the first question by a postulate, namely that we see the world as we do, or as it is, because we have the required capabilities to do so.

Never mind, let us with Gibson assume that the sensory system is capable of picking up the information in the optic array, which his analysis has revealed. To do so, however, would leave us with yet another problem. Thus, we could well ask how the sensory system can determine that the information in the stimulus, as described by Gibson, is only information about gradients and high order variables in a *projection* of the world, and not the world itself? For the sensory system or perceiver to be able to infer the existence of the world from the stimulus thus determined, the sensory system or perceiver would have to be able to recognize the *true* nature of the stimulus. The question then becomes: how is it that the sensory system or the perceiver comes to determine the true nature of the stimulus as invariant structures in the *optic array*, which correspond to *structures* of the world, and which *specify* the world? This question needs to be answered, if we can agree that optico-geometrical patterns of light in the optic array corresponding to the structures of the shape of a chair is not a *chair*; nor does a description of such optico-geometrical information in the optic array amount to a description of a *perception* of a chair.

Now, in order for the perceiver or sensory system to recognise the stimulus for what it is and, furthermore, to make sensible use of and inferences from the structures in the stimulus to the structures of the world, would require not only that the perceiver, or his sensory system, had knowledge about the true nature of the stimulus, but *also* that it had knowledge about the world itself. Thus, the answer to the question by postulating capabilities of the perceiver or his sensory system, is an answer and a postulate to the effect that the perceiver is capable of picking up information in the stimulus and of interpreting it as information about the world – because he or his sensory system has *the required knowledge about the world to do so.*

This is of course tantamount to saying that in his explanation of how it comes about that we perceive the world as it is, and thus acquire knowledge about the world, Gibson is presupposing that which he set out to explain. The problem with Gibson's theory and its attempt at explaining perception by way of functional relations between perception and the world, may be illustrated this way. On the one hand we have that

stimulus is a function of the world (1)

and, on the other hand, that

perception is a function of stimulus (2)

From this it follows by simple deduction, according to Gibson, that

perception is a function of the world (3)

Now, if perception of the world is a function of the world, we may as well eliminate the troublesome intermediate link of the stimulus. In all our further speculations we may answer the question of why we see the world and objects as we do simply by saying that it is because the world and objects *are* as we see them. Why do we see a triangle as a triangle? Because it *is* a triangle.

Obviously, Gibson intends to say more than just that. In his theory of vision, perception of the world is a function of the world *in virtue of* being the product of processes by which information emitted from the world, which specifies the world, is picked up by the perceiver or his sensory system. However, granted that his theory is thought to have universal validity, it must, of course, also apply to his own perception, and thus to his own description of the world with rooms, tables and chairs, pictures on the walls etc. This being the case, serious, though classical, difficulties in his theory will arise. For, then we have that the description of the world, which Gibson himself uses as his point of departure, and on which rests, necessarily, his technical description and determination of the optico-geometrical patterns in optic array corresponding to structures of things in the world, can only be the world *as perceived*, or a "percept" of the world. This may be expressed thus:

the world is a function of perception (4)

So, now we have *both* that

perception is a function of the world (3)

and that

the world is a function of perception (4)

from which we have by a simple reduction that

perception is a function of perception (5)

That is, we end in a circle into which, I believe, we will always fall if we try to construct causal or functional theories of perception, which rest on the assumption of a dualistic or polar opposition between perception of the world and the world itself.

Gibson's theory of perception was thought to provide a guarantee of the objectivity and veridicality of our perceptions. And, we may ask, how could it be otherwise, how could there be any question of error, when our perception is a function of that which we perceive? (See also Hamlyn, 1957, for a similar argument.)[3]

The reason why in theories of perception like Gibson's we have to presuppose that which we want to explain, becomes apparent when we consider how the various stages in the chain of events between W → S → E → P are determined. Thus, *W* is described in terms of walls, rooms, pictures on the walls, chairs, tables etc. *S* is described in terms of geometrical perspectives, i.e. it is a description of W such as it may be described in a plan-geometrical description. That is, the description of S may be considered just *another* way of describing W, i.e. a way of technically *representing* particular features of W, which has already been described in ordinary everyday terms. Indeed, the description of S *rests* on and *presupposes* this ordinary everyday description of W. In this sense the description of S rests on an abstraction, i.e. in the sense that it presupposes another description of the world. Now, let us with Gibson disregard the next stages in the process, and go directly to *P*. How is P described? P is described in terms of walls, rooms, chairs, tables etc., so

[3] As an experimenter trying to test his hypothesis, Gibson must almost of necessity find himself in the position of having to disavow his own theoretical postulate; in fact he cannot set up an experiment without doing so. When, for example, he wishes to test the hypothesis that certain gradients or high order variables are the gradients or high order variables we use when we perceive a tunnel (Gibson *et al.*, 1955), he cannot, of course, do so by showing the subject a real tunnel. He must construct an artificial tunnel, a so called "optical tunnel", which is not a tunnel at all, but is constructed in such a way that the gradients correspond to those of a real tunnel. An "optical tunnel" consists of sheets of plastic with a hole in each, black and white sheets being mounted alternately in parallel planes perpendicular to the line of sight of the subject. The gradients of such a tunnel correspond to those of a real tunnel, and the subject, when looking at the optical tunnel through a hole in a screen with one eye, reports seeing a tunnel. The hypothesis is verified, but the theoretical postulate is thereby disavowed. For if it were really the case, as Gibson claims, that the objectivity of our perceptions is guaranteed by the stimulus, the subject should certainly not have said that he saw a tunnel, but that he saw a series of black and white plastic sheets with a hole in them.

now we have the same circularity as before. For the description of the initial conditions of the events is identical with the description of the causal outcome of the event.

Now, it could be objected that the analysis so far of the problems to which Gibson's account of perception gives rise, does not take into account the differences between his earlier and later theories of direct perception (as presented in his book, *The Ecological Approach to Perception*, 1979). So, although the critical points made here about Gibson's concepts and his attempts to explain perception may well apply to his earlier work, they do not apply to his later theory of perception.

To this objection it must be said that a significant part of the conceptual apparatus developed by Gibson in his earlier work, e.g. "invariant optico-geometrical structure of the light specifying the layout of the world", reappears and plays a central role in his later theory of perceptions and its concepts about "information in the ambient optic array specifying objects, events, and their places in the environment". However, there are new concepts and explanatory terms as well, among them the key-concepts *affordances, information,* and *environment.* To begin with the latter, Gibson no longer talks about The World, but rather about the Environment (of an animal or a human being), which is defined as the "part" of reality to which the *sensory-behavioural system* of the perceiver relates, and towards which his or her acts are directed (e.g. acts that the perceiver *needs* to carry out in order to survive). Perception of this environment, Gibson stresses, is an *on-going* process, and what goes on in this process is the detection by the perceiver's sensory system of *information* in the ambient array in the form of *patterns of high-order structures* that specify the objects, places, and events of the perceiver's environment. In his later work as in his earlier, Gibson contends that rigorous geometrical analysis of these patters both provides an appropriate and solid foundation for determining the information available to the perceiver, or his or her sensory system. Indeed, to determine the visual information available to a perceiver, says Gibson, involves identifying a geometry which is appropriate for such a determination. (See e.g. Michaels and Carello, 1981; Reed and Jones, 1982; Reed, 1988).

The theory of *affordances* is a theory of what information informs about, and just like the notion "environment", affordances are determined relative to a perceiver and his sensory-behavioural system. Affordances are what the objects and their properties *mean* to the perceiver - whether they are good or bad, dangerous or harmless - as well as the functions they have or purposes they may serve. Furthermore, these affordances of objects and events are described in terms of the behaviour they *invite* or *permit* by virtue of their structure, composition, position, and so forth - *and* by virtue of the perceiver's sensory-behavioural capabilities.

The latest version of Gibson's theory of *direct perception* involves the claim, firstly, that affordance meaning of things and events in the perceiver's environment, as well as information about the place, shape and size of things, can be perceived directly, when information specifying them is available; i.e. this information requires no (further) processing by the perceiver or his or her perceptual system. Secondly, it involves the claim that the perceiver's perceptual system is sensitive (or has become sensitive in the course of the perceiver's ontogenetic development) to the structured energy, which invariantly specifies properties of the environment of significance for the perceiver. However, before venturing further into Gibson's theory of direct perception and affordances, let us first attempt a conceptual analysis of what visually perceiving a "thing" implies, expressed in ordinary everyday terms.

Part of visually perceiving a thing is to see that the thing has a particular shape, size and place in the environment. And part of having perceptually identified the thing as a particular thing is that the thing has particular functions and may serve particular purposes; that we may do various things with it, and that various things happen to the thing - or to ourselves - as a consequence of so doing. For example, the perceptual identification of the object lying on my desk as being my *cigarette lighter* implies that it is cylinder shaped and hollow, although I cannot see it all the way round or look through it. And it implies knowing that if pressed at a given point, a flame will be emitted from another part of it. Thus, to have perceptually identified the object as this particular object implies having identified a whole range of properties and possibilities of action that I may carry out with or upon it - some of the properties being directly visible and others being non-visible at the moment of observation - all of which may be said to form part of the meaning and implication of the perception of the object as being my cigarette lighter.

It is worth noticing at this point that a description of my *perception* of the thing would not be different from a description of the *thing*. Although I can certainly distinguish between seeing a thing and describing it, I cannot distinguish between *what* I see (i.e. the percept of the thing) without describing it as *that* which I see (i.e. the thing). (I shall come back to this point later.) And it is worth noticing, furthermore, that if part of perceiving what things are, implies perceiving what we may do to them - and knowing what happens to things or to ourselves as a consequence - then part of being someone having this knowledge, must be someone who knows that one may *designate* oneself as an agent, who may *initiate* actions with things, which lead to changes in one's future possibilities of action and observation. (For example, by pressing at some particular part of my cigarette lighter, I subsequently have the opportunity to observe a flame, and to light my cigarette.)

Now, according to Gibson's Ecological Theory of Perception, we directly perceive the properties of things and the possibilities of action we may have with or towards them - as well as the *projected* consequences of such acts - because information specifying these affordances is there in the ambient array to be picked up directly by the perceiver. It is this information about the affordances of things which is the *basis* for our "epistemic contact" with the environment, and which ensures that we acquire (true) knowledge of things and events in the world. Indeed, that this information may picked up *directly* by the perceiver's sensory system means not only that it needs no processing by his perceptual system, but also that no "capabilities" on the part of the perceiver are required - apart from the ability to move around in the environment with the right kind of sensory system.

However, I think that Gibson would have to agree that for his theory of direct perception and affordances to be a theory which "completely" explains how picking up information in the optical array gives rise to perception, would require that a description could be given of this information and these affordances, which corresponds completely to how we actually perceive things and events in the world, their meaning, functions, values, the possibilities of action we have with or towards them, and *all* the rest. But he would certainly also have to agree that no such description, in whatever technical terms it be expressed, can be made independently of, but, rather, must rely on a description of *what* we actually perceive and *how* we perceive it. That is, Gibson would have to agree that the set of explanatory terms used in his theory of that which *gives rise* to perception, logically depends on the set of terms used in the description of the *outcome* of the perceptual process. Indeed, without this inter-dependency of terms, we would not know to what the technical terms apply. And this brings us right back to the circularity discussed earlier.

In *The Ecological Approach to Perception* Gibson argues that if perceptual systems had evolved so as to detect the information specifying the environment and its affordances, then the very act of this detection *in and of itself* would constitute some kind of awareness of what it specifies. But this argument does not take his theory of direct perception out of the circle. For if it is a precondition for this theory of direct perception that information exists in the ambient array about objects and events, which itself is describable in terms of how we actually see them, then his argument amounts to saying that the detection of this information causes perceptual awareness of what we see - because our sensory system happens to have the sensitivity required to make us perceive objects and events as we see them. - And now I think it is obvious that Gibson's latest version of direct perception suffers from the same kind of problems as did his earlier attempts to solve the problem of how perception of things in the world comes about.

The difference - if there is any - is that in his ecological theory of perception the determination of the information in terms of a technical (optico-geometrical) description of the structured optic array is supplied with a description of the affordances of things in the environment, *which is indistinguishable from a normal everyday description of those same things and their properties* - and thus indistinguishable from a description of what we actually perceive, and what perceiving things implies and amounts to (cf. the conceptual analysis above). At best, his later version of direct perception is as circular as its predecessors. At worst it is a theory of perception which relegates *to things and events in the world* what rightly belong to the perceivers and their knowledge about the world and of themselves as agents, and their ability to initiate and evaluate their possibilities of action to things and events in the world.

Now, it may be objected that it is wrong to read into Gibson's later theory of perception an intention to *explain* perception, let alone solve the problems precipitated by the Cartesian Mind-Matter dichotomy. Rather, he intends to shows how these problems may be "resolved" or "overcome". However, this view is not shared by followers of Gibson who, supposedly, would understand the implication and meaning of Gibson's theory of direct perception and affordances. Edward S. Reed, for example, writes in his biography, *James J. Gibson and the Psychology of Perception* (1988):

> "Gibson was convinced that the theory of affordances, in conjunction with the concept of information, persistence, and change, would enable him to transcend the ancient debate between subjectivity and objectivity and to resolve the mind-body problem ... and (to expose and eliminate) many of the scientific and philosophical confusion of earlier theories." (ibid. p. 208, 282.)

However, in his *direct-perception-cum-affordances* theory of perception Gibson accomplishes this transcendence by allowing himself - quite uncharacteristically - to be rather imprecise in his use of terms. For example, he insists that affordances are "neither" subjective "nor" objective, but "both", i.e. they are a "mixture of subjective and objective".[4] Well, if we allow ourselves in this way to muddle up the concepts we are using, we may of course transcend the ancient debate between subjectivity and objectivity, indeed, we may quite easily overcome all sorts of problems, including the problems of perception to which the mind-body dichotomy gives rise.

[4] This "mixture", however, is the unavoidable effect of the inter-dependency between the set of explanatory terms used to describe, on the one hand, that which gives rise to perception and, on the other, the set of terms used to describe the outcome thereof.

To extract from Gibson's work its lasting contribution to perception psychology, one has, in my view, to distinguish his theory of perception from his experimental and methodological approach to studying *concrete* perceptual problems. This approach includes both a precise outline and re-formulation of the kind and range of perceptual problems which ought to be addressed by perception psychology, and methods by which such problems may be analysed; herein lies in my view the originality and usefulness of Gibson's work. However, his later version of a *general* theory to explain how perception comes about, is neither useful, nor does it "resolve" or "transcend" the problem arising from the mind-body dichotomy any better than any other theory of perception put forward within the framework of the assumed possibility of a polar opposition between "mind" and "reality". And it is this assumption and problem which has been the issue of my discussion.

But are the problems to which Gibson's theory give rise inescapable in theories which attempt to explain how perception of the world comes about? Researchers advocating a more recent alternative, the so called computational approach to visual perception, claim that their models not only avoid, but *solve* such problems; in effect, solve the "mind-body" problem, and thus the problem of circularity and infinite regression which hamper Gibson's theory. One of the most influential of these more recent theories of visual perception is the model proposed by David Marr (1982), which will be discussed next.

3.4 Marr's computational model of vision

As Gibson did, Marr insisted that a complete explanation of perception will have to be able to account for how people perceive realistic everyday situations during normal conditions of observation. What is needed is a general theory, i.e. a theory general enough to account not only for vision of drastically limited scenes, so often displayed in A.I. and contemporary psychological research on visual perception, but also for the infinitely more complex realistic perceptual situations and tasks encountered in normal everyday perception.

To Marr as to Gibson the stimulus or input to the visual system is the information in the optic array which, granted optimal and realistic conditions, is rich in information about structures corresponding to features to be found in the world. But unlike Gibson who assumes that this information can be picked up directly by the perceiver, Marr believes that several complicated stages of processing by the visual system are required in order to detect this information - and to organize it in such a way that *representations* of features of the surroundings of the system will occur. In Marr's own words:

"Our overall goal is to understand vision completely, that is, to understand how descriptions of the world may efficiently and reliably be obtained from images of it. The human system is a working example of a machine that can make such descriptions, and as we have seen, one of our aims is to understand it thoroughly, at all levels. What kind of information does the human visual system represent, what kind of computations does it perform to obtain this information, and why? How does it represent this information and how are the computations performed and with what algorithms? Once these questions have been answered, we can finally ask, How are these specific representations and algorithms implemented in neural machinery?" (Marr, 1982, s. 99.)

In the discussion of Marr's model of vision which follows, only one of the three levels of explanation which the model encompasses, i.e. the *computational level*, will be addressed.[5] At this level a *computational theory* is developed which describes the kind of computations necessary for solving the problems of computing information about light intensity changes in the retinal image into outputs of ever more reliable and relevant representations of the environment from which the light is emitted. In Marr's view, a whole sequence of processes in various modules of the sensory system are successively extracting visual information from one representation, organizing it, and making it explicit in another representation to be used by other modules to compute other representations, etc, until the final representation of the surroundings is reached - and appears as "pure" perception.

The sequence of representations computed are grouped into three broad classes according to the type of description they produce, namely *image description* (Primal Sketch), *scene surface description* (2½D sketch), and *volume description* (3D object models). The first two modules constitute the first stages in the visual process; they describe the so called low-level processing of image structures from light intensities, and are claimed to be purely data-driven, i.e. they are described by way of bottom-up processing, which does not rely on information or procedures from higher levels. The third module constitutes the third - and final - stage in vision in which information from the two low-level modules are processed, the output of which is perception of objects and scenes in the "outside" world.

In my discussion of Gibson's theory of perception presented earlier, I argued that by a re-description of the world in a plan projection of the world corresponding to the retinal image, Gibson had only managed to move "the problem of perception" a step nearer to the organism - i.e. to the level of the retinal receptors. At this stage the question would be similarly transformed into the question of "how it is that the receptors are capable of picking up

[5] The two other and contrasting levels are the *algorithmic level*, and the implementation or *hardware level* respectively.

and *interpreting* the structures in the retinal image as information of structures of the world which specifies or represents the world?". It could be said that where Gibson leaves of, Marr takes over. He believes that his computational account of the process of constructing representations will explain not only how *descriptions are produced of the scene which produced the image in the first place* - but that it will also explain "the problem of perception" as it has now been transformed. Marr writes:

> "The true heart of visual perception is the inference from the structure of an image about the structure of the real world outside. The theory of vision is exactly the theory of how to do this, and its central concern is with the physical constrains and assumptions that make this inference possible." (ibid., s. 68.)

To illustrate Marr's procedure, examples will be given in what follows of some of the structures being computed at the various stages, and how, according to Marr, inferences may be made about the structures of the real world - as well as examples of the assumptions which make these inferences possible. After the presentations of each of these examples, Marr's procedure and argument will be discussed.

3.4.1 The Primal Sketch

In an early stage of the production of the primal sketch, a representation is computed which describes and locates the light intensity changes which exist all over the (retinal) image. Moreover, different types of light intensity changes are identified - and given names such as "edge-points", "small blobs", "edge segments", and "terminations of edge segments", etc. The light intensity changes described in this representation (called the Raw Primal Sketch) reflect the intensity changes in the light arising at, say, the boundaries of objects (hence the label "edge points"), or shadows, or surface structure markings. However, the problem of determining *what*, precisely, they arise from can only be solved by careful analyses of the nature of the world being viewed and the image structures it is likely to give rise to. As Marr puts it,

> "As we have seen, in each case the surface structure is strictly underdetermined from the information in images alone, and the secret of formulating the processes accurately lies in discovering precisely what additional information can safely be assumed about the world that provides powerful enough constrains for the process to run." (ibid., p. 265.)

So, in order to solve the ambiguity problems and make the process of *coherent* surface structure computations run (i.e. computations of coherent structures from "dots", which corresponds to edges of objects, surfaces, etc.),

Marr builds into the algorithm *grouping rules* of combining edge-points, which are based on the assumption that, say, edges of objects are usually continuous. Therefore, it makes sense to choose edge point groupings which display maximum continuity, as well as proximity. Moreover, object edges or surface markings are usually similar in their characteristics, which justifies grouping together edge points of similar types. In other words, the computational problems of grouping together local dots that belong to the same "thing" is solved by implementing the Gestalt principles of *proximity, figural similarity, continuity, closure* etc., in the algorithm. That is, rules and assumptions which, like the Gestalt principles, are based on our knowledge of objects in the world and how we perceive these objects visually.

Discussion of example: It is the strategy and argument of Marr that in order for *proper representations of features and structures in the world to be computed*, his model has to be supplied with knowledge of what amounts to *proper* representations of such features - knowledge which in its turn rests on knowledge about *what the world looks like*.

Now, the full primal sketch which describes the location, boundaries, shapes, and texture properties of large sized regions of the image, does not have any notion of "things", only of "regions" or "lines". But, needless to say, a representation equipped with "edge-points", "regions" etc. could not have been computed from the light intensity changes on the retina unless the system had been provided with rules of how to group these light intensity changes in such a way that a *correlation* would occur between structures in the representation and features of physical objects. Therefore, it is misleading to claim that knowledge of such a correlation is *inferred* by the system. This would amount to claiming *both* that the necessary conditions for the system to compute proper representations in which structures correlate with features in the world (e.g. objects having delimited surfaces) are built-in rules which *ensure* this correlation, and *also* to hold that this correlation is *inferred* by the system.

3.4.2 The 2½D Sketch

The 2½D sketch serves the dual function of providing information about the layout of surfaces and, in so doing, implicitly solves the "segmentation" problem by making explicit the discontinuities between different surfaces and objects.

Now, no object was identified in the Primal Sketch, so *surfaces* cannot be identified in terms of the objects of which they are part. How, then, is the 2½D sketch constructed from the Primal sketch? Marr suggests that a variety of computations are performed by independent modules, and that their output descriptions are combined in the 2½D sketch, which acts as a short-term memory while the computations are carried out. *Discontinuity in depth* is assumed to be computed partly by *stereopsis mechanisms* and partly by the presence of occlusion. Occlusion may be specified by discontinuities in the pattern of motions present. And information about *surface orientation* is given by stereopsis, by surface and texture contours, and by an analysis of structure from motion.[6] In what follows I shall present Marr's account of how stereopsis mechanisms contributes to compute a 2½D sketch representation of distance and segmentation of surfaces.[7]

Because of the distance between the two eyes, slightly different images will be projected on the retina under normal viewing conditions. The image of the object which the eyes are focusing on lies at the same point on the two retinas. The images of other objects, however, would be displaced to a greater or lesser extent relative to one another if the images on the two retinas were superimposed. The different displacements are straightforwardly related to the distance of the objects from the one in focus, so those distances can be computed from the differences between the two images. However, the primal sketch does not identify *objects* in the scene being represented, so how to compute distance from the information in two slightly different primal sketches of the scene?

It is a well studied phenomenon that so called random-dot images presented as stereograms will produce stereoscopic identification of surfaces at different depths, that is, without any objects having been recognised (Julesz, 1971). Therefore, Marr argues, it must be possible to produce what is called a stereoscopic *depth map* from information available in the Primal sketch images presented in the two eyes - i.e. from images which do not identify any objects in the scene. Indeed, it is possible so to do. But, again, it cannot be done without providing the model with general assumptions of what the world looks like. This is due to the almost inexhaustible ambiguities involved in solving, first of all, the *alignment problem*, i.e. the problem of ensuring that the focus of the two images is on the same "fixation point", and that they have the correct orientation with respect to one another and the scene being viewed. And, secondly, it is due to the ambiguities involved in deciding which elements or primitives in one image are to be matched with what primitives or elements of the other, so that their disparities can be measured.

[6] Marr called the representation resulting from the computation of this information the 2½D sketch because the representation of the third dimension is incomplete. Even though information about *relative* distance of different surfaces is accurate, information about *absolute* distances of the surfaces from the viewer is not.

[7] Stereopsis is the effect that two images of a scene produced from slightly different angles, when shown each to the left and the right eye respectively, will merge into one perceived scene presenting surfaces at different (3D) depths.

The rule which helps to solve the latter ambiguity problem is based on the following general assumption: because the aim of the stereo combination process is to give information about entities in the scene, the primitives extracted from each of the images to be used for matching purposes must correspond to well-defined locations on the physical surfaces being imaged. From this a rule of binocular combination is arrived at which says that elements are allowed to form a potential match if they could have arisen from the same surface marking (i.e. edge points having roughly the same orientation, same scale or direction of intensity change etc.). Another rule is derived from the fact that a given point on a physical surface has a unique position in space at any one time. So, matches should be selected in such a way that any given matching primitive derived from one image should be allowed to participate in one and only one of the selected matches. To break this rule would be tantamount to allowing one and the same surface marking to be in two places at the same time.

A third example of a general assumption encoded as a rule in the stereo fusion algorithm, is that continuous surfaces make up a much greater proportion of the visual field than boundaries. Because matter is cohesive it appears in fairly large chunks, so surfaces of objects in our normal visual world are generally smooth. So, if what you see looks like a continuous surface it almost certainly is one - rather than *two* different surfaces at different distances, which are being illuminated in such a way as to give an impression of (only) one and the same surface. In other words, neighbouring points on a surface are likely to be similar in their depths away from the viewer. From this assumption another rule of binocular combination is arrived at, namely the rule that preference should be given to matches that have neighbouring matches with similar disparity. That is, matches should be chosen such that, as far as possible, disparities vary smoothly across them.

Discussion of example. Marr's strategy of building in rules and assumptions about the world in the algorithms producing the 2½D leads to the same general conclusion as before, namely that in order to make the process run in such a way that proper *representations* be computed of structures corresponding to spatial structures in the world, the system has to be supplied with knowledge of what amounts to proper representations of such features. The major difference seems to be that the higher the level of the so called low-level processing, the more extensive and specific is the knowledge about the world and objects required in order to solve the computational problem. Thus, in the example described above, in order to make possible the computation of distance and segmentation of surfaces from the primal sketch, assumptions about the effect of illumination for identifying an object relative to others will have to be taken into account. However, this requires knowledge of how the light intensity emitted from surfaces of *objects in depths* are affected by the position of the source and distribution of illumination.

We may contend that Marr's model of vision presented so far demonstrates how it is possible to develop successive representations of structures of things from information in light, by employing rules of how to extract information in such a way that a correspondence exists between these representations and the properties of that which is represented. However, the whole enterprise of developing representations hinges on the possibility *at each stage* in the process of identifying a correspondence between the primitives of the representations (edge-points, regions, etc.), and the features and properties of objects *as we visually perceive them* (boundaries, surfaces, etc.). That is, at every stage of these low-level computations, production rules are implemented which are based on knowledge about what objects and scenes look like - for which reason it is obviously misleading to claim that the computations involved in the model are purely data-driven, i.e. occur *before* or *without* any knowledge of what objects or scenes in the world look like. Conversely, granted the claim were admitted, it is hard to see how it could be maintained that the output of the computations were *representations* of precisely these objects or scenes.

3.4.3 The 3D Model

The 2½D representation does not yet represent any objects - only shapes and relative distance. Therefore, in order to get to the final stage which represents the objects and scene from which the image originated, the 2½D representation has to be transformed into another representation suitable for identifying and recognizing the shapes as being shapes of *particular* objects taking up space in a 3D world. However, the shape of a particular object may look different not only from other objects, but from different points of view as well; indeed it may even change it's shape when moving (e.g. like birds in flight). This means that the shape of an object could be described by almost an infinite number of different representations - each of which would have to be recognizable by the system as one and the same object. One way of solving the problems of this manifold of representations as well as the ambiguities which go with it, Marr argues, is to get at a representation which - as far as possible - will describe the shape of an object in an *observer independent* fashion. This will prevent the representation having to be replaced or changed whenever the observer moves - or whenever the shape of the object is changed as a result of its own movements.

In order to construct or compute such an *object centred* model, a formal system is needed for the description of shapes of objects - plus rules for its application by the system. However, no such general system exists. We do not have a system which will adequately specify the myriads of different shapes which we observe - from balls to crumpled newspapers or items of clothing waving in the wind. But it appears to Marr that "generalised cylinders" seem adequate to model a whole range of objects, so he suggests that the formal system for modelling the 2½D representations into a 3D model of objects could consist of a library of exactly such cylinders - plus rules for how they should be attached to each other. Figure 1 shows how the shape of a human body could be modelled using generalized cylinders as primitives.

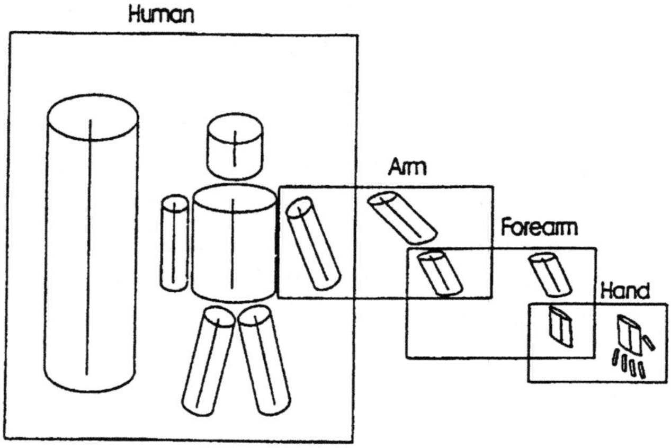

Figure1. (From VISION by Marr © 1982 by W.H. Freeman and Company. Used with permission)

The representation is *volumetric* and *hierarchical* as described in this quotation from Marr and Nishihara (1978):

> "First the overall form - the 'body' - is given an axis. This yields an object-centred coordinate system which can then be used to specify the arrangement of the 'arms', 'legs', torso', and 'head'. The position of each of these is specified by an axis of its own, which in turn serves to define a coordinate system for specifying the arrangement of further subsidiary parts. This gives us a hierarchy of 3D models; we show it extending downward as far as the fingers. The shapes in the figure are drawn as if they were cylindrical, but that is purely for illustrative convenience: it is the axes alone that stand for the volumetric qualities of the shape, much as pipe-cleaner models can serve to describe various animals."[8]

[8] Incidentally, to show an image of the 3D model as in figure 1 is misleading. An image will always show the model from a specific angle - whereas the 3D model is an object-centred representation. It is a representation of the shapes of objects described in abstract, symbolic (volumetric) terms, which does not lend itself to be illustrated as an image.

According to Marr the recognition and identity of an object in a scene can only be established using already stored information about what particular objects look like. This means that at this stage in the visual process Marr assumes the existence of top-down procedures and information. Thus, to identify or recognise a 3D model of cylinders making up two arms, two legs, a body and a head as a man *as opposed* to a horse with 4 legs, a body and a head, the 3D model will have to be matched against a catalogue or database of 3D models already stored in the visual system. Once the best-fitting model has been selected, moreover, it may supply information for further analysis of the 3D model. For example, it may suggest interpretations for parts of the image that are unclear or obscured, such as an arm partly hidden behind a door-post, or the like.

Discussion of example. First, a "thing" with the same *shape* or *volume* as a horse - or of a match of it in the database - may conceivably have been computed and identified during the processes described by Marr. But it is misleading to claim that by these processes an object has been identified as being a *particular* object, i.e. a horse. To identify an object as being a horse requires a *concept* of a horse, i.e. knowledge of what a horse is, which goes far beyond the knowledge of what shape it has or what "volume" it occupies in 3D space. No amount of computation of the kind suggested by Marr would enable identification of an object in this sense.

Next, the question of how it is that the visual system is capable of constructing representations of the world and its objects which make us recognize and identify these objects as they are, is answered by Marr by positing an already existing database of representation of those objects in the visual system. However, this does not answer the question at hand any more than does the answer to the original question, i.e. that we see the world as it is, because we happen to have the required knowledge to do so.

Furthermore, the assumptions about the nature of the world being encoded in the model at the various computational stages were introduced by Marr in order to cut down the number of ambiguity problems - and they did so considerably. But at the cost, of course, that these assumption cannot be explained by his theory - and therefore, the theory can hardly be said to be a *complete*. In summing up the main points of the computational or information processing approach to vision, Marr writes:

> "The critical act in formulating computational theories for processes [for recovering the various aspects of the physical characteristics of a scene from images of it] is the discovery of valid constraints on the way the world behaves that provide sufficient additional information to allow recovery of the desired characteristic. The power of this type of analysis resides in the fact that the discovery of valid, sufficient universal constraints leads to conclusions about vision that have the same permanence as conclusions in other branches of science.

> Furthermore, once a computational theory for a process has been formulated, algorithms for implementing it may be designed, and their performance compared with that of the human visual processor. This allows two kinds of results. First, if performance is essentially identical, we have good evidence that the constraints of the underlying computational theory are valid and may be implicit in the human processor; second, if a process matches human performance, it is probably sufficiently powerful to form part of a general purpose vision machine." (Marr, 1982, p. 330-31).

Well, it is perfectly alright to introduce all kinds of assumptions of what the world looks like if our aim is to construct computational models which may solve problems of a "visual" nature, say, in artificial systems like robots. The enormity of the complications involved, and the ambiguities to be resolved in creating computational model which can account for or simulate even extremely simplified forms of vision, are probably extremely enlightening and useful for researchers of human vision as well. But if the aim of a computational account of vision is to provide a complete explanation of how human vision comes about, it will not do. To the extent that such accounts or models of human visual perception are provided with assumptions and knowledge of how the world is visually perceived by us, they are inevitably models in which is assumed that which the model set out to explain.

Moreover, to introduce such assumptions and knowledge only leaves us with the problem of explaining how the visual system got this knowledge in the first place. And, as we have seen, there will be a lot of knowledge to account for. Indeed, one of the major problems for the computational approach is the sheer amount of implicit constraints and assumptions about the world necessary for a model of vision to run. In the case of a model of normal everyday human perception the amount will have to be virtually infinite - thereby making such a model devoid of theoretical interest.

In the next chapter I shall discuss some further problems of a principle nature concerning Gibson's and Marr's theories of perception.

Chapter 4

Consequences for perception psychology and epistemology

4.1 Similarities between Gibson's and Marr's theories of perception

Despite arguments to the contrary in the ongoing controversy between exponents of the Gibsonian and the computational representational theory of perception, the analysis in the preceding chapter of Gibson's and Marr's account of vision shows that the problems to which they give rise are *in principle* the same. Given the similarities of conditions which apply for the two accounts - rather than the differences most often focused on - this is not surprising. Thus, it applies to both Gibson's theory and Marr's computational model of vision that substantial assumptions and presuppositions are made about the "capabilities" and "knowledge" on the part *either* of the sensory system picking up information in the ambient array, *or* of our visual system to process representations from information in light images. And it applies to both Gibson's theory and Marr's model of vision that the descriptions of "high order variables in the optic array" or of "computed structures in representations from light images", rest on an abstraction in the sense that they rely on ordinary everyday descriptions of things in the world and how we perceive them. In this respect it makes no difference in principle whether a technical description of the information extracted in patterns of light projected on to the retina is reached by an analysis of a Gibsonian kind, and expressed in terms of gradients, high order variables, and affordances, or whether the description of virtually the same information is the outcome of complicated computations of representation, expressed in some other technical language. It applies to Gibson's as it does to Marr's technical descriptions that they are ways of representing reality and the objects of which the experimental scenes are made up, which rely on ordinary everyday descriptions of these objects and scenes, i.e. rooms with walls and pictures on walls, or various shaped blocks put on a surface. If not, it would not be possible to determine whether, or to what extent, or for what purpose, their technical descriptions were *adequate* or *relevant* descriptions of - precisely - these scenes.

Because of the assumed knowledge about the structures of the world on the part of the sensory system, introduced by Gibson in his theory of direct perception, and of virtually the same knowledge being implemented in the production rules of Marr's computational model of vision, none of the theories can claim completely to explain how visual perception and, thus, knowledge about the world comes about. And because a technical description of both the structures in the stimulus or in the computed representations of them depend on an ordinary everyday description of how we see the world, both theories will have to be theories which assume that which they claim to explain.

As we have seen, both Gibson and Marr places great emphasis on understanding the relationships between the structures in the world and ways in which they structure light. In Gibson's approach the emphasis is on optico-geometrical structures in the optic array and the corresponding spatial structures of the world; in Marr's approach algorithms are developed producing representations of structures of light from light images, which correspond to the spatial structures of the world. The dispute between the Gibsonian ecological view and the computational representational view, however, is not whether a complete account of visual perception is possible by referring to the existence of such corresponding structures. Rather, the dispute concerns whether the structures in the optic flow can be picked up *directly* by the perceiver's sensory system (the Gibsonian view), or whether such structures have to be computed (mediated) from "light images" *before* it can be picked up by modules at higher levels of the visual system, and used as input for further computations of representations at those levels (the computational view).

However, whether we endorse the Gibsonian or the computational view, we are still left with the question discussed earlier of how it is that sensory or visual systems are capable of discerning the true nature of what is picked up - i.e. whether it is mere formal symbolic descriptions of structures in a 2½D "sketch" of the world, or whether it is virtually the same structures as they appear in the optic array projection on the retina. As neither a technical geometrical description of structures in the optical array corresponding to spatial structures in the world, nor a formal symbolic description of structures in 2½D representation corresponding to the same structures, may tell the *whole* story about what the world looks like when visually experienced by human beings, we are in both cases left with the additional question of how the sensory or visual system respectively completes the story. How, for example, does the sensory or visual system manages to *interpret* mere formal, technical descriptions of structures in the optic array or computed images as being structures which *correspond* to spatial structures to be found in the world and its objects? And how in the process of such interpretations or inferences does it happen that visual experience of the world and objects "out there" emerges out of this interpretative and inferential process?

Both positions agree that visual perception is mediated by light reflected from surfaces and objects in the world, and both agree that some kind of physiological system is needed for handling information available in the light. For the Gibsonians, however, once invariant information in the light necessary for perceiving spatial features has been described, there is no need to inquire further into how the nervous system handles this information. That is the task for physiologist to unravel. Besides, our knowledge of how our nervous system handles this information does not shed any light on, nor does it add to our knowledge about the "ecological" changes of the optic array, which provides information to perceivers about the layout of his or her environment.

These views of the Gibsonians would seem reasonable enough. Where they go wrong is in assuming that by describing the optic array as a "medium" which not only carries information about the size, shape, slants and distance of objects, but also about their meaning, identity, function, value, what they are good for, and all the rest, a theory has been provided which *completely* explains the epistemic "contact" of perceivers with their environment. For, this theory of information pick-up and affordance no more explains how perception of the world comes about than does a mere conceptual analysis of what visually seeing the world amounts to, and on which rests *any* description of the information and affordance in the optical array. As pointed out by critics from the computational camp, the consequences of Gibson's theory of direct perception is that both the terms 'invariant' and 'directly detected' are left so unconstrained as to be trivial and devoid of meaning. (See e.g. Fodor and Pylyshyn, 1981.)

But the solution offered by the computationalists runs into quite similar problems and consequences. Thus, to explain visual perception as a process whereby structures of light from light images are extracted and representations computed by an algorithmic model implemented with rules about the layout of the spatial structures of things in the world, which, eventually, will be matched with other, already stored representations, carrying all the information worth knowing about things in the world,[1] amounts to a similar trivialization of the concept of 'representation'. This way of explaining how the visual system solves the problem of computing representations of structures in the light corresponding to spatial structures in the world, and of "matching" these representations with a database of already existing representations containing knowledge of what things in the world looks like, does not add anything to a mere conceptual analysis of what visually perceiving things amount to, on which rest the rules for computing these representations.

Thus, the fact that the implications and consequences of Gibsonian and computational theories of vision are *in principle* the same, may be summarized thus.

To the Gibsonians the explanation of how perception occurs is that all information is there in the optic array to be picked up directly by the perceiver. Invariant structures adequate and sufficient for the perception of shape, meaning and function of objects may be found and detected - provided our sensory systems know in advance *what amounts to a proper correspondence between spatial structures of objects in the world and the structures in the optic array* - and know what inferences from these structures may be made about objects in the world.

To the computationalists the explanation of how perception occurs is that it is all there in the light images to be computed by the visual system into representations of "pure" perception. Structures of light adequate and sufficient for computing representations of shapes, meaning and function of objects may be found and detected - provided the computing visual system knows in advance *what amounts to a proper correspondence between spatial structures of objects in the world and structures in the representations* - and knows what inferences from these structures may be made about objects in the world.

4.2 Conditions for carrying out investigations in perception psychology

The analysis of Gibson's and Marr's general theories of perception has shown that neither of them solve or overcome the philosophical problem of perception to which the Cartesian Mind-Matter dualism gives rise. It has done so by showing that the assumptions on which both theories build of a partition of "reality" and "perception of reality" into *independently determinable* parts, does not hold – and, as a consequence, both their general, causal theory of how it comes about that we perceive reality as it is, or the

[1] The very same information which Gibson in his theory of direct perception and affordances attribute to the optic array.

way we do, will necessarily become circular. The reason for this circularity is twofold: First, in the case of Gibson's theory of perception it has to be presupposed that the visual system, in order to pick up information in the stimulus emitted from the environment *as* information about things in the world, is already equipped with a considerable amount of knowledge about how we actually perceive features and properties of things in our familiar world – and therefore, the theory has to presuppose what it set out to explain. In the case of Marr's computational model of vision, in order for the model to determine the true nature of the information and solve the ambiguity problems of the representations computed at the various levels of processing, it has to be provided with rules and a data base of knowledge about how we actually perceive features and properties of things in the familiar world – and thus, as in the case of Gibson's theory, the model has to presuppose what it aimed to explain.

Secondly, the circularity of Gibson's and Marr's theories is inevitable due to the fact that their technical description of the stimuli (or the computed representations of it), thought to underlie and cause our perception, rely on an ordinary everyday description and perception of the world from which the stimuli are emitted. Thus, in an experimental test of the explanations and prediction of Gibson's and Marr's theories – as well as in general of any causal account of perception – the technical descriptions of the stimulus situation (or computed representations of it) may be conceived of as ways of representing the reality and objects of which the experimental scenes are made up. However, as pointed out in the previous section, such a technical description of the stimulus situation cannot be given independently of but, on the contrary, presupposes and rests on an ordinary everyday description and perception of these experimental scenes. If not, it is not possible to apply the technical descriptions to these scenes, nor to determine whether, or to what extent, or for what purpose, the technical descriptions are adequate or relevant descriptions of precisely these scenes. In this respect a technical description of the stimulus situation (or computed representations of it) rests on an *abstraction*, i.e. in that it depends on an ordinary everyday description of our perception of the world.

However, this dependency of a technical description of the stimulus situation (or computed representations of them) on an ordinary everyday description and perception of the world, is a condition which applies *in general* for the implementation and empirical test of any accounts and theories of the processes of perception – and not only for Gibson's and Marr's. Consequently, to the extent that such accounts and theories aim to provide general and causal explanations to solve empirically "the problem of perception" inflicted by the Cartesian dichotomy, they will be just as circular as the theories of Gibson and Marr. More importantly, granted this condition for a technical description of the stimulus, this Cartesian problem of how it comes about that we perceive reality as it is – or the way we do – cannot be stated consistently, neither scientifically nor conceptually, philosophically. Why this is so, becomes obvious when we consider the impossibility *in either case* of talking consistently about and describing *reality* and *things existing in reality* without or independently of referring to our perception of them, *and* without presupposing that reality and these things exist as things that we may perceive and talk correctly about. And, conversely, consider the impossibility of talking consistently about and describing our *perception* of reality and things in reality, without or independently of referring to and describing reality and these things. Indeed, it would seem that any discussion and problems concerning the nature and status of our perception of reality, just as in general any discussion and problems about reality and things existing in reality, will have to presuppose that our perception of reality and these things do in fact concern reality and the things which exist in reality. Otherwise, neither our notions of 'reality', 'things in reality', nor of 'perceptions of things in reality', could have any well defined meaning – nor could such discussions and problems be well defined.

Now, it has been objected that the above conditions and presuppositions, which are just as necessary for consistent implementations of psychological theories of the process of perception as in general for talking consistently about reality and our perception of it, does not mean that *causal functional theories* of perception are not possible. Nor do these conditions and presuppositions invalidate the explanations of such theories concerning the processes in the "machinery" of the perceptual system, and the stimulus conditions which cause them. It is only, so the objection goes, if we "mix up" or confuse philosophical *epistemological* issues about the nature and status of perception with *psychological* issues about the causal functioning of the perceptual system that the problems (among them the problem of circularity) arises. However, once Gibson's and Marr's theories are freed of their undeniable epistemological pretensions, and instead interpreted as *functional theories*, so the objection continues, the critique I have raised and problems I have pointed out in are not longer valid.

To this I think we shall have to agree - provided it is understood that it would be just as problematic to maintain that our perception of reality may be accounted for adequately, let alone exhaustively, by functional theories concerning the causal processes taking place between stimuli in the world and the "machinery" of the perceptual system. In particular, I contest that the assumption that the outcome of any *concrete* investigation of the causal functioning of the perceptual system in which human observers take part, can be the uncovering of perceptual processes of "cognitively impenetrable" functions, i.e. processes and functions not influenced by or dependent on "higher level" knowledge and expectations. The reason, as I shall argue next, is that no experiment in perception psychology can be carried out, which do not presuppose a substantial amount of background knowledge and perceptual capabilities on the part of the experimental subjects – indeed, background knowledge and perceptual capabilities, which form a substantial part of the subjects' perception during the experiment, and which *therefore* preclude that the perceptual phenomena experienced by the subjects during the experiment be accounted for adequately – let alone exhaustively - in terms merely of processes of some or other 'local' functions in the "machinery" of their perceptual systems, and their causal link with some or other 'local' stimuli in the experimental set up. To see why this is so, and, in particular, to see *why it makes no sense to distinguish causally between the subjects' percepts and the stimuli presented to them during the experiment,* we shall first have to consider the conditions and circumstances under which investigations and experiments on human visual perception are carried out.

To these conditions belong, as already noted, that the experimenter himself is able correctly to perceive and describe the situation in which his experiment takes place, the experimental set up, etc. Equally importantly, the experimenter must take for granted that the *subjects* taking part in the experiment are also capable of correctly perceiving the experimental situation and that, being users of the same language as he himself, they are capable of correctly describing this situation and the phenomena they perceive during the experiment. To this must be added the condition that the subjects understand the experimenter's instruction about what - according to the purpose of the experiment - are considered relevant descriptions of the phenomena to which they are supposed to attend. If these conditions were not met, the experimenter could not ask his subjects to sit down on *that* chair and look at the "lines", "figures" or "dots" on *this* monitor, or through the little hole in *that* screen, and report what they see on the monitor or through the screen. - In this respect an experimental situation resembles a perfectly ordinary everyday situation in which two persons may come to agree on the correct descriptions of the things they perceive in the situation in which they find themselves. And just as in such everyday situations, the communication between the experimenter and his subjects relies on the fact that as users of the same language, they know correct description of the things in those situations, and that they can use them correctly to describe what they perceive.

Well, it could be objected that although subjects taking part in an experiment on visual perception may well be able correctly to describe what they perceive, they rarely if ever describe correctly that which the experimenter has presented to them. To take an example, in a study of the illusion created by the chevrons attached to the lines of the so called Müller-Lyer figures, the two lines presented are equally long. However, when presented with these figures and asked what they perceive, the subjects will often say that the lines connecting the inward and outward going arrows are of unequal length. Therefore, it makes no sense to say that the subjects are able correctly to describe what is presented to them in the experimental situation. Conversely, it makes good sense to distinguish causally between the subjects' percept and the stimuli presented to them. However, this objection has no bearing on the point I am arguing. For one thing, in an experiment in which the subjects are given the task of determining the relative length of two lines, the investigator must necessarily be able to count upon the subjects' knowledge of the correct implications and application of *description of lines,* and *description of length*. Thus, he must be able to take for granted that his subjects know the implications of describing the two lines as of *unequal length*, implications which could be expressed in statements such as: "If I measure the distance with a ruler, the values will turn out to be different". Equally, in the Gibson-experiment mentioned earlier, the experimenter must be able to count upon the subject's knowledge of a correct description of a tunnel. If the investigator could not count upon such knowledge of concepts and correct descriptions on the part of his subjects, and take for granted that they are capable of describing what they see correctly, it would make no sense - let alone be possible - to conduct psychological experiments on visual perception.

However, it has to be remembered that in most, if not all investigation on perception, the subjects' condition for observing and describing the experimental set-up, and their opportunities for checking whether the implications of their descriptions of it hold true, is severely limited compared to those of the investigator's. Given better opportunities of observation and description - for example those available to the investigator - the subjects may no longer abide with the original description of the stimulus situation; but, for the reasons given above, it has to be presupposed that also during the restricted opportunities of observation and description prevailing in an experimental situation, the subjects still know correct descriptions of *what they see*, and know how to apply them correctly. Now, the reason why this means, secondly, that it makes no sense to distinguish causally between the subjects' percepts and the stimuli presented to them, is this:

The result of an experiment or of a series of experiments in visual perception are often formulated in law-like statements of the form 'if A is presented to the subjects, they will perceive B', where A is a description of the stimuli presented to the subjects, and B is a description of what they perceive. It is a fundamental rule of scientific method that if one wishes experimentally to verify a law of the form 'if A then B' one must be able to give an unambiguous description of A. If not, it is impossible to define, set up, or reproduce the experimental situation in which the law is claimed to hold.

To give a correct, unambiguous description of the stimulus may in some experiments require that the experimenter knows a physical or technical, formal description of the stimulus, and thus a description of the stimulus may involve a series of measurements using various measuring instruments. However, in order to know that any such physical or technical description is a description of precisely this particular stimulus set-up, the experimenter must know that he has used the correct measuring instruments in the correct way and at the correct places. This presupposed knowledge, however, is *not* itself described physically. On the contrary, it is part of his ordinary everyday knowledge and description of the situation, measuring instruments etc., and *on which the physical description of the stimulus situation rest*. Thus, a physical description may be considered in some respects and for some purposes a more accurate account of aspects of the stimulus situation, *which the experimenter has already identified and described in an everyday manner*. Therefore, the stimulus situation is primarily an everyday situation in which certain measurements have been carried out.

The experimenter must, moreover, be able to identify the descriptions or reactions of the subject, considered as the result of the experiment, as descriptions of the experimental situation, or rather of that part of it to which the subjects have access. Otherwise the experimenter would be unable to attribute the subject's descriptions of the experiment, or claim that the answers or reactions of the subject could be described or identified as the dependent variable related to the independent variable of the experiment. This means that it must be a condition for a psychological investigation on perception that the subjects' as well as the experimenter's descriptions must be descriptions of the *same* thing. That is, they are descriptions of the experimental set-up and its objects, as observed under two different conditions of observation. The two conditions of observation differ in that the investigator always has unlimited access to the experimental set-up and can make any measurements he pleases. In contrast, the conditions of observation of the subjects, as already mentioned, are limited relative to those of the experimenter. Indeed, if this were not the case, *the descriptions of the subjects and the experimenter would not be different, and the result of the experiment would be given by logic, and therefore devoid of empirical content.*

Beside the description which is necessary to make a reproduction of the investigations possible, the investigator may choose to describe the stimulus situation in a number of different ways. (Cf. Gibson's and Marr's descriptions of information in the optic array - or of computed representations of it, respectively). However, the subjects' descriptions of the stimulus situation, precisely because of their limited opportunities of observing and describing it, will always be that of ordinary everyday language.

Let us now look at the causal, functional model of perception and its interpretation of the result of an investigation on perception. In this model it is assumed that as a result of some physical stimulation (or computation of it) a percept or representation is generated in the subject. According to this view the experimenter's and the subject's descriptions refer to *two different things*, i.e. to the physical stimulus and to a percept or representation of it in the mind of the subject respectively - because the same thing cannot be both e.g. a "series of real plastic sheets" and a "percept or mental representation of a tunnel".

Now, the experiment could in fact be carried out and be described in causal terms, i.e. just like and in the same way as are experiments in classical physics - *if* the experimenter were able unambiguously to identify what the subject is describing. In that case, the experiment would result in two descriptions of two different things - stimulus and percept or representation - of which the former is the cause of the latter. But how can the investigator unambiguously identify what the subject is describing? There is no possibility for the experimenter and the subject to identify together this "something", which the subject describes as something *other* than, or something *different* from, the object which the subject looks at, and the experimenter has placed in front of him. They simply have no independent procedure for determining in advance whether they are talking about the subject's *representation* or *percept* of the object, or about the object itself. Not even the subject himself would know what he is talking about. For one cannot point to the representation or percept, or take it in one's hand and show it to someone. But neither can it be *unambiguously* identified relative to the tangible object, nor be identified independently of it. An identification of the percept *relative* to the object would require that to any given object there corresponds one, and only one percept or representation, which is false, whilst an identification of the percept *independently* of the object would require that it be possible to describe *what* we see (the perception of the object) without describing it as *that* which we see (the object) which, to put it mildly, would be an impossible requirement.

To conclude, then, the subject's description cannot be anything other than a description of what the experimenter has placed in front of him and asked him to describe; for if it were anything else, the experiment, as argued, would have failed. Thus, in a psychological law of the form 'if A is presented to the subjects, then they will perceive B', A and B cannot be descriptions of different things or objects. On the contrary, we have to presuppose that they are descriptions of the same thing, namely, *the experimental situation, as it can be unambiguously identified and correctly described under two different conditions of observation and - possibly - by different means of description.*

The interpretation of the laws of perception psychology presented above renders the application of a causal terminology to express the relation between stimulus and perception rather artificial. This would be immediately obvious if the subjects taking part in experiments on perception had the same opportunities of observing and describing the stimulus situation as the investigator. In this case the subject's and the experimenter's description would be identical. But if perception is to be considered explainable in terms of a causal process at all, it must also apply in this case, and we would be compelled to say that the stimulus is what the subject sees, and also the *cause* of what he sees. Such a formulation can hardly be said to have any meaning.

However, this case is special. Indeed, it characterizes psychological investigation on perception that the subject's opportunities for observing, investigating and describing the stimulus situation is severely restricted - as opposed to the experimenter's, who has free and unlimited access to the experimental set-up. However, to the extent that the investigation requires both parties to agree on the *general description* of the situation, we can naturally no more than in the previously special case give an account for this fact in causal terms. There remains only the possibility of saying that the information which is available *only* to the experimenter indicates the cause, while the information given by the subject in *addition* to the general presupposed description of the experimental situation indicates the effect. But as it is more or less arbitrary, and varies from experiment to experiment, which procedures and types of investigations and descriptions the experimenter chooses to reserve for himself and denies his subjects, there can hardly be any doubt that this would be a rather unusual and inconvenient employment of the concepts of cause and effect.

Let me sum up the consequences of the conditions and presuppositions for carrying out investigations and experiments in perception psychology.

First, the data or results of such investigations and experiments will always be two different description of the *same* thing, namely the experimental set up as it may be described under two different conditions of observation by experimenter and subjects respectively. The existence of causal processes between the stimuli presented and the visual system during perception notwithstanding, this means that the result of such experiments cannot be given straightforward causal interpretations. Secondly, since any experiment on perception, as we have seen, rely on and presuppose a substantial amount of background knowledge and perceptual capabilities of the subjects - indeed, background knowledge and capabilities forming part of the subjects' perception during the experiment - it is misleading to account for the subjects' perception during the experiment in terms of a causal connection between some particular isolated stimuli of the experimental set-up and processes in the "machinery" of some particular isolated function of their perceptual systems. Indeed, neither the perceptual phenomena reported by the subjects, nor the processes of such perceptual functions may be *determined* and *accounted* for adequately, *independently* of referring to the subjects' *general* knowledge and perception of the experimental situation; nor without referring to the purpose and conditions of observing and describing the things presented to them in those situations. For this reason, the perception reported by the subjects during the experiment cannot be accounted for adequately, let alone exhaustively, merely in terms of functional theories derived from experiments investigating local processes in the "machinery" and function of the perceptual system caused by local stimuli in the experimental set-up.

Now, whatever the results may be of investigations carried out by perception psychologists concerning the causal functions underlying human perception, they are immaterial to the points I have tried to make here. What is not, however, is that attempts by psychologists to develop theories of perception *in terms* of such causal functions, inevitably ignore the conditions and implicit presuppositions about the perceiver and his perception during the investigations, on which their investigations rest. In so doing, causal functional theories of perception are just as inevitably at variance with the presupposition and assumptions on which they rest.

In conclusion, *because* of the existence of presupposition of an epistemological nature for carrying out experiments in perception psychology, no accounts of the causal processes of the "machinery" of the perceptual system underlying perception may be given *without* referring to these presuppositions, i.e. without referring to our knowledge and perception of the concrete situations in which we find ourselves - including experimental situations; nor without referring to the purposes and conditions of observing the things in those situations. Therefore, logically, perception cannot be accounted for adequately in terms of causal functional theories of perception.[2]

The mistake of Gibson and Marr is not that they committed the error of "mixing up" what may rightly be interpreted as *causal functional theories* of perception with epistemological theories concerning the nature and status of perception. Rather, the mistake is that they failed to realise - as did and still do perceptions psychologists from Helmholtz, Hering, Wund, Koffka,

[2] As we shall see in Chapters 9, 11, and 14, the conditions and presuppositions for implementing and testing causal functional theories of perception, apply just as well for the implementation and test of causal functional theories of cognition and language, among them the so called *Syntactic Theory of Mind* by Stich (1986), and the Causal Theory of Content by Fodor (1985, 1994).

Koehler and onwards - that any *causal functional theory* just as any *general* theory of perception necessarily rests on presuppositions of an epistemological nature and thus, in this sense, inevitably *is* "mixed up" with epistemological assumptions. In the section which follows, I shall summarize the epistemological presuppositions on which such theories of perception must necessarily build, just as I shall outline the implications of these presuppositions.

4.3 General epistemological consequences and implications

The most important general result of the discussions so far is that the epistemological "problem of perception", inflicted by the Cartesian dichotomy between "reality" and "our perception of reality", cannot be solved nor overcome empirically by perception psychology. The reason for this is fundamental in that the very assumption behind this dichotomy and problem, i.e. the assumption that "reality" and "our perception of reality" are *independently determinable*, simply does not hold – and, as a consequence, the problem itself cannot be consistently stated, nor therefore solved or overcome. That this assumption is untenable, is quite obvious. As already pointed out, we only have to consider just how impossible it is to talk consistently about and describe *reality* and *things existing in reality* without or independently presupposing that reality and these things exist as things we may perceive and talk correctly about. And consider, conversely, how impossible it is to talk about our *perception* of reality and things in reality, without or independently of referring to reality and these things – i.e. to talk about and describe *what* we perceive (our perception of things), independently of and without at the same time talking about and describing *that* which we perceive, i.e. the things. Thus, it seems that for any discussions of and problems posed about both reality and the nature and status of our perception of it to be well defined, it is necessary to assume that an interdependency exists between *reality* and *our perception of reality*. Indeed, without so assuming, neither our concepts of 'reality' nor of 'perception of reality', would be well defined, nor therefore would any discussion of or problems posed about either reality or our perception of it. Given these presuppositions and conditions for talking consistently and meaningfully about both reality and our perception of reality, it ought to be obvious why the assumptions of a Cartesian dichotomy between "reality" and "our perception of reality" as independently determinable parts, is untenable, and hence why the "problem of perception" which relies on this assumption, cannot be consistently stated.

In view of the inter-dependency between our notions of 'reality' and 'perception of reality', it would seem that the only consistent alternative to the assumption of Cartesian dualism is to assume that a *necessary relation* exists between reality and our perception of it. The implication of this assumption is that no *general* questions can be asked about the relation between reality and our perception of it, nor of whether our perception of reality and things in reality does in fact concern reality and things existing in reality. On the contrary, this has to be taken for granted and be the point of departure for any *further* concrete questions and investigations, not only into reality and facts of reality, but also into our perception of reality and these facts - just as it must be taken for granted and be the basis on which build the implementation and scientific tests of any theories of the processes of perception. This means that no causal account of the perceptual processes can be used to explain scientifically the relation which exists between our perception of reality and reality itself, nor be used to justify or 'give proof' of the correctness of our perception of reality. But neither can the relation between reality and our perception of it be questioned, in the sense of doubted or denied, nor can the correctness of our perception of reality. For such doubt or denial to be meaningful, would require that we can indeed talk correctly about both reality and our perception of it, and, hence, would have to presuppose *both* that we may indeed perceive reality correctly, *and* that reality exists as something we may perceive. Because they have so to presuppose, such doubt and denial would be absurd.

However, the existence of an inter-dependency or necessary relation between reality and our perception of it does *not* mean that we cannot distinguish between mental phenomena or processes, such as perceiving things in reality, and material things existing in reality. Nor does it mean that reality and things in reality do not exist independently of being perceived. To start with the first point, we can certainly distinguish between being aware of objects "our there" and the object "out there", just as we can distinguish between, on the one hand, perceiving an object (seeing, smelling, touching, or hearing it), or verbally describing it, thinking of it, imagining it, walking around it, and, on the other, the tangible object to which we relate in these different ways. However, we cannot describe our perception of the object or any of the acts we carry out with it, without referring to the tangible objects we perceive and act with, i.e. we cannot describe *what* we see and act with, without describing it as *that* which we see and act with, namely the tangible object.[3] Indeed, it is precisely if we maintain that perceiving things in the

[3] To give a concrete example. I cannot talk about my perception of the cup on the table in front of me without at the same time talking about the cup; but neither can I talk about the cup on the table without or independently of presupposing that the cup exists as something I may perceive and talk correctly about.

world amounts to having percepts or mental representations of them that a distinction is *not* possible between perceiving a thing and the actual thing being perceived, and thus between *a representation* of a thing and *the thing*. However, secondly, it is necessary to assume that things in the world we perceive, and with which we may carry out acts, exist independently of being perceived and acted with. Indeed, this would seem just as necessary to assume, as it is to assume that our bodies and sense organs exist independently of the things we perceive and act with. So, although there is an interdependency or necessary relation between reality and our perception of it, we shall also have to assume a crucial *asymmetry* between things in the world and our perception of them. The asymmetry concerns the fact that although perception of things does not exist independently of or without things, things we perceive do exist without or independently of being perceived.

Furthermore, although the assumption of a necessary relation between reality and our perception of it implies that no *general* questions can be asked about this relation, this does not mean that no *concrete* problems of perception may exist and be investigated. However, such problems can only exist and be investigated on the condition that no *general* problem of perception and this relation exist. Thus, as shown in the previous section, such investigations – e.g. of the kind carried out within perception psychology - rely on the presupposition that in normal everyday situations under normal conditions of observation, both the investigator and the experimental subjects are able correctly to perceive and observe reality and things in reality, and that as users of the same language, they are both able correctly to describe what they perceive and observe. Because of this general presupposition for carrying out investigations in perception psychology, *and* of the dependency of the investigator's technical description of the stimulus situation on an ordinary everyday perception and description of the experimental set up, it would be misleading to say that the technical description is *more* concerned with what exists in reality than is the subjects' descriptions, which are *more* concerned with some or other phenomena going on in their minds. Both descriptions must necessarily be descriptions of the *same* thing, namely the experimental set up as it may be described under different conditions of observation, and using different means or apparatus of description. In this sense both a technical (e.g. physical or geometrical) description and an ordinary everyday description of reality depend on an abstraction, i.e. in that they both depend on specific conditions of observation, and descriptive apparatus which permit *correct* description of what is being observed and perceived. And again, the causal processes involved in perception notwithstanding, because of this condition, together with the other general condition for implementing and experimentally testing any theories about the perceptual processes, there is no way that such theories, be they causal functional or otherwise, will ever amount to explaining how perception of the world comes about, nor to justifying that our perception of the world does indeed concern the world and things, which exist in the world.

The epistemological points presented here has far reaching consequence, not only for theories of perception, but in general for theories of cognition and language. Thus, it seems that for any investigation of and problems posed concerning our perception, cognition and language to be well defined, be those investigation scientific or philosophical, they must necessarily, logically be based on the presupposition that as persons and language users, we may both perceive, have knowledge about and use language correctly to describe ourselves and the material world of which we are part – and, conversely, that we ourselves and the material world exist as something, we may perceive, and of which we may have knowledge and put forward propositions in language which are true. This is something which we cannot explain, neither scientifically nor philosophically, but on the contrary, it is something we have to take for granted and which must be the point of departure for any further investigation, *both* into the world *and* our perception, knowledge and linguistic description of it.

In later chapters of this book on theories of mind, cognition and language, I shall further discuss and develop these general epistemological conditions and presupposition – just as I shall discuss their consequences for a variety of recent so called causal computational theories of mind and cognition, and for similar models and theories of language, and the acquisition and use of language. A thorough analysis of these theories will show that they suffer from problems and inconsistencies which are quite similar, if not identical to the problems and inconsistencies of current theories of perception. This is not surprising since these theories and models of language are based on the same fallacious assumptions of a Mind-Matter dichotomy, as are these current theories of perception. In these chapter I shall also try to make clear the implications and consequence of the alternative assumption of a necessary relation between Mind and Matter presented in this book, for developing theories of mind, cognition and language, which are both consistent and meaningful.

4.4 Assumptions and aims for a psychological science of perception

If the points above are correct perception psychology will have to accept that everyday public and communicable perception under favourable, i.e. free and unlimited conditions of observations cannot be given an empirical, psychological explanation. On the contrary, the correctness or "veridicality" of this perception is something which we shall have to take for granted, and which must form the point of departure for all further investigations of our perception. On this epistemological assumption must rest any research and theorizing in perception psychology.

It is worth noticing that this assumption does not imply that errors of perception, or mis-perceptions of things, do not occur in everyday situations. Rather, it means that such errors or mis-perceptions could not be identified, determined as such, unless it is presupposed that we know of, or may come to know of, correct perceptions of the things being mis-perceived. Nor does the fact that any research on perception rests on the assumption of the veridicality of our everyday perception mean that there is nothing at all about our perception, which we may investigate and come to know about by psychological investigations. Rather, given that the assumptions and presuppositions on which psychological investigations rest are correct and made explicit, they may provide perception psychology with invaluable insights into the kinds of questions and problems which may be consistently and meaningfully addressed by perception psychology - and which not.

Thus, the conditions for perception psychology outlined above suggest that instead of discussing which of the current theories of perception makes for the better candidate for a general, causal theory of perception (Ullman, 1980, Reed, 1988), perception psychology should ask questions of the form: under what conditions do we perceive - or mis-perceive - things in particular ways, and what are the descriptions of those conditions, which reliably and consistently may predict such perceptions - or mis-perceptions. In this pursuit, the research by Marr and Gibson on functional aspects of the visual system will undoubtedly contribute to our understanding of what these conditions are. However, in view of the presuppositions for investigating such questions, accounts of the relation between the perceptual phenomena being studied and the stimulus conditions under which they occur, cannot be of a causal, but only of a *correlational* kind.

What we *can* do in psychological investigations of vision, for example, is to arrange situations in which the subjects of the experiments are asked, with particular conditions of observation and possibilities of description, to describe a situation which the investigator has been able to investigate more thoroughly. We can thereafter compare the subject's description of the situation with that of the experimenter and note agreements and disagreements. We can vary such situations and investigate systematic relations between the two sets of descriptions. In this way we may discover certain sets of conditions for "veridical perception" and certain sets of conditions for particular deviations from "veridical perception". Now, to give a description of the stimulus condition which reliably and consistently will predict the outcome of the subjects' perception may require that the investigator has carried out physical measurements of features of the stimulus situation. However, as already mentioned, physical measurements will be of interests to the experimenter *only* in so far as variations in physical measurements of the situation make a difference to the subjects' descriptions. Secondly, it would be misleading to maintain that a physical description - or for that matter any technical description - of the stimulus situation is more concerned with what exists in this situation than is an ordinary everyday description, which is more concerned with our perception or other phenomena of our minds. To the extent that a technical description always rests on an everyday description and determination of the stimulus situations, the stimulus situation can only be an ordinary everyday situation in which certain measurements have been carried out. And it is worth reminding ourselves, thirdly, that both an everyday and a technical description of the stimulus situation *depends* on an abstraction, i.e. in the sense that they both depend on specific conditions of observation, descriptive apparatus, which permit different correct descriptions of it. And they rest on an abstraction in the sense that they are both *partial* (in the sense of limited) descriptions of the things and features of the stimulus situations - i.e. descriptions of them with regard to some particular purposes, interests, points of view, etc.

For these reasons I shall contend that the general purpose and aim of research in perception psychology can only be to solve problems of the form: what problems of a perceptual nature do people who, under normal conditions of observation are generally capable of correctly perceiving and describing things in the world, nevertheless encounter in situations in which the conditions are *not* normal, or in which the limits of normal perception are exceeded, or hampered by e.g. their moods, emotions, particular interests, and so forth; how do people try to understand and solve these problems; and with what means do we best describe the condition under which these problems occur, which may suggest ways in which to optimized the conditions, so that the problems may be overcome. To this end, perception psychology would be free to use whatever means of describing the stimulus conditions; it is not bound to using descriptions fitting any particular physical, mathematical model or theory about the functional "machinery" of the perceptual system, but only bound to finding the descriptions of the stimulus conditions which reliably and consistently predict the perceptual phenomena being investigated, and which serve the purpose of the investigation. However, given the observed correlations between stimulus condition and perceptual phenomena are correct, the outcome of research in perception psychology may be the development of technologies and means to solve problems of a *practical* nature - just as is the case in other sciences.

Now, to acquire relevant psychological knowledge about our visual perception is complicated by the fact that very few - if any - problems are exclusively "visual" in nature; they are rarely problems about how we avoid banging into things - or how we manage to reach out and get hold of things in our environment, i.e. to form a correct estimate of the distance to things in the world. Neither are our difficulties in attending to and recognizing what we see due to the fact that exposure time is in the order of a few hundred milliseconds. The problems we envisage are much more often - and insepa-rably - related to what *meaning* the objects and the environment have for us, what *sense* we make of it in the particular situations in which we find our-selves, (Bruner 1990). In order to gain knowledge of such problems, experi-ments in artificially simplified laboratory settings using meaningless dots - whether static or moving - are of extremely limited value. It may be a lot more scientifically gratifying to restrict one's investigations of "visual phe-nomena" to those occurring in controllable experimental settings without the remotest resemblance to real life situations - and the results of such experi-ments may *seem* rigorous and well-founded as well. But the complexity of visual perception does not disappear because we simplify the experimental situations and cut down the size of variables to a minimum, as is so custom-ary in most current laboratory research on visual perception. Neither is there any great hope that models and theories of "local" functional processes being generated by this kind of research, will ever be generalizable to or find its place in more global attempts at understanding human visual problems. In particular, the risk to perception psychology of re-defining perceptual prob-lems in the terms of "computational vision" is that perception psychology continues to commit itself to problems which are not of a human, psychological nature - but problems which rightly belongs to disciplines such as applied mathematics and robotics. The danger is that perception psychology will be reduced to the pursuit of delivering so called psycho-physical "evidence" to models of artificial perception or "robotic" vision developed within these disciplines.

But there are lots of problems "out there" of a genuinely psychological nature - not least in a visual environment being increasingly complex and difficult to understand and cope with, due to the development of ever more sophisticated and complicated technologies, which dominate our everyday life and work situations. These problems are of a perfectly practical nature which await solutions by imaginative and courageous psychological researchers. The number of examples is probably infinite - but to give a hint of what I refer to let me mention just two examples:

What problems of a visual, cognitive nature do operators of hazardous and highly complicated production plants encounter in controlling such systems; in diagnosing faults - and in correcting them? How do operators' themselves try to solve these problems? And how could we help them do this, by e.g. displaying and representing the processes of such systems in ways which increase the operators' understanding of their dynamics? And how can we help them retrieve and choose the relevant information about the system and its states, which is available in those representations. (See e.g. Praetorius and Duncan, 1991). Or, what problems of a visual and cognitive nature do people encounter when driving cars? How do we design better traffic signs, cars, and lay out of roads in order to solve these problems? (See e.g. Ebbeson, Parker and Konëcni, 1977, Lee, 1976, Lee and Lisham, 1977.)

For the solutions of these problems of a genuinely psychological nature, current perception psychology, in my experience, has precious little to offer. If, for once, we did concentrate our research efforts on *observations* of perceptual problems facing us in real life situations, there would be hope for perception psychology to be a science acquiring relevant knowledge of variables influencing our perception. That is, knowledge on which to develop hypotheses that may be tested experimentally, and thus for experimental studies in perception psychology to be both scientific and relevant. As I have tried to argue, however, our understanding of what are relevant *psychological* observations, as well as theoretical and practical problems to be pursued by psychological research on human perception, must go hand in hand with an understanding of the condition and presupposition of an epistemological nature on which rest such observations and research. Once these conditions and presuppositions are understood, there is hope that perception psychologists may come to agree on a set of basic concepts, assumptions and principles for describing phenomena and issues of perception of a *psychological* nature, and to develop theories which may appropriately account for such phenomena and problems. If so, we may begin to have a situation within perception psychology resembling the situation in the natural sciences, who already have a common basis of concepts, assumptions and principles about the nature and properties of the phenomena falling within their subject area.

PART II

THE RELATION BETWEEN LANGUAGE, COGNITION AND REALITY

Chapter 5

The relation between language and reality

5.1 Introduction

In previous centuries philosophers were preoccupied with questions of how our Minds come to be filled with experience and knowledge of the world (or rather with "perceptions" and "ideas" about the world), and also whether this experience and knowledge were really in accordance with a reality existing independently of our experience and cognition of it. These discussions left philosophers with a number of unsolvable problems. The most prominent was that since all we have access to is our experience and cognition of reality as it appears in our Minds, then how can we possibly settle the question of whether our knowledge is in accordance with reality existing independently of and "external" to our Minds? Indeed, the assumption of a reality existing independently of our experience of it would be an assumption of a reality beyond our experience and, therefore, unknowable to us. It seemed inevitable that our notion of what exists in reality depends on our experience and cognition of it - and hence that reality does not exist independently of our cognition, nor independently of being experienced.

In this century philosophers have been preoccupied with questions about the relation between language and reality, and the possibility that speakers of a language may say anything about reality which is true. This swing of interest to language is not surprising since discussions about our experience and cognition of reality[1]are carried out in language. Hence, the possibility of developing consistent and tenable theories of mind and cognition, would seem to hinge on whether we get our assumptions right about the relation

[1]Or of the world, things in the world, or whatever we experience and have knowledge about.

between language and reality, and the conditions for using language to talk correctly about reality. If we are mistaken about these assumptions there is little hope that we shall be able to say anything of significance about our minds and cognition - be it in philosophical or scientific contexts. I shall argue in this Part, however, that far from recasting the unsolvable epistemological problems of previous centuries, philosophers taking part in the current debate within the domain of "Philosophy of Language" have only managed to contaminate the language-reality issue with problems of a quite similar nature - and to make the original problems even more obscure. In one of his recent books, to which I shall refer repeatedly in this introductory chapter, Searle puts his finger right on the problem:

> "Twentieth-century philosophy has been obsessed with language and meaning, and that is why it is perhaps inevitable that somebody would come up with the idea that nothing at all exists apart from language and meaning. Earlier centuries were obsessed with experience and knowledge, and correspondingly, philosophers came up with the idea that there is no reality independently of experience and knowledge." (Searle, 1995, p. 168).

In my view, the major problems with both ideas are that it is assumed it makes sense to talk about language, cognition and reality without presupposing, first, that we may have knowledge of and use language to *refer to* and say something which is true - or false - *about* reality, and therefore, secondly, without presupposing that reality exists independently of and as something of which we may have knowledge and talk correctly about.

Nevertheless, these assumptions in various forms lie behind the recent theories of language and cognition to be presented in the chapters which follow. Here, I shall argue why these assumptions are wrong, and propose alternative assumptions which, I argue, we shall have to take for granted in any discussion and investigation of language, cognition and reality. In the chapters to come I shall discuss the consequences of these alternative assumptions for developing theories of language and cognition which are consistent, and for what issues about our cognition and use of language such theories can and cannot explain.

5.2 Basic assumptions

Let me begin my argument of what I think we shall have to presuppose about the relation between language and reality with the following observation: Two language users, sitting opposite to each other in a room, may discuss what it implies to be language users sitting opposite to each other in a room. Indeed, they may spend the rest of their lives discussing the implications of the statements they may put forward on this subject - without ever being able to come up with an exhaustive list of what these implications are. However, they cannot at the same time question or subject to discussion whether they are in fact language users sitting opposite to each other in a room discussing this subject, nor doubt the correctness of statements to this effect, *without removing the very basis on which their discussion is conducted.*

To grant this is to grant that although it may never be possible for the two speakers to exhaust the implications of what it is to be language users being in some particular situation, no discussion between them as to what may count as implications can be *groundless*. Put differently, any such discussion would rest on presuppositions which cannot be called into question. Among the most obvious are that the two language users are speakers of a *public language* finding themselves in some particular situation in a *public world.* And it would rest on the presupposition that the speakers may put forward statements to this effect which are correct, and therefore that the speakers may use language to say something about themselves and the situation in which they find themselves. Furthermore, the discussion rests on the assumption that these statements forming part of the basis on which the discussion is conducted, have *determinate* and *determinable* implications and that, as users of language, the speakers know what they are. That such statements have determinate implications means that neither speaker can say. "I am a language user, who may use language to discuss with another speaker the implication of being language users sitting opposite to each other in a room, but I cannot use language to refer to myself or say anything about myself or the speaker sitting opposite to me, nor about the room in which we find ourselves" (for obviously, "I" just did and at the same time tried to question or deny that "I" could.) Nor can either of them say, "in the language I and my companion speak no true statements exist about myself, my companion or the room in which we find ourselves", nor say, "I, my companion and the room do not exist as things about which true statements may be made". In this sense statements put forward by the speakers about themselves and the situations in which they carry out their discussion have logical implications, i.e. the speakers not only know and may determine their implications, but also determine to *what* they apply. That statements forming part of the basis for the discussion have *logical* implications means that these statements cannot be used arbitrarily.

Now, should the speakers maintain that they do not know the correct implications and use of statements about themselves and other things forming part of the basis of their discussion, it would be immediately obvious that they were using language to maintain that they could not use language to talk correctly about anything. That is, it would be immediately obvious that they were contradicting themselves, or quite simply that they were talking nonsense. For, how could they even contradict themselves without assuming that the statements in which they deny that they can use language to talk correctly about anything do have logical implications, i.e. they know correct implications of such statement and how to use them correctly. So, if they do not want to contradict themselves or to talk nonsense, they shall have to assume that they can put forward statements which are true about themselves and the situations in which they are, and thus that, indeed, language may be used to say something about that which they may talk. If so, they shall necessarily also have to assume that that about which they talk exists as something about which true - or false - statements may be put forward in language. It is worth noticing at this point that this certainly rules out linguistic scepticism; for although we may be in doubt about the implications and correct use of any *particular* statement, then, to the extent that this doubt would have to be formulated in language, no *general* scepticism can be formulated which rules out the possibility of using language to talk correctly about that of which we may express doubt.

It is not difficult to show that what applies to the discussion between the speakers in the example above, applies *mutatis mutandis* to discussions carried out of any statement about anything in the world, i.e. such discussions necessarily rest on and presuppose the existence of statements about these things, which have logical implications. The speakers taking part in such discussions know what the implications of those statements are, and know to what they apply, i.e. they have a basis on which agreement or consensus may be reached as to *what* they talk about, and of statements which may be used correctly about it. If they did not, or disagreed on what those statements are, it can be shown that any discussion aiming at clarifying the disagreement, necessarily rests on a presupposed set of statements, the correct implications and applications of which they *do* know, and do *not* disagree about.

Let me give a concrete example. That the statement, "the thing on the table is an apple", has logical implications, means that a speaker making it cannot at the same time say, "but there is no apple on the table", or, "it is transparent and may be used for drinking beer out off", or, "it is sharp, and may be used for cutting". Conversely there will be a whole series of implications in the form of other statements to which the speaker has *committed* himself by putting it forward. These will be statements such as, "it is a fruit", or, "it is round in shape and has a stalk", or, "it grows on such and such kinds of tree", or, "if I cut it into halves I will see that it has a core" - or other statements of the type, "if I do such and such with it, this or that will happen, or this or that will prove to be the case"; or statements of the type, "it can be used in such and such a way" - all of which form part of the implications or *meaning* of the statement in question.

Now, the range of implications of a statement in the form of other statements to which a speaker has committed himself by putting it forward may of course vary. The range may vary in what may be called *intentional depth*, or *breadth* as well as in *intentional direction*, which, in its turn, may vary e.g. with the purpose, interest, or knowledge of the speaker, or the situation or context in which he finds himself. What is meant by this may be illustrated with the following example. The statement, "this is an apple", put forward by me, i.e. with my general everyday knowledge, experience and interests in apples,[2] and opportunities for observing and relating to apples in ordinary everyday situations, will certainly not have the same intentional depth as if put forward by a plant biologist, whose knowledge of, special interests in, and opportunities for observing and studying apples far exceed mine.[3] However, in order that he and I be able to discuss the implications of the statement, "this is an apple", to which each of us have committed ourselves by putting it forward, requires not only that we both know to what the statement may be applied, but also that, as users of the same language, we know and may agree on some general everyday implications of the statement. That is, *statements on which his and my further investigations of apples, as well as implications of statements about apples, must necessarily rest.* Without such a basis of agreed implications of the statements we use, our discussion could not get off the ground. And in the event that I and the plant biologist should disagree on a common basis of implications for the statement, "this is an apple", if for example I maintain that it does not imply, "it has a stalk", nor, "it is a fruit", the disagreement can only be cleared up if we both know the implication of the statements, "this is stalk", and, "this is a fruit" - and thus that a common basis of correct implications and use of "stalk" and "fruit" exists, and so on.

5.2.1 The reflexivity of natural language

Within philosophy of language and linguistics it is widely agreed that language is reflexive, i.e. we can use language to talk about language. This should come as no surprise since all investigations carried out on language within linguistics (i.e. on the constituents and structure of linguistic expressions) and philosophy (i.e. on the conditions for language and use of language) rely on and presuppose this fact. Indeed, there would be no sense in,

[2] i.e. that it is a fruit which grows on trees, that it is edible and has a particular taste, that when cut into pieces and boiled in a small amount of water to which sugar is added, makes a nice filling in a pie, and so on.

[3] Opportunities of observation and interest which for example may have provided him with knowledge of what family of trees apple trees belong to, what carbohydrates and proteins apples contain etc.

nor any justification for the existence of either disciplines without this assumption. In this respect philosophers and linguists are in accordance with ordinary language users, who not only find it *sensible*, but from time to time *necessary* to ask questions such as 'what does that mean' and 'what is that called', and who expect that there are quite straightforward sensible answers to such questions. For example, 'What does it mean that Peter is a bachelor?' - 'It means that he is not married'. Or, 'What is the thing I am pointing at called?' - 'It is called an apple'. The reflexive potential is inherent in natural languages, indeed it is so fundamental to natural languages that if we fail to grasp the significance of this fact, "we could no more be participating members of any linguistic community than we could play chess without understanding our role as players" (Harris, p. 163, 1996).

Not all communication systems have this reflexive potential of natural languages, and to-day it is agreed by philosophers and linguists alike to be a key feature which distinguishes human language from all other forms of animal communication (Harris, op.cit.). Animals may use signs of various sorts to communicate the presence of predators, but so far there is no evidence to suggest that they may communicate about the signals they use - for example, discuss whether different signs would be more suitable or appropriate as indicators of the presence of predators. In this respect the communication systems of signs among animals resemble man-made communication systems, such as the systems of traffic signs and traffic lights having signals for "stop" and "go". However, such systems do not include descriptions of the signs or signals themselves, nor of what they are meant to signify. Rather, the design of signals and signs, as well as the description of their significance, is something which has been decided and agreed on from *outside* the systems. Indeed, man-made communication systems of this kind rely on a language in which both the features of signs and their significance may be determined and discussed.

5.2.2 The concept of truth of natural language

There is yet another feature which crucially distinguishes both animal and man-made communication systems from natural language, and that is the existence in natural language of the notion of 'truth', i.e. of linguistic expressions to be true or false. This notion does not exist in mere "signal-systems" - not even implicitly - nor may the signs or signals of such communication systems be true or false. For example, an alarm call may be a signal indicating the presence of a predator, but it cannot *fail* to be a signal indicating the presence of a predator, nor be false about the presence of the predator any more than the signal "walking green man", once it has been determined to signal "go", can fail to indicate *go*, or be false about indicating go. In this respect the signals of animal and man-made communication systems are similar to "natural signs", such as smoke or footprints. Smoke may be a sign of fire, however, as a *sign* it cannot fail or be wrong about fire. Likewise, footprints in the sand may be a sign of someone having walked in the sand, however, as a *sign* it cannot be wrong or false about someone having walked in the sand.

In contrast, for someone (human or animal) to have a notion of 'truth', and thus of 'true', requires not only the notion that something matches or is in accordance with, or may be right or the case, but also that something may fail to match or be in accordance with, or may be wrong or not the case. In other words, to have a notion of truth requires that one may be mistaken and realize that one is mistaken. However, to realize that, say, one is mistaken in taking the "thing" over there to be a predator or an apple requires that one may realize that something is the case or *true* about predators and apples which is *not* the case or *false* about the "thing" over there. Conversely, it requires that something is the case or true about the "thing" over there, which is not the case or false about predators or apples. In other words, to be mistaken about predators or apples requires the notion that predators or apples are things about which something is the case or true and something (else) is not the case or false - indeed, infinitely many things may be false about them.

The consequences of this argument may be put in this way: If in natural language, just like in animal and man-made communication systems, we could only put forward propositions which were in accordance with the thing they represent, then we would have no notion of 'false', and therefore, no notion of 'truth', and none of 'true'. If so, we could not even begin to discuss what the implications of linguistic expressions are and are not, nor to what they may be correctly applied, and thus not use language to talk about language and the conditions for using language. (This point and its consequences will be further argued in Chapter 14 and in Part III Chapter 16.)

5.2.3 Consequences for Subjective Idealism

Now, although philosophical questions of the form 'What does that mean?' may be seen as extensions of everyday questions having the same form, and to which straightforward sensible answers may be expected, they differ from such question in that they invite us to reflect on the use of a certain term or terms. What makes questions such as 'What is justice?' or 'What is virtue?' philosophical is that they ask us "to ponder, for example, 'What is it that is called virtue?', 'What kind of conduct is called virtuous?' Or at least to begin by pondering such questions" (Harris, op.cit. p. xv). But this is not all.

> "To take such questions seriously is already, albeit implicitly, to endorse certain assumptions about how language functions. Behind the question 'What is virtue?' lies the assumption that the words virtue, virtuous, etc. are not randomly applied to human behaviour, any more than the words oak and ash are randomly applied to trees. Otherwise there would be no point in raising the issue in that form in the first place," (Harris, op.cit p. xvi).

As Roy Harris so succinctly points out in his short, but immensely important book, the problem is that philosophers (and linguists) take the reflexivity of language to mean that we are free to ask all sorts of questions about language which *violate* the implicit assumptions about how language functions. It is as if the mere fact that we may use language to ask these questions guarantees that they make sense, and also that sensible answers to them may be expected. Despite the fact that no linguistic expression may have a meaning and be understood by anyone independently of referring to communicative contexts or situations that speakers may be in and have knowledge about,[4] it is assumed by both philosophers and linguists that language and linguistic expressions may be investigated independently of how language and linguistic expressions are used by speakers in such contexts, and independently of the knowledge they have about these contexts and the things being talked about. Hence, however incomprehensibly, the assumption has formed in the minds of philosophers and linguists that because we may use language to talk about language, then language is something "inherent" which exists and may be investigated *sui generis*.

According to some theories language exists as a system of signs, the meaning and implications of which are entirely dependent on their relations to other signs of the system, and of the structure of linguistic expressions in which they occur (cf. Chapter 6 on Saussure's theory of the language system). Other theories have it that language is a system of syntactic forms which come ready-made in the minds or brains of speakers from birth, and which give speakers the competence to generate and understand infinitely many different propositions - independently of the experience and knowledge that speaker have of that which they speak, or of the context in which the propositions are generated and understood (cf. Chomsky's theory of a universal grammar (1990), and Fodor's argument for an innate language of thought (1976). Both theories and their assumptions will be discussed - and refuted - in Chapter 15). Now, for these theories, and many others which have in common with them the view that language may be separated from its communicative functions and contexts, and hence be investigated as a

[4]Anybody doubting this fact only needs to consider the substantial amount of background knowledge it is necessary to feed into a computer model of language understanding in order for the model to decode even quite simple everyday conversations.

system existing "in and of" itself, the problem inevitably arises that although we may use language to talk correctly about language and propositions put forward in language, we may not necessarily use language to talk correctly (i.e. in non-arbitrary ways) about anything else. And with this problem arises just as inevitably the problem of how to explain that language "hooks onto" or may "correspond" with that of which we speak - if indeed it does - i.e. the world, reality, things in reality, ourselves, the situations in which we find ourselves, etc. That is, we are back to the traditional Cartesian problem, now in the guise of a language-reality dualism and its assumption of the possibility of a polar opposition between language and reality - and with it the problem of whether anything exists in reality independently of or outside language.

In what follows I shall argue, first, that the existence of language is *incompatible* with the assumption that reality and things in reality do not exist independently of or outside language.[5] Later I shall argue why it is not possible to explain how language or linguistic expressions "hook onto" reality and things existing in reality, let alone possible to prove that a correspondence exists between statements about these things and the things "themselves". On the contrary, we shall have to take it for granted. That this is so, would seem immediately clear if we consider that the alternative would be to assume that we may use language to talk correctly about language and propositions put forward in language, and even use language to discuss and talk correctly about how, whether, or on what conditions propositions put forward in language "hook onto" or may be true or false about reality and things existing in reality, and at the same time assume that we may not necessarily use language to talk correctly about that to which the these conditions apply, i.e. reality and these things.

The former assumption, i.e. that reality and things in reality do not exist independently or outside language, has a long history, which in its current formulation in terms of *Language-Reality Idealism* may be traced back to Berkeley and his argument in favour of what is now called *Subjective Idealism*. Not surprisingly, the rejection presented below of Berkeley's arguments for Subjective Idealism will do just as well as a rejection of Language-Reality idealism. Now, the central tenet of Berkeleyan idealism - put in modern terms - is that because we can only come to know about things in reality via our perception of them, then things in reality do not exist

[5]Needless to say, I would not waste my efforts and time - nor that of my reader - to refute the assumptions of what I shall call *Language-Reality Idealism*, had it not been for the fact that this -ism and its assumptions have recently become rather fashionable - not only within philosophy and linguistics, but also within so diverse fields as psychology, literature, philosophy of science, computer science, biology and even physics. In a section which follows I shall discuss the two most frequently used arguments in favour of language-reality idealism.

independently of how we perceive them: indeed, things in reality only exist in virtue of being perceived. The quotations from Berkeley below shows the line of arguments by which he reaches this conclusion.

> "That neither our thoughts, nor passions, nor ideas formed by the imagination, exist without the mind, is what everybody will allow. And it seems no less evident that the various sensations or ideas imprinted on the senses, however blended or combined together (that is, whatever object they compose), cannot exist otherwise than in a mind perceiving them. - I think an intuitive knowledge may be obtained of this by any one that shall attend to what is meant by the term exists, when applied to sensible things. The table I write on I say exists, that is, I see and feel it; and if I were out of my study I should say it existed - meaning thereby that if I was in my study I might perceive it, or that some other spirit actually does perceive it. [...] This is all that I can understand by these and the like expressions. For as to what is said of the absolute existence of unthinking things without any relation to their being perceived, that seems perfectly unintelligible. Their esse is percipi, nor is it possible they should have any existence out of the minds or thinking things which perceive them. [...] In short, if there were external bodies, it is impossible we should ever come to know it; and if there were not, we might have the very same reasons to think there were that we have now." (Berkeley, 1930, V, Section III, p. 30-31, and p. 40)

Well, we shall have to agree with Berkeley that no ideas or imaginations may exist without the mind, and agree that sensations and perceptions of things do not exist "otherwise than in the mind". And we shall also have to agree that we cannot talk about the existence of things, which we have not perceived, and of which, therefore, we have no knowledge. The notion of the existence of such things is quite simply unintelligible, as Berkeley points out.

Now, following Locke, Berkeley believed that some of the ideas in our minds were *perceptions* of objects as opposed to mere ideas of imagination, and that perceptions of objects occurred as a consequence of sensations being imprinted on our senses. For example, sensations imprinted on our eyes would give rise to visual perceptions of objects, and sensations imprinted on tactile senses in our bodies and hands would give rise to perceptions of how things we touch with our bodies and hands "feel". However, so to believe implies granting a whole series of assumptions. Thus, Berkeley could not talk sensibly about perception of objects caused by sensations imprinted on senses without assuming that *senses* (e.g. eyes or tactile senses in hands) exist "external" to and thus independently of the mind in which perceptions or sensations occur. Nor could he talk sensibly about senses being part of our bodies and on which sensations may be imprinted, without assuming that something exists external to and independently of senses, which may cause sensations to be imprinted on senses. Indeed, Berkeley cannot tell the story of how sensations may be imprinted on senses in his hand and eyes without referring to actual objects - e.g. his table - which exist independently of the senses in his hand and eyes.

So, far from drawing the conclusion that the things he perceives (such as the table he writes on and the chair he sits on) only exist *in virtue* of being perceived and thus conclude, in effect, that things only exist *in virtue* of the existence of senses on which they may make imprints, his argument stands - and falls - with the assumption that he can refer to and talk meaningfully about the existence of things (tables and chairs), which may cause imprints on his senses, and thus that he can refer to things which necessarily exist *independently* of and external to senses and bodies of which they are part. What may now seem so curious is that Berkeley and other idealists could believe that it made sense to assume, on the one hand, that their bodies and senses existed external to their minds as things on which imprints may be made by things in the world, such as tables and chairs, and thus assume that senses and the bodies of which they are part exist independently of other bodies and things in the world, and, on the other, also assume that those other things and bodies may not be distinguished from and do not exist independently of their own bodies and senses.

So, although we have to agree with Berkeley that we cannot talk sensibly about things in the world which we have not perceived and of which, therefore, we have no knowledge, we shall have to say that the things we perceive necessarily exist independently of being perceived. Hence, the fact that we can only come to know things by perceiving them does not warrant the conclusion that then things do not exist independently of being perceived, nor the conclusion that things we perceive do not exist "external" to our minds. On the contrary, if we take into account Barkeley's assumptions of how perception - as opposed to mere ideas of our imagination - is caused by imprints made on senses by things in the world, quite the opposite conclusion is warranted, namely, that perceptions of things residing in our minds concern things which exist independently of and external to our minds. That is, it warrants the conclusion of an external reality existing independently of our perception of it.

The mistakes and confusions of assumptions in Berkeley's argument are two-fold. First, we have that senses are *means* of perception but also that senses, just like anything else existing outside our minds (including the things which may leave imprints on our senses) only come into existence as a consequence of being perceived. From this, secondly, we have the derived identity of 'to exist' and 'to perceive'. The absurdities which follow from this confusion become clear when we consider that it implies the claim that things which we perceive are indistinguishably identical with our perception of them. That is, it implies the claim *either* that the only things which really exist are perceptions or ideas in our minds of things such as tables and chairs. *Or* it implies that tables and chairs existing external to our minds only exist as perceptions or ideas. For example, when Berkeley is sitting on a chair, he is merely sitting on a perception or an idea of a chair, or rather, his perception or idea of his body is sitting on a perception or an idea of a chair, and so on.

More seriously, if things we perceive did not exist independently of being perceived, but only *in virtue* of being perceived, there could be no false perceptions or mis-perception of things, but only perceptions which were in accordance with or corresponded with the things being perceived. Apart from being patently wrong, this consequence leaves Subjective Idealism with serious epistemological problems. For, if there could be no false perceptions of things, and thus no notion of 'false', there could be no notions of 'truth', and none of true perceptions of things. If so, not only would the notion of the existence of un-perceived things be unintelligibly, but so also would be the notion of things which we perceive.

It is perhaps understandable that the mistakes of Subjective Idealism may occur and go unnoticed when our discussions about the existence of things in external reality are conducted in terms of our perception or ideas of them, i.e. in terms of phenomena residing in our minds which are not public, in the sense of not immediately available to others. However, if instead we analyse Berkeley's arguments from the point of view of the language and statements involved in his arguments - and in particular the presuppositions on which they rely, as well as the conditions on which any of them may have a meaning which may be understood - it becomes obvious what the mistakes of his arguments are, and why they are invalid. First of all, Berkeley would not have embarked on the project of writing a book and publishing it for others to read unless he assumed that he and his readers were users of a public language, and thus that the statements used in his argument have the same implications and correct use for other speakers as they have for himself. Nor, secondly, without assuming that that which his statements concern - i.e. rooms with tables and chairs, and persons with bodies, who may sit on chairs and write on tables, and who have eyes and hands with which they may see and feel these things - are things which exist in a public world of which both he and his readers are part, and about which he and they have knowledge. Indeed, they exist as things which are different from and may be distinguished from statements about things, and, therefore, are things existing independently of the statements he and other speakers put forward about them. And in order to talk about his mind and the perceptions and ideas of things residing in his mind, Berkeley necessarily has to assume that it makes sense to talk about his mind and his perceptions and ideas of things as something being distinguishable from and different from *statements* about his mind and his perception and ideas of things - just as the things and objects of his perception are different from and may be distinguished from the statements he puts forward about them. Thus, a statement about a chair is *not* a chair, nor is a chair a *statement* about a chair. Neither would anyone mistake a statement about a chair for a chair, for example think that one may sit on a statement about chair, or the like. Likewise, statements about chairs are not perceptions of chairs, nor would anyone mistake his perception of a chair for a statement about the chair.

If, for the reasons given above, Berkeley and everyone else has to assume that we can use language to talk about things existing in a public world, and also assume that the things about which we talk exist as something different from, and as something existing independently of the statements and description we use when taking about them, then I think it would be far less obvious to make the idealist mistake of identifying 'to describe' and 'to exist'. For, in the public world of things of which speakers are part and find themselves with other persons and things, no one would confuse a description of a chair with a chair, nor hold that it made sense to say that a description of a chair is identical with a chair, i.e. that a chair is something linguistical.

If we did talk in this way about things, i.e. maintained that the things about which we may talk do not exist independently of statements we put forward about them, but only in virtue of such statements, there could be no false statements about things, but only true ones. However, as argued before, without a notion of false statements, and thus of 'false', there can be no notion of 'truth', and none of 'true'. That is, we would be doing away with one of the key features of language and use of language, without which we could not begin to use language to talk correctly about anything - including language and the conditions for using language to say anything about anything, which may be true or false. But it would certainly also be obvious that we were using language to do away with things in the world as something which we may talk about - and therefore at the same time would be doing away with language as something we may use to talk sensibly about these things and this world - including ourselves, other speakers and the situations in which we find ourselves.

The discussion so far may be concluded thus: although we cannot talk about e.g. chairs without using statements about chairs, nor put forward descriptions of chairs without *referring* to and *describing* chairs, we shall have to assume that chairs exist as something different from and independently of the description we may put forward about them. Indeed, we cannot even begin to use language to talk about chairs and other things without assuming both that these things exist independently of the statements and description we put forward about them, and that those things necessarily exist as things which we may talk *correctly* about. That is, we shall have to assume that things in the world exist as things about which something is the case or true - and something else is not the case or false - and that we may use language to put forward statements about that which is true and false about them. To grant these assumptions is to grant the independent existence of external reality, i.e. of a reality existing external to and independently of language and the statements we may put forward in language about it. Indeed, the existence of language is incompatible with Language-Reality Idealism.

5.3 The principle of the general correctness of language

The discussion in the previous section has shown, first, that the use of language by speakers to communicate about themselves and other things existing in the reality in which they find themselves, rests on the presupposition that language may in fact be used by its speaker to talk about themselves and these things and situations. This presupposition implies that an *interdependency* exists between, on the one hand, statements put forward about reality and those things and, on the other, reality and those things. That is, as speakers of the language in which we talk, we know and may determine the correct implications of statements concerning that which we talk about, and know and may determine to what they may be correctly applied. And, secondly, it implies that things about which we may talk exist as things we may talk *correctly* about, i.e. they are things about which something is the case or true - and something else is not the case or false - and that as users of the language in which we speak, we may put forward statements about that which is true and false about them. Without granting both implications, if, say, we denied or doubted their validity, we would be using language to deny or doubt that we can use language to say anything correctly about anything; and we would be denying or doubting that there *is* anything which is the case or true, or not the case or false, concerning that which we talk about. In effect, we would be denying and doubting that we could use language to talk correctly about anything – and at the same time assume that we could get away with using language so to assert.

It seems that for reasons similar to the reasons why a polar opposition between perception and reality as two independently determinable "entities" does not work, such dichotomizing of language and reality does not work either. Just as it was shown in the previous chapters that we cannot talk about our perception of things in reality independently of or without referring to the things being perceived, we cannot in any well-defined way talk about or characterize the content of statements put forward about reality and things in reality, independently of or without at the same time referring to and describing reality and these things. Conversely, just as we cannot characterize reality and things in reality without referring to our perception of them, nor say anything sensible about what exists in reality without presupposing that we may perceive reality and things in reality correctly, we cannot talk in any well-defined way about reality and things in reality, i.e. talk about *what* reality and these things *are* and are *not*, without presupposing that we may use language to say something about them which is true and false, nor without presupposing that reality and these things exist as things about which true and false statements may be made.

Because the assumption of this inter-dependence between language and reality (or whatever we may talk about) is necessary for any of the notions we use to characterize both language and reality to be well-defined, I shall call the relation which exists between language and reality a *necessary relation*. However, as is the case for perception, we shall also have to assume that an *asymmetry* exists between language and reality. Thus, although true and false statements about reality and things in reality do not exist independently of language and speakers who put them forward, we shall necessarily have to assume that reality and those things do exist independently and as something different from the statements that speakers may put forward about them. The inter-dependency and asymmetry may be expressed in this way: no statement about things in reality exist without language and speakers who may put them forward, nor may any statement about these things have a content and be true or false independently of or without referring to things which exist *independently* of language, and as things about which true and false statements may be made in language.

The necessary relation which exists between language and reality I shall formulate in a principle which I call *The principle of the general correctness of language* (for short: the *Correctness Principle*). It is a principle which implies that, generally, language may be used by its speakers to say something about reality - and whatever they may talk about - which is correct, true or false. It is a principle, moreover, which has to be taken for granted and must form the point of departure for any discussion about language and reality, as well as for discussions about the correctness or truth of any *particular* propositions or statement put forward in language about reality. Indeed, without the presupposed *general* correctness of language, the correctness or truth of any *particular* proposition and statement about reality could not be discussed, let alone be determined. For, no such discussion or determination would make sense without presupposing that, generally, the language in which the discussion and determination is conducted may be used to say something which is correct, true or false, about that which we talk, and thus is a language to which the Correctness Principle applies.

It is not difficult to show that the Correctness Principle, and the necessary relation between language and reality which it expresses, has to be taken for granted. Nor is it difficult to show that the presuppositions entailed in this principle, just as the principles of classical logic, have to be taken for granted in order to talk consistently about anything. However, it is much more difficult to make clear what the consequences of this principle are, because there is a long tradition within philosophy of asking nonsensical questions about the relation between language and reality. We are so accustomed to thinking that, somehow, it must be possible to prove, justify or give evidence of the fact that language and propositions put forward in language may be correct or true of reality or that which we talk about, and even to thinking that it must be possible to explain how it comes about that a correspondence exists between our linguistic statements and non-linguistic reality. However, the claim of the Correctness Principle cannot be proven or justified, nor may it be explained how it comes about that we may use language to say something which is correct about reality or that about which we may talk. For, any such proof, justification or explanations would inevitably involve characterizing, and thus describing reality and, therefore, would have to presuppose that we can indeed use language and statements put forward in language to talk correctly about reality, etc. In other words, such proofs, justifications and explanations would be circular in that they would have to presuppose that which was to be proven, justified or explained.

Nor, for the same reason, may the claim of the Correctness Principle be disproved, denied or doubted. For any such disproof, denial or doubt to be well-defined, it would have to be presupposed that it could be formulated in language, and thus that we can indeed use language to talk correctly about that which the disproof, denial or doubt concern, i.e. reality and the possibility of using language to talk correctly about reality. In other word, such disproof, denial or doubt would be self-contradictory in that they would be attempts to disprove, deny or doubt the very presupposition for formulating the disproof, denial or doubt.

Hence, as in the case of principles of formal logic and axioms and theorems in mathematics, we shall have to accept that the only way to argue for the necessity of the Correctness Principle is to show that any proof of it would be circular, and that any disproof would lead to contradictions and absurdities. Now, if this will do to substantiate the principles of logic and axioms and theorems of mathematics, it will certainly do just as well to substantiate presuppositions and assumptions for theories of language and epistemology.

5.3.1 Consequences for Correspondence Theories of truth

According to the *Correspondence Theory of Truth* first proposed by Austin (1950), 'true' may be defined as correspondence to facts. Thus, the statement, 'the cat is on the mat', is true if and only if it corresponds to the fact that the cat is on the mat. However, as pointed out by Strawson in his famous debate with Austin about truth and fact (Strawson, 1950), this way of defining 'true' is problematic for at least two reasons. First, that a statement may correspond to some matter of fact would seem to imply that two independent things exist, i.e. a statement and a fact, and that it would be possible somehow to compare elements of the statement with elements of the fact, and come to the conclusion that a relation of correspondence exists between them. However, that 'true' is a matter of isomorphism or match between elements of statements and elements of fact, and thus between something linguistical and something non-linguistical, is absurd. And, secondly, if the only way of specifying a fact is to make a true statement, then 'facts' are not extralinguistic things, but already have the notion of 'statements' and 'truth' built into them. Or, as Strawson puts it, "facts are what statements (when true) state; they are not what statements are about". (ibid., p. 41).

I think we shall have to agree on Strawson's objections to the Correspondence Theory of Truth - and, for that matter, to any theory which undertakes to show that statements put forward in language may be true (or false) *in virtue* of a special relation, isomorphism, or fit between statements and some matters of fact. What is wrong with such attempts to define truth is the assumption that facts may be specified independently of true statements, and conversely, that statements may be specified independently of referring to facts - and thus independently of having already built into them the notion of facts.

Well, it is almost a triviality that one cannot compare a statement, i.e. a string of words, with some matter of fact, and that no straightforward relation exists between linguistic elements of a statement, say, 'the apple is lying on the table' and elements of the apple and table in question. Nevertheless, attempts have been made to determine properties of statements which are claimed to correspond to properties of facts in such a way as to make statements having these properties true - and thus in effect explain the notion of truth, and of language and statements put forward in language to be true about that which they are concerned, i.e. reality and its facts. The first and to date most elaborate attempt of this kind was Wittgenstein's "picture theory" of meaning and truth (of which I shall have a lot to say in part IV). But similar ideas may be found in theories espousing the view that statements put forward in language may be conceived of as *representations*, i.e. statements about facts are linguistic ways of representing facts. Accordingly, what makes statements true is that they represent that of which they are representations, and - by implication - what makes language "hook onto" reality and its facts is the property of language to represent reality and its facts. As has been pointed out by others (see e.g. Searle, 1995), the problem with both Wittgenstein's picture theory of language and the far less sophisticated theories which view the relation between statements and facts as a *representational relation*, is that they can only account for true statements about that of which they are "pictures" or representations, but not for false statements. For, as already mentioned, a representation of particular things or facts, as we normally understand the notion of representation, cannot be false about the things or facts, nor be a denial of the existence of things or facts, i.e. represent non-existing facts. So, if language and statements put forward in language really were representations or "pictures" of facts, neither the statement 'the cat is not on the mat', nor the statement 'the cow has five legs' would be false, but quite simply meaningless. Indeed, it would not be possible to distinguish between false statements about some particular thing or fact, and statements about non-existing things or facts, or statements which were devoid of meaning. However if, according to representational theories of language, no such distinctions could be made, then I think we shall have to agree that they are theories which do away with one of the key features of language, i.e. its notion of truth. A feature, that is, which crucially distinguishes language from representational systems and systems of signs - be they natural or man made.

And this leads us right to another problem from which such theories suffer, namely the problem that a "picture" theory of language as well as a theory of language as a representational system stand in need of an other language in which the representational relation between statements and the facts they represent are determined. For, as already noticed by Wittgenstein, such a ("pictorial" or representational) language cannot be reflexive,[6] i.e. it cannot be used for talking about the language itself, even less for taking about the conditions for how language and statements put forward in language may be used correctly. The problem may become clear if we consider what it would take to substantiate the claim that language and statements put forward in language stands in a representational relation to that which they represent, i.e. reality and its facts, and thus that language may be used to talk

[6] cf. Russell's comments in his introduction to *Tractatus* on Wittgenstein's "doctrine in pure logic, according to which the logical proposition in a picture (true or false) of the fact, [...] has in common with the fact a certain structure. It is this common structure which makes it capable of being a picture of the fact, but the structure cannot itself be put into words, since it is a structure of words, as well as of the facts to which they refer. Everything, therefore, which is involved in the very idea of the expressiveness of language must remain incapable of being expressed in language, and is, therefore, inexpressible in a perfectly precise sense" (*Tractatus*, 1922/1974, p. xx - xxi).

correctly about reality and its facts. So, for the sake of argument, let us suppose that it were possible to argue that statements put forward in language are in fact representations of reality and its facts. The first requirement would be that we have criteria and rules for determining whether the statements were *valid* representations of reality and its facts, and thus for determining what may count as an appropriate correspondence between the representations and the represented. But that would not be enough; it would also require that we have access to some other and *independent* determination and description of that which is represented, i.e. reality and its facts, and thus that we have access to a language about which it is presupposed that true *and* false description may be made of reality and its facts, and, moreover, that in this language we may discuss and determine the criteria for a correspondence between reality and its facts and statements being representations of reality and its facts.

Now, this presupposed independent language and its descriptions could not, of course, themselves be *representational*, i.e. descriptions to which the same kind of questions about their correspondence with reality could be asked; as said above, it would have to be a language of which it is *taken for granted* that it can indeed be used to talk correctly about reality and its facts, as well as about statements about reality and its facts. However, it is not at all clear how a distinction could be made between this presupposed independent language in which true statements may be made about reality and its facts, and the language we speak and which is being investigated as to its representational correspondence with reality and its facts. Anyway, why make do with the notion of the language we speak as being *representational*, if any substantiation of claims about the existence of a representational correspondence between this language and reality would have to presuppose that we have a language in which we can talk correctly about both reality and statements representing reality, and thus have a language which has the defining features of a natural language being reflexive and having a notion of truth? Indeed, why not assume that the language we speak is just this very language of which we have to take for granted that we may use it to say something about that which we speak, i.e. reality and its facts?

Let me summarize the point above in this way. The relation between language and reality cannot be characterized by an analogy to some matter of fact and a representation of it, as e.g. a face and a photo of a face. For, if we do try to talk in this way about reality we are, necessarily and inevitably, at the same time involved in describing reality. Although Strawson was right in pointing out that *facts* cannot be specified independently of true statements, and that in this sense 'facts' have built into them the notion of 'statements' and 'true', the conclusion he draws that facts, therefore, are something linguistical, was wrong. What he and virtually every other philosopher in this century overlooked and continue to overlook is that, just as importantly, nor can *statements* be specified independently of referring to facts, and thus in this sense *statements have already built into them the notion of 'facts'* - i.e. a notion referring to things existing independently of statements, and to which statements may refer and be true and false about. If so, it is obvious that we do not *need* a Correspondence theory of truth, and also why we end up in circularity and nonsense when we attempt to define true statements, and thus truth, in terms of a correspondence between statements and facts.

Because of this inter-dependence between statements and facts, and thus the necessary relation between language and reality and its facts, we shall have to assume that a *logical relation* exists between the notions 'statements', 'facts' and 'true', as well as 'language' and 'reality', i.e. none of these notions have well defined meanings independently of or without referring to well defined meanings of the others, nor may any of them be reduced to or explained in terms of the others. This means that the relation which exists between language and reality is *unanalyzable* - and this is precisely what is implied in the principle of the general correctness of language and statements put forward in language. It is a principle which renders impossible any *general* question of whether a correspondence exists between our linguistic statements and non-linguistic reality and facts, because neither notions of 'reality', 'statements', nor 'true' would be well defined independently of presupposing that this is the case.

However, that the relation between language and reality is unanalyzable, and that facts of reality cannot be specified independently of true statements, and thus that *truth conditions* for statements cannot be specified without using language, does not mean that truth conditions must therefore be something linguistical. But nor does it mean that the determination of the truth of any particular statement does not involve confrontation or checking with non-linguistic facts. What it does mean is that such confrontation or checking would not make sense without presupposing that, generally, we may use language to talk correctly about reality and its non-linguistic facts.

Let me illustrate this point with the following example of when and under what conditions it would make sense to determine the truth of the statements, 'there are pyramids in Egypt', by going to Egypt and checking. Let us suppose I am having an argument with someone as to whether such things as pyramids exist in Egypt, and hence whether the statement, 'there are pyramids in Egypt', is true. As far as I can see, there are two kinds of disagreement concerning the truth of a particular statement or description, which could be determined by means of a confrontation. Using the example in question, either there is a disagreement about

1. whether there exist such things in Egypt, to which a pyramid description may be correctly *applied*.

Or, there is disagreement about

2. what would be considered a correct description of pyramids, - i.e. disagreement about what the *implications* of a pyramid description are.

In the first case of disagreement a determination for or against the truth of the statement, 'there are pyramids in Egypt', necessarily presupposes that both I and my companion know the correct description of a pyramid, i.e., we know the implications in the form of other descriptions, to which one has committed oneself by describing something as a pyramid. If there is no agreement here, no amount of confrontation with "reality" will settle the dispute. If, for example, my companion holds that the implications of describing a thing as a pyramid are that pyramids are cubic and five feet high, no amount of confrontation with the specimens of Egyptian architecture will force him to change his opinion. Nothing would prevents him from insisting on his description of pyramids, or from maintaining that what he is looking at are not pyramids at all, and thus that the statement, 'there are pyramids in Egypt', is false.

Thus, a determination as to whether or not it is true to say that there are pyramids in Egypt can only be made by means of a confrontation, if I and my companion already know what it implies to describe something as a pyramid, i.e. provided we both know the correct implication of a pyramid description.

In the second case of disagreement, i.e. whether a description of the burial chambers to be found at a certain place in the Egyptian desert, must imply one meaning or another in the form of other descriptions, the disagreement may only be settled by means of a confrontation, if both I and my companion agree that the burial chambers in question are pyramids. If now it happens that my companion believes that a pyramid description implies that pyramids are cubic and five feet high, then a confrontation must lead to his necessarily withdrawing his original "pyramid description" - together with his previous opinion about the meaning or implication of a pyramid description. That is, a settlement of the argument will imply a revision of the original description according to the consequences that he on personal inspection has observed to be part of a correct description of a pyramid.

In neither case of disagreement however, whether when discussing the correct *application* of a particular statement, or when discussing the correct *implications* of a statement about some particular object, will it be reality itself which *unconditionally* terminates the discussion. This is to be understood in the following sense. In both instances of disagreement or discussion a settlement may only be reached, and the truth of the statement be determined because, first, the discussants are able to identify what they are talking about. And secondly, because in the language they speak, true - and false - statements may be made about that which they talk, i.e. they know their correct implications and applications. Without these conditions granted, and thus without presupposing that, in general, language may be used to talk correctly about reality and its facts, attempts to determine the truth of any particular statement about reality and its facts would not make sense.

This could also be said thus. Whereas the determination of the truth of any particular statement may be an empirical matter involving confrontation and checking with facts, it cannot be an empirical matter, nor a matter of confrontation or match that language may be used to talk correctly about facts. On the contrary, it is because it is taken for granted that language may be used to talk correctly about reality and its facts that we may determine the truth of any particular statement about reality and its facts.

I shall contend that to have a language and to be a language user is to know correct, true and false, statements about reality and its facts. Indeed it is to know and to be able to put forward infinitely many true - and false - statements about reality, oneself, other persons, and the situations in which one finds oneself. If this does not seem immediately obvious, we only have to consider the number of true statements and descriptions that I and my companion would have to know in order to settle the dispute in the example above - such as statements about Egypt, what Egypt is, where we find it and how we get there; or about shapes or sizes of objects; or about ourselves being persons with bodies who, like material objects, may be in or move to particular places at particular times, and so on. To determine the correct application or implications of the statement, 'there are pyramids in Egypt', both implies and presupposes countless other descriptions and statements, the correct application and implication of which we know and take for granted.

Now, if and when the description, 'this is a pyramid', proves to be true, it is so because it is correct in *this* situation, in *this* place, under *these* circumstances, to describe that which we have identified and are talking about as a pyramid. No other grounds or justification can be given. Put in general terms, the existence of the inter-dependence or logical relation between notions such as 'statements', 'facts' and 'true' means that we cannot ask how it comes about that we know true statements, nor how we know that true statements are true. That is, we cannot ask, nor explain how it is that we know correct uses of the word 'true'. On the contrary, notions such as 'true' and 'correct' have to be key-concepts or “primitives” of any theory of language and cognition (as they are in logic and mathematics), which we have to know correct uses of in order to talk about language and cognition and what language and cognition are about, i.e. reality and its facts. If we could not or doubted that we could use 'true' or 'correct' in a correct way, we would quite simply be cut off from using language for talking about anything.[7] If, for example, we say, "it is true that an apple is on the table, but there is no apple on the table", we would immediately know that the word 'true' was used in an incorrect way. However, that we and every child - maybe even philosophers - do know this suffices, indeed it *has* to suffice. For, there are no other, let alone non-question begging ways of explaining how we know correct uses of 'true' and 'correct'.

Nevertheless, it has been suggested, notably by adherents to logical positivism, that the reason we know true statements is that facts exist in reality about which true statements[8] may be put forward - and that we so to speak come to know true statements, and thus 'true', by investigating those facts of reality. This view implies that our notions of 'true statements' and thus of 'true' are *functions* of facts, i.e. something which we may acquire rigorous definitions of by analysing facts, or which we may derive from facts. In other words, this view implies that it is possible to investigate how it is that we can say something true or correct about reality - by empirically exploring the facts of reality about which true statements may be made. Now, this can hardly be the case since, by logical necessity, such investigations can only take place in any well defined sense on the condition that, as the point of departure for these investigations, we already have notions of 'true' and 'correct', and that we can correctly identify and determine that which the investigations concern - namely reality and its facts. Another way of saying the same thing would be that we cannot say how it comes about that we have notions of 'true' and 'facts' by analyzing facts - since we cannot begin to analyze facts and talk sensibly about the facts being analyzed, independently of knowing how to use 'true' correctly. In other words, 'true statements' and thus 'true' cannot be made into functions of that about which true statements may be made.

A view which is in principle quite similar to the view presented above may be found in the attempts by Computational Functionalist Theories to account for the meaning, intentionality and truth of beliefs and propositions. According to one version of this theory (i.e. Fodor's Causal Theory of Content which I shall discuss thoroughly in Chapter 12), true propositions

[7] Needless to say, we could not express doubt or deny the we know how to use 'true' and 'correct' correctly, nor express doubt or deny that we know what these notions mean, without having and using a notion of 'truth' and 'correct' when expressing such doubts or denial of which we presuppose that we know correct uses of and thus the meaning of. For the same reason it could not be proven that we know how to use the notions of 'true' and 'correctly' correctly, nor what these notions mean.

[8] So-called "protocol statements".

about facts in the world, i.e. about tables and chairs, molecular structures and states of the brain, may be explained in terms of a causal link between propositions and facts, and thus the notion of 'truth' may be causally derived from things which we can make true propositions about. Taken to their extreme conclusion both the attempts by Logical Positivism and Computational Functionalism to explain and derive the notion of truth from the things we may talk correctly about, would seem to imply *either* that things such as tables, chairs and molecular structures may be true in the *same* sense as statements about these things may be. However, so to say would hardly make sense. A chair or a molecule structure may exist, and it may be true that it exists, just as statements about their existence may be true; but it would be nonsense to say that a chair or molecule structure can be true. *Or* it would imply that statements about chairs and molecule structure, just like chairs and molecule structures, are not really the sorts of things which may be true. In either case it is not difficult to see that we would have done away with *both* language as something in which we may put forward true statements about reality and its facts, *and* with facts as things which exist in reality.

Where both logical positivism and a causal-*cum*-computational theory go *fundamentally* wrong, however, is by assuming that the notions of 'truth' and, by implication, the notion of 'reference', may be reduced to, deduced from or explained in terms of something more elementary or basic, which does not imply the existence of the notion of 'truth' and 'reference'. The untenability of this assumption is obvious if we consider the impossibility of accounting for "this" more elementary or basic, without *referring* to it, and without using the notion of 'truth' *when* referring to it.[9]

Let me conclude this discussion of the consequences of the Correctness Principle for Correspondence Theories of Truth by saying that this principle does not, of course, imply that everything we say about reality and its facts is correct or true, or that we cannot be mistaken. We can certainly be mistaken, indeed we may very often discover that we are, and determine how we are mistaken, i.e. specify quite precisely what the mistakes or faults consist in. However, it would seem obvious that a condition for doing so necessarily would be that in the language we speak true or correct descriptions or statements do exist about that which false or mistaken statements are put forward. Again, any questions or doubts as to the implications and truth of statements

[9] It could be suggested, indeed, it has been suggested (i.e. by Churchland, 1981), that the possibility remains of a future development of science in which accounts could be given of "this" elementary something in terms of which the notions of 'truth' and 'reference' (along with other notions of cognition and language), could be explained, and to which they could be reduced. That is to say, *reduced* in such a way as completely to eliminate and render superfluous these notions. In chapter 7 I shall discuss - and refute - this suggestion.

about reality and its facts will have to be carried out in a language in which we may put forward true statements about that which we speak. Otherwise, such questions or doubts would make no sense.

5.3.2 Consequences for Language-Reality Relativism

The term *Conceptual Relativity* refers to the fact that the same reality and things may be described from different perspectives or points of view, and using different conceptual systems or vocabularies. Indeed, nothing in reality puts a limit to *what* points of view may be developed or from *how many* we may observe, act with or investigate things existing in reality, nor limits to the conceptual systems and vocabularies we may use to describe what we observe. In this respect the points of view and conceptual systems we choose to observe and describe things are *arbitrary*, and will vary with or be relative to the conditions of observation available to us, as well as to the interests or purposes we have with reality and the things in the situations in which we find ourselves. This is fairly common place, just as it is - or ought to be - common place that both the interests, purposes and the possibilities of observations and description available to us to a very large extent are influenced by and depend on social, cultural and historical circumstances.

That the development of conceptual systems and conditions of observation in this sense depend on us, has been taken to mean that reality and the things and facts being observed and described by us are therefore mere *products* of how we observe and describe them - and that reality and those things and facts do not exist independently of, nor "outside" the conceptual systems with which we describe them. One of the arguments in favour of this view which, in some form or another, has been used by a wide variety of thinkers,[10] is that two conflicting descriptions of the same thing or event put forward within different conceptual systems may be true and, conversely, that one and the same statement about a thing or event may be true when put forward within one conceptual system and false when put forward within another. This shows, so the arguments goes, that the ways in which we "carve up" reality are dependent on the conceptual schemes we happen to have developed, and therefore that the truth of descriptions of reality is entirely a matter of how reality and things in reality are represented in those schemes. And, the argument continues, because conceptual schemes depend on *social convention* and *consensus*, which, in their turn, vary with or are relative to purposes and interests in the "consensual domain", the concept of

[10] In his book quoted earlier, Searle lists the following: Michael Dummett, Nelson Goodman, Thomas Kuhn, Paul Feyerabend, Hilary Putnam, Richard Rorty, Jacques Derrida, Humberto Maturana, Francesco Varela, and Terry Vinograd. However, the list could be extended to include Kenneth Gergen, Rom Harré, Jerome Bruner - and many, many others.

'true descriptions', and thus of 'truth', are themselves arbitrary or "relative" concepts depending on social conventions and consensus.

However, conceptual or linguist relativity cannot be used as an argument for the view that there are no non-arbitrary or *unambiguous* ways of determining what exists in reality, nor used for the view that reality and things in reality do not exist independently of language and conceptual schemes. Indeed, it is not difficult to show that any of the (precious few) examples in the literature in support of this view warrant the exact opposite conclusion. Let us take the example given by Putnam (1990, p. 97). He invites us to imagine that a part of the world exists as shown in the figure below.

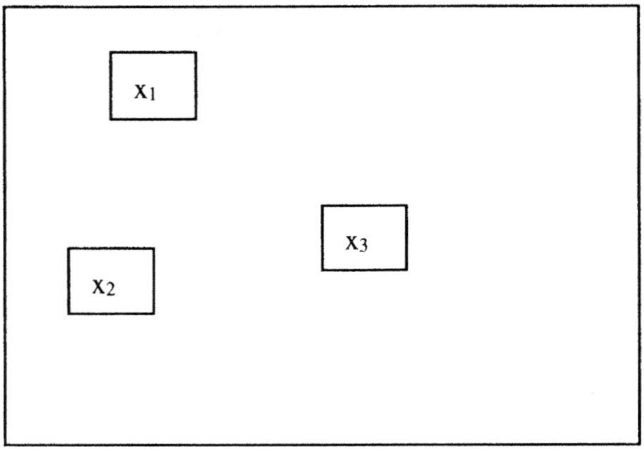

He then asks us to tell how many objects there are in this "mini-world". According to Carnap's system of arithmetic - and to the ordinary everyday scheme of things - there are three objects, namely x_1, x_2, and x_3. However, according to the mereological calculus invented by the Polish logician Lesniewsky there are 7 objects, namely

$$1 = x_1 \qquad 5 = x_1 + x_3$$
$$2 = x_2 \qquad 6 = x_2 + x_3$$
$$3 = x_3 \qquad 7 = x_1 + x_2 + x_3$$
$$4 = x_1 + x_2$$

So, how many objects are there really in the imagined world - are there 3 or 7? To this question, says Putnam, there is no absolute or unambiguous answer. The answer we may give will always be relative to an arbitrary choice of conceptual schemes. According to one of the schemes in the example, the same sentence 'There are exactly three objects', will be true, whereas it will be false according to the other. This seems to show that

Realism, i.e. the assumption that reality and things in reality exist independently of conceptual schemes and vocabularies, leads to inconsistencies, because it allows for conflicting descriptions of the supposedly independently existing reality and objects existing in reality. However, as pointed out by Searle in his convincing attack on this line of argument,

> "if conceptual relativity is to be used as an argument against realism, it seems to presuppose realism, because it presupposes a language-independent reality that can be carved up or divided up in different ways, by different vocabularies. Think of the example of alternative arithmetics: Putnam points out that one way to describe the miniworld is to say there are three objects, another way is to say that there are seven objects. But notice that this very claim presupposes something there to be described prior to the application of the description; otherwise there is no way we could understand the example." (op.cit. p. 165).

The discussion in later chapters of this book of other and more complex examples by Putnam (in Chapter 14) and Nelson Goodman (in chapter 17 of Part IV) will show that, for precisely the reason pointed out by Searle, linguistic relativity cannot be used as an argument *against* Realism, let alone as an argument *for* Anti-Realism. On the contrary, all examples showing that the same thing may be described in different ways according to different conceptual schemes, and conversely, that the same statement may be true according to one conceptual scheme and false according to another, necessarily presuppose that things exist independently of the schemes to which the descriptions and statements may be applied. That is, they presuppose that there is something which these things are or are not independently of (or prior to) our descriptions and statements about them, and that true or false statements and descriptions may be put forward by us concerning what those things are and are not. Inconsistencies and conflicts only arise when linguistic relativity is used as an argument against Realism and for Anti-Realism. For, such an argument inevitably involves *both* the claim that we can make different true statements about the same thing, just as the same statement may be both true and false, *and* the claim that nothing exists in reality about which the true statements may be true, and the false statements may be false, i.e. nothing exists to which those true and false statements may be applied. Even worse, however, if nothing exists "outside" language and conceptual schemes, then there would be nothing to which different descriptions belonging to such schemes could be applied, let alone be applied correctly, and thus no difference to determine of ways in which the same things may be correctly described. Indeed both *things* and *different description of them* would disappear.

Arguments to the effect that linguistic relativity supports Anti-Realism or, for that matter, any other claim that reality and things in reality do not exist independently or "outside" language and conceptual systems, have built into them consequences which are not only conflicting, but also self-defeating. For, such arguments imply, on the one hand, that we may indeed use language to talk correctly about and determine how different conceptual systems and vocabularies are used to describe the same reality, i.e. how, according to conventions and consensus determined by social, cultural and historical circumstances, statements and descriptions may be applied to this same reality. On the other hand, such arguments also imply that this *same* reality, to which these different conceptual systems and descriptions may be applied, is itself purely "conceptual" and "linguistic", and just as "arbitrary" and "relative" as are all other notions of socially, culturally and historically determined conceptual systems and vocabularies. In other words, if Anti-realism and the view about the "arbitrariness" and "relativism" of concepts and the truth of statements and descriptions were correct, it would also have to apply to the descriptions of examples used in the arguments for Anti-realism and this view. Neither those examples, nor the notion of *the same reality* would be "immune" to this relativism and arbitrariness. In the final analysis, the very statements used to formulate this view, can also only be "arbitrary" statements, which can only be "relatively" true.

Let me conclude the discussion so far by saying that any investigation into how different conceptual systems and vocabularies may be used to describe reality and things and events in reality in different ways, presupposes that we have a language in which we may talk correctly, not only about such conceptual systems and vocabularies, but also about the reality and things and events in reality to which statements put forward in these conceptual systems and vocabularies may be applied. Moreover, any such investigation presupposes that reality may be "carved up" into things and events about which different things may be the case or true, and about which different true statements may be put forward from within different conceptual systems. In other words, the fact of linguistic relativity, in order to be a *well defined* fact, presupposes that there is nothing arbitrary or relative about the use and implication of statements put forward within such conceptual systems or vocabularies. Just as it has to be presupposed *in general* that a necessary relation exists between language and reality, it has to be presupposed for *any part* of language and any conceptual system developed within language. To presuppose this necessary relation is to presuppose that statements about reality put forward in language, or any conceptual system being part of language, has logical implications, i.e. they have determinable implications and correct applications. That statements have logical implications means that they cannot be used arbitrarily; nor can we give reasons for their correctness without using other statement about which it is presupposed that they have logical implications. Hence, it is *because* statements have logical implications that they may be true or false, i.e. *that their truth may be determined.*

For these reasons it is wrong to say that the truth of statements, and thus of truth, relies entirely on social conventions agreed among language users, and wrong to say that correct use of language is *fundamentally* a matter of consensus. Following the arguments above it can always be shown that for language users to develop conventions of how to describe things, presupposes that they have identified the things into which reality may be carved up as things about which something is the case or true, and something else is not the case or false, and that in the language in which they speak and formulate the convention, a notion of 'true' and 'false' already exists of which the speakers know both their correct implications and applications. If the implications and uses of 'true' and 'false' were *arbitrary*, and merely a matter of what speakers could come to agree on, then discussions of the truth and falsehood of statements would be discussions in which the very conditions governing the discussions would be arbitrary and could be questioned. That is, one would be taking part in a discussion of how to use 'true' and 'false', independently of or without presupposing that such discussion was conducted in a language in which true and false statements could be put forward. Put differently, if 'true' and 'false' were mere conventional notions, i.e. notions the use and implications of which language users come to agree on, we might well ask in what language and with what concepts of 'true' and 'false' such conventions are being established.

5.3.3 The logical space of descriptions

As said in the previous section, reality and things in reality may be described correctly in different ways. How, in an actual situation we describe things, will vary with our aims, purposes and interests, as well as with the conditions of observation and conceptual systems available to us in the situation. Now, the meaning and truth of any particular statement or description put forward in a particular situation will be determined by a limited set of relevant, correct descriptions or statements. Such a limited set of descriptions or statements I shall, following Mortensen (1973), call *the logical space* of the description or statement, and I shall be talking about the meaning and truth of a particular statement as being determined by its relation to other descriptions within a given logical space.

Let me give an example. Take the description, 'This is O'. In a previous section I said that some of the implications of a description to which a speaker has committed himself by putting it forward, will be statements of the type ‘If one does H with O, it will be observed that P’. H is a description of some act or other, whereas P is a statement about the consequences of carrying out H with O, which belongs within the limited set of correct statements, or logical space, which determines the meaning and truth of the description, 'This is O'. Now, what will be the implications of a given description in the form of other descriptions to which a speaker has committed himself by putting it forward, will depend on the situation, i.e. it will depend on the conceptual system and means available to the speaker for observing and acting with O. If, for example in an everyday situation with everyday means of observation and description I say 'This object is red', I may imply 'If I compare its colour to the colour of a British letter box, I will notice that it has the same colour as the letter box'. Or if I am a physicist studying the physical properties of colours, I may imply 'If I measure the composition of the reflected light, it will be observed that the greater part has wavelengths exceeding 600 mmy'.

A given description, then, will always belong to a particular logical space of possible relevant descriptions and concepts within which its meaning may be understood, as well as procedures by which its truth may be determined. If, say, the truth of a physical description is to be determined, the investigation will always be based on previously defined, relevant physical descriptions and arguments, i.e. with regard to the logical space of correct descriptions of physics, and in accordance with their correct use in well-defined physical situations. Similarly, a determination of the truth of an everyday description will be made with regard to the logical space of everyday descriptions which define its meaning, and according to everyday procedures for determining its correct applications.

But note: this does not mean that a description is only true in a situation in which its meaning is defined and its truth may be determined. Thus, the statements, 'This is a cup', and, 'This is a particular structure of molecules', put forward about the cup on the table, may both be true - and are undoubtedly true throughout all situations in which the cup exists. However the statement, 'This is a particular structure of molecules', put forward about the cup, does not belong within the logical space of possible everyday descriptions of the cup. This only means that the meaning and truth of this statement cannot be determined within an ordinary every day situation. What it *means* to describe the cup as 'a particular structure of molecules' is defined within a logical space of physics, and with regard to the possibilities of observation, action and descriptions belonging to situations in physics. That the meaning and truth of the statement, therefore, can only be *determined* within such situations of physics, and not within normal everyday situations, does *not* mean that the statement is false in an everyday situation - in the sense that in an everyday situation cups are not built up from molecules. If we did say that, we would be saying that the properties of reality and objects, as described within physics, only *exist* within such situations of description, not in everyday situations; for example, that the law of gravity only applies in situations in which the terms of the law are well defined, and in which the precise effect of the law may be observed according to some particular and well defined procedures. However, if we do say this, we would be saying that physics and the descriptions and laws of physics do not concern the very same material reality and objects, which in other situations may be correctly described as 'cups', 'tables' and 'chairs', or 'mountains' etc. - Conversely, I think it is obvious that we shall have to say that the statement, 'This is a cup', meaning, 'something out of which we may drink tea', is true in all situations - regardless of what other (e.g. physical) properties of the cup are being investigated and described in those situations.

So, let me finish this chapter as I began it by saying that it takes a language, which may be used by its speakers to talk *about* language and the conditions for using language, to make conventions of how language may be applied to that which it is about, i.e. reality and things existing in reality. But it also takes a language of which it is presupposed that it can be used to talk in non-arbitrary and correct ways about that to which the conventions may be applied - and thus that the things in reality exist as things, which we may use language to talk about correctly. Therefore, language and use of language cannot be *fundamentally* a matter of social convention.

Chapter 6

Language, concepts and reality

6.1 Introduction

Natural languages do not exist without people who speak them, and a world of things they speak about and of which they themselves are part. It is essential to natural languages that they may be used by their speakers to express and communicate what they know and think about this world, what sense they make of it, of themselves, others, the situations in which they find themselves, as well as the acts they perform in those situations. And it is this communicative fact about languages that makes it possible for its users to develop social institutions, societies and cultures. It is through language that persons may come to agree and make conventions and rules for how to organise and carry out their lives in societies, and determine what acts to perform - and not to perform - within their institutions. And it is through language that people establish and come to learn about the values, significance and purposes of the institutions which make up the societies and cultures in which they live.[1]

However, in significant respects, language may itself be considered as a social institution. Thus, the way we use language and what we say in particular situations, and how we understand what is being said to us, all depend on our knowledge of the rules and conventions that govern how things may be talked about in these situations. Indeed, to be capable of carrying out any of the wide range of quite basic speech acts, such as making descriptions,

[1] For the most consistent and comprehensive account to date of the role played by language and linguistic symbols in the "Construction of Social Reality", I refer the reader to Searle's book of 1995.

asking questions, putting forward propositions, requests, promises and giving orders, expressing feelings and emotions, rely to a large extent on our knowledge of rules and conventions for how to perform such acts. But although language is part of the reality in which persons exist, as are other social institutions, language is not an "entity" which may be observed at particular places, nor a "substance" or "phenomenon" which may be found or discovered like explorers may find or discover the sources of the river Nile (Mortensen, 1972, p. 17). What may be found and observed are *people who speak*, and *texts being produced in language*.

Unlike a thing or an object, a linguistic occurrence has implications or a meaning, and may be used to refer to something else; indeed, a linguistic expression has implications or meaning *only* in virtue of being about or referring to something else; that is to say, it has a meaning *for* someone, *about* something, *in* some particular situation or context, and the meaning of a linguistic expression is understood by a language user in virtue of his knowledge of *what* it refers to, and thus knowledge about *that* to which it refers. Take away the existence of apples in the world as things we may refer to and talk about, and the expression, 'The apple is lying on the table', would have no meaning, nor would it, as argued in the previous chapter, be a statement which could be true or false. Hence, it is in virtue of the inter-dependence between the meaning or implications of a linguistic expression and its reference to non-linguistic reality that language users may use language to communicate about things in the world. Without this inter-dependence between language and what language is about there *is* no language. In this sense the relation between language and what language is about is a necessary relation.[2]

To this it may be objected, firstly, that we not only use language to make statements or descriptions of things in material reality; we may also use language to talk about other things, for example use it to talk about language and assertions put forward in language. Indeed, it is a defining property of natural language that we can use it to talk about language. As argued in the previous chapter, however, I do not think that language users can make assertions about anything, including language and assertions put forward in language, without presupposing that they can make assertions about themselves and the situations in which they find themselves in material reality - including the situations in which they talk about language; that is, without presupposing that users of language can talk correctly about themselves as persons, who may be in particular places at particular times in material reality. *Any* language, which any language user may make use of

[2] Natural language may of course also be used to talk about things which do not exist in reality, e.g. about fictitious things and circumstances. The relation between the meaning of terms and what they refer to in such circumstances will be discussed in Chapter 10.

must of necessity be, or rely on, a language which has concepts about matters of space and time that serve to locate not only the physical surroundings, but also the persons making assertions about them. The alternative assumption, that we may use language to talk correctly about language, but not use language to talk correctly about ourselves or other things existing "outside" and independently of language, would be utterly nonsensical.

And it may be objected, secondly, that the inter-dependence between meaning and reference only applies to the type of linguistic occurrences quoted above, i.e. to speech acts such as descriptions, assertions or propositions, but not to others such as orders, requests, exclamations, or promises. However, here I think we shall have to agree with Austin (1962) that descriptions, propositions or assertions are fundamental in that they are implied in all other speech acts, and thus must exist in any natural language. For example, the order, 'Give me the axe', could hardly exist as a speech act in a language in which the assertion, 'This is an axe', did not exist. But nor would anyone understand what the order involved, unless a description of the axe being referred to was implied, i.e. that it is a physical object of some particular kind, having a particular shape, being made of some particular material and having some particular function.

Thirdly, as pointed out by the Swiss linguist Ferdinand de Saussure, to whose theory of language I shall turn next, it is not only linguistic expressions that have content or meaning; a whole range of other social phenomena may function as *signs*, and thus are phenomena which belong within the domain of what he calls *semiology*. According to Saussure a linguistic sign - as well as signs in general - may be characterised as a union of a *signifiant* or signifier (in the case of a linguistic sign, patterns of sound), and an idea or a concept signified, which he calls the *signifié* or signified. (Saussure, 1960, p. 158). Furthermore, in order for signs to communicate ideas they must be part of *a system of conventions* - a system of signs. In this respect the signs of natural languages, according to Saussure, are comparable to signs belonging within other conventional systems being used to express ideas - such as "the system of writing, the alphabet of deaf-mutes, symbolic rituals, forms of etiquette, military signals, etc." (Saussure, 1983, p. 15, [33]).

How Saussure came to see language as a system of signs, and in particular how he came to view this system as a methodological unity within the domain of linguistics, will be the topic of the next few paragraphs. As will become clear, however, Saussure's assumptions about the properties and principles by which the language system may be characterised - and distinguished from its use by speakers - disagree with almost all the general points about language and use of language for which I have argued up to this point. According to Saussure, the language system - by which he means that part of language that constitutes *the body of necessary conventions and grammatical forms* adopted by speech communities to enable their members to talk - is a *self-contained* system which can and should be studied by linguists independently of its use by speakers in actual situations of communication, and independently of that which language is used to refer to. Thus, far from assuming that signs of a language system acquire their meaning and implications in virtue of referring to things and events outside this system, Saussure insists that the meaning or "value" of words and other linguistic elements is inherently given within the language system itself, namely by the differential relations and oppositions of those elements to other elements within the sets or categories to which they belong.

After introducing Saussure's arguments for the view of how the language system may be conceived of as a methodological unity, as well the principles applying to this system (i.e. *the principle of the arbitrary nature of the sign*, and *the principle of the differential and relational value and identity of signs*), there will follow a discussion of both the methodological and epistemological problems to which these views and principles give rise. Problems, which show themselves most acutely in attempts to give concrete examples by which to demonstrate their feasibility.

Views and assumptions which are in principle similar to, and thus have consequences which are correspondingly similar to those of Saussure's, may be found in a wide range of current theories of language and cognition within the humanities, anthropology and social psychology - and not only the so called *structuralist* and *deconstructionist* theories which are claimed to derive from the views and principles developed by Saussure. Although these theories are dressed up in different and more contemporary idioms, they have in common with Saussure's view that language may be seen as an independent, self-contained and holistic system of conventions and categories, and also that language - whether it be seen as a social construction negotiated among members of a speech community, or as an innate syntactic structure built into our genes - somehow organises and determines our cognition of the world and its properties. And they have in common the view, conversely, that language and its structures and categories somehow "structure" or "organize" the world of things and their properties - and thus "emerges" as products of the structure and categories that language imposes on them. In the present chapter I shall give only a few examples of such theories, and only briefly. In later chapters, other examples will be more extensively discussed, including models which attempt to give so called *computational* accounts of language and use of language.

However different these various models or theories of language as an independent, self-contained system may appear, they all without exception have one further thing in common: they ignore the presuppositions on which rely the determination of the principles and rules, which the theories claim to structure language and govern its use. This neglect, I shall argue, causes problems which are not unlike the problems which hampered the attempts by Marr and Gibson to develop so called complete theories of perception (cf. Chapter 3). Thus, Marr and Gibson, in their attempts to account for how it comes about that we perceive things in the world as they are, had to rely on or supplement their models of perception with prior knowledge about what the world looks like. In the same way, linguists and psychologists, in their attempts to account for language and its structure independently of its use, have to presuppose and rely on knowledge about what language and expressions put forward in language are used to refer to and are about.

The recurring neglect of the presuppositions on which are based theories and models of language from Saussure onwards, is seen no clearer than in Saussure's own arguments for the status of the language system as a self-contained or *immanent* system - nor are the problems to which these arguments gives rise; this justifies an extended discussion of Saussure's view of language and the assumptions on which they are based.

In the discussion - and refutation - of the notion of a self-contained language system, I shall use turns of phrase such as, "linguistic occurrences considered as acts that people carry out in situations in which they find themselves", and I shall be talking about "the knowledge (linguistic or non-linguistic) of language users about such situations", however without first having defined or made clear what I mean by the terms 'situations', 'act', and 'linguistic vs. non-linguistic knowledge'. I shall make up for this in the chapter which follows (Chapter 7) in which an account is given of the formal implications of these terms.

6.2 Saussure's delimitation of the language form as an independent object of linguistic research

After having made considerable methodological and theoretical contributions to the analysis of Indo-European languages and their development, Saussure was for many years intensely preoccupied with fundamental problems concerning the general nature of language, and with how language could be conceived of as a coherent organised structure, amenable to scientific study. In this pursuit, he attempted to develop methods and terms which would provide linguistics with logical classifications of linguistic facts, and to determine the points of view from which they could be treated. Sadly, Saussure's untimely death at the age of 56 prevented him from completing this work, and even more sadly, he only left very few notes about his thoughts and work, and never published any of its results. The book, *Cours de Linguistique Generale*, which for the most part consists of an edition by his colleagues of notes taken by students attending his introductory lectures in linguistics during 1908 - 11 (and published in 1916, three years after Saussure's death) may be seen as Saussure's attempt to give a coherent account of his thoughts and views thus far about what sort of an object language is, and, as he puts it, 'to show the linguist *what he is doing*'.[3]

Right from the start of *Cours* Saussure points out the dilemma which confronts linguistics: language, taken in its entirety is immensely heterogeneous and complex, and "no one object of linguistic study emerges on its own account". The range of facts involved in just the most simple speech act, and the points of view from which these facts may be considered, is virtually endless. One may consider how the sounds are produced by the speaker and perceived by the listener, or how the meaning or idea is communicated by the speaker and understood by the listener; or consider the intention of the speaker, the circumstances or context in which the speech act is being performed, or consider what in the situation motivated the speaker to perform that particular speech act, etc. All these aspects of language, as well as the points of view from which they may be considered, situate the study of language within a variety of different research areas - such as psychology, sociology, physiology and philology - each with their own sets of terms and methodologies. However, unlike other sciences, which are provided with an object of study given in advance, and which is then examined from different points of view, the object of linguistics is not given in advance of the points from which it may be viewed; rather, "*it is the viewpoints adopted which create the object*". (*Course*, p. 8, [23]. Italics added.) Saussure suggests the following solution to the dilemma of linguistics, a solution in his opinion, which resolves all the difficulties:

> "*The linguist must take the study of linguistic structure as his primary concern, and relate all other manifestations of language to it.* Indeed, amid so many dualities, linguistic structure seems to be the only thing that is independently definable and provides something our minds can satisfactorily grasp." (*Course*, p. 9, [25].)

This *linguistic structure*

> "is a set of forms, a fund accumulated by the members of the community through the practice of speech, a grammatical system existing potentially in

[3] Unless otherwise indicated, the quotations in English from *Cours de linguistique generale* (1960), is from *Course in general linguistics* (1983), translated by Roy Harris. Although the book was not written by Saussure himself, quotations and passages referred to will be attributed to Saussure - as is the usual practice. The book will be referred to by the shorthand *Cours*, and in references made to quotations from the English translation, by the shorthand *Course*. References are made to pagination of both the English translation and the French original; the latter appear in square brackets.

every brain, or more exactly in the brains of a group of individuals." (*Course*, p. 13, [30])

And it is the *system of signs* and their union of meaning (or ideas) with patterns of sound, which exists independently of the utterances that the individual language user happen to express in order to communicate his own thoughts. (*Course*, p. 14, [32]) This part of language, the social and collective linguistic structure of language, for which Saussure reserves the term *langue*, he distinguishes from "the execution of language", *parole*, which are individual acts of speech carried out by speakers. For linguistics the primary concern must be *langue* - in so far as it must be the primary interest of linguistics to determine the units and rules of combination which makes up the linguistic system, but not to describe speech acts. To study language as a *form*, a system of signs, Saussure says, is to study the *essential features of language*:

"By distinguishing between language itself and speech, we distinguish at the same time: (1) what is social from what is individual, and (2) what is essential from what is ancillary and more or less accidental." (*Course*, p. 14, [30].)

By this distinction and delimitation of the language system from its use, linguistics is provided with a well defined object of its own. A science which studies linguistic structure, says Saussure, "is not only able to dispense with all other elements of language, but is possible only if those other elements are kept separate". (*Course*, p. 14, [31].) Indeed,

"A language as a structured system is both a self-contained whole and a principle of classification. As soon as we give linguistic structure pride of place among the facts of language, we introduce a natural order into an aggregate which lends itself to no other classification." (*Course*, p. 10, [25]).

Well, a distinction between, on the one hand, language considered as a linguistic structure or system and, on the other, speech acts performed in language, would seem perfectly reasonable. For if use of language is to be considered as *acts* performed by speakers in language, and thus that any linguistic utterance expressed may be seen as a *choice* made by a language user, there must be *possibilities of choice*, i.e. some kind of system or "repository" which defines the possible choices - a "repository", that is, which is shared by users of the same language. The distinction between "system" and "speech act", or between *langue* and *parole* as defined by Saussure, has become axiomatic within modern linguistics, and it is generally agreed among linguists to-day that a linguistic occurrence, be it a literary text, a shout in the street, or a military order, may be described in two ways which are different in principle, namely

1. as an element or component defined by its position in a linguistic system,

2. as a speech act, i.e. a linguistic choice made by a speaker in a particular situation of communication.

From the perspective of a *linguistic system* a linguistic occurrence may be characterised as a sentence in a particular mood, e.g. indicative, imperative, interrogative etc., whereas the same linguistic occurrence according to the *speech act description* may be characterised as a particular utterance, e.g. a proposition, an order, or a question. The speech act description indicates what a particular linguistic occurrence is *used for*, whereas the systematic description is about the *structural characteristics* of the occurrence.

In the next paragraphs I shall present what, according to Saussure, are the main features and principles of the language system.

6.2.1 The Principle of the arbitrary nature of the linguistic sign

The first principle to be mentioned in *Cours* concerns the arbitrary nature of the linguistic sign. This principle is based on the observation that no natural or inevitable link exists between the sound pattern and the idea (or concept) with which the sign is united. In English, for example, the signifier 'dog' represents the notion or idea of an animal of a particular species, but this sound pattern is no better suited to serve this function than are other sound patters used in other languages, such as 'chien', 'hund', or 'cane'. There is no intrinsic reason why any one of these signifiers rather than others should be united with the idea or concept of a *dog*, and, therefore, no inevitable or natural link between signifier and signified.

That in this sense linguistic signs are arbitrary may seem a rather trivial fact and, as pointed out by Culler (1986), on this interpretation alone, the principle of the arbitrary nature of the sign could hardly have the paramount consequences that Saussure assigns to it. However,

> "it is often easier to discover a truth than to assign it to its correct place. The principle stated above *is the organising principle for the whole of linguistics*, considered as a science of language structure. The consequences which flow from this principle are innumerable. It is true that they do not all appear at first sight equally evident. One discovers them after many circuitous deviations, and so realises the fundamental importance of the principle." (*Course*, p. 68, [100]. Italics added.)

One of these deviations leads one to the observation that not only is the choice of *signifier* arbitrary, but even the *signified* itself is arbitrary. Just as there are no universal or fixed signifiers being intrinsically linked with particular signifieds, no universal or "pre-existing" concepts or signifieds exist with which signifiers may be linked. Language, says Saussure, is not a *nomenclature* for a set of universally existing concepts. The arbitrary nature of signifieds, i.e. the concepts or ideas contained in linguistic signs, is evidenced by a series of observations.

First, if languages consisted of universal or fixed concepts, the signs of languages would never undergo changes, and in the historical evolution of a language its concepts would remain stable. However, it is well known that the concepts of a language do change as the language develops, and that the boundaries of its concepts also change.

Moreover, if language were just a nomenclature for a set of universal concepts, there would be no problem in translating from one language to another or in learning a new language. One would only have to replace a name for a concept in one language with the corresponding name in another. However, as is well know to translators and to learners of a new language, the concepts of one language may differ quite a lot from those of another. Both German and French have two different concepts for *knowing*, i.e. 'erkennen' or 'wissen', and 'savoir' or 'connaître', each with distinct meanings, whereas English only has one, which must somehow cover both the French and the German concepts. Conversely, English has two words, 'sheep' or 'mutton' to represent a concept of an animal or its meat when it has been prepared for a meal, while French only has one, 'mouton', to represent both concepts. To give another example, English has two concepts for *flowing water*, i.e. 'river' and 'stream', and so does French, 'rivière' and 'fleuve', however, they do not cover comparable concepts. What distinguishes the English signifieds 'river' and 'stream' is size, whereas in French 'fleuve' does not necessarily differ from 'rivière' because it is larger, but because it flows into the sea, which a *rivière* does not.[4]

Thus, the fact that different languages may operate perfectly well with different conceptual distinctions, and that each of them may divide up different spectrums of conceptual possibilities, means, according to Saussure, that a language does not simply assign arbitrary names to a set of "full-blown" and "pre-existing concepts". (*Cours*, p. 97.) The signs of a language are unions of arbitrary signifiers and signifieds which are just as arbitrary. In this respect, one may say, it is the nature of the sign to be *doubly* arbitrary. Not only do different languages produce different sets of signifiers, but also different sets of signifieds - that is, each language has distinct and completely arbitrary ways of "organizing" its concepts or categories. Moreover, to Saussure the notion that language is a nomenclature, i.e. "a list of terms cor-

[4] This example, and a others which will be used later in this text, is taken from Jonathan Culler's book, *Saussure* (Culler, 1986).

responding to just as many things", is wrong, because it entails the assumption that concepts of things exist prior to words.

6.2.2 The differential identity and relational value of the linguistic sign

Well, granted the arbitrary nature of the sign, according to which we have to think of a linguistic sign "*not* as a union of a thing and a concept, but of a concept and a sound pattern" (*Cours*, p. 97), how then do concepts, terms or signs acquire their meaning? And what is it that defines the identity of a signifier and its signified? These two questions are not unrelated. On the contrary, the first question may, according to Saussure, be answered in terms of an answer to the second. This question Saussure answers by suggesting a new principle, namely a principle according to which both signifier and signified are purely *relational* or *differential* entities. What this principle implies is, firstly, that neither signifiers, nor signifieds, are autonomous entities each of which have their own "identity" or "essence", which may be defined in isolations from other signifiers or signifieds. Both may only be defined - and thus have a *value* and *identity* - by their relation to and difference from other members belonging within the linguistic system as a whole.

To illustrate the notions of value and identity, Saussure uses as an example a game of chess, which, he says, "is like an artificial form of what languages present in a natural form". In a game of chess the value of the chess pieces depends on their position on the chess board, just as in language each term has it value through its contrast with all other terms. (*Course*, p 88, [125 - 126]). However, the individual pieces of the game are not in themselves part of the game. Thus,

> "Consider a knight in chess. Is the piece itself an element of the game? Certainly not. For as a material object, separated from its square on the board and the other conditions of play, it is of no significance for the player. It becomes a real, concrete element only when it takes on or becomes identified with its value in the game. Suppose that during a game this piece gets destroyed or lost. Can it be replaced? Of course it can. Not only by some other knight, but even by an object of quite a different shape, which can be counted as a knight, provided it is assigned the same value as the missing piece. Thus it can be seen that in semiological systems, such as languages, where the elements keep one another in a state of equilibrium in accordance with fixed rules, the notion of identity and value merges." (*Course*, p. 108 - 109, [153 - 154]).

It is in virtue of the fact that a sign belongs within the language system that it acquires a value and an identity, i.e. an identity and meaning which is determined entirely by its internal relations to and differences from other signs in the system. In his comments on the general principle according to which linguistic signs acquire identity and 'value', Saussure says,

> "what we find, instead of *ideas* given in advance, are *values* emanating from a linguistic system. If we say that these values correspond to certain concepts it must be understood that the concepts in question are purely differential. That is to say they are concepts defined not positively, in terms of their content, but negatively by contrast with other items in the same system. What characterises each most exactly is being whatever the others are not." (*Course*, p. 115 [162]).

So far, two of Saussure's principles of language have been presented: (1) the principle of the arbitrary nature of the sign, and (2) the principle that both signifiers and signifieds are relational and differential elements, the identity and value of which depends on the simultaneous coexistence of all other elements in the language system. For Saussure these principles are both fundamental and general in the sense that they apply to elements at all levels of language, i.e. from the levels of signifiers and signifieds, to the levels of signs or words, and to the level of grammatical and syntactic combination of sequences of signs into sentences. Indeed the entire linguistic system can be explained in terms of bipolarities or differences between elements at the same level as well as their relations to elements at others. I shall turn next to the difficulties entailed in these two principles applying to the language system, and in particular the difficulties in giving consistent accounts of how the meaning of a sign is determined entirely by features and relations internal to the linguistic system.

6.2.3 Problems and consequences of the twin-principles of the arbitrary and relational nature of the sign

In the accounts so far of Saussure's principles of the language system, the terms 'value' and 'meaning' have been used interchangeably. But, according to Saussure, 'value' and 'meaning' are not synonymous terms. Although value, in its conceptual aspect, is undoubtedly part of meaning, he says, it is necessary that a distinction be drawn - "*if a language is not to be reduced to a mere nomenclature*". (*Course*, p. 112, [158]. Italics added.) His explanation of how this distinction is made, or rather what is being distinguished, goes as follows:

According to the preliminary definition of the sign as a union of a signifier and signified, the *meaning* of a sign is simply the *counterpart* of a sound pattern. The relevant relation of a sign thus defined is one between a sound pattern and a concept, "within the limits of the word" [...] "which is for this purpose treated as a self-contained unit, existing independently". (*Course*, p. 112, [159].) However, this linguistic sign itself has counterparts, i.e. the other signs in the language, and the value of its constituents depends on the simultaneous coexistence of all the other signs and their constituents.

But how, then, does it come about that *value*, defined by the relation between a sign and other signs of the system, can be assimilated with *meaning*, i.e. with the counterpart of the sound pattern within a sign? In answering this question, says Saussure,

> "it is relevant to point out that even in non-linguistic cases, values of any kind seem to be governed by a paradoxical principle. Values always involve:
>
> 1. something *dissimilar* which can be exchanged for the item whose value is under consideration, and
>
> 2. *similar* things which can be compared with the item whose value is under consideration.
>
> These two features are necessary for the existence of any value. To determine the value of a five-franc coin, for instance, what must be known is: (1) that the coin can be exchanged for a certain quantity of something different, e.g. bread, and (2) that its value can be compared with another value in the same system, e.g. that of a one-franc coin, or a coin belonging to another system (e.g. a dollar). Similarly, *a word can be substituted for something dissimilar: an idea*. At the same time, it can be compared to something of like nature: another word." (*Course*, p. 113-114, [159 - 160]. Italics added.)

And that is all, as far as Saussure is concerned. But how could it be? How can a word, defined as a union between a sound pattern and an idea, be substituted with an *idea*, i.e. with something being *part* of itself, its constituents? And how can an *idea* be characterized as something *dissimilar* to the sign, given that one of the constituents of a sign, i.e. its signified, is an idea? It seems that the categorical distinction made between (1) *value* in terms of exchangeability with something dissimilar, and (2) *value* in terms of comparability with something similar, does not work in the case of linguistic signs as conceived of by Saussure. For according to this conception of the sign (1) must be void. So, given that the value of a linguistic sign has been accounted for in terms of the relations that its signifier and signified have to signifiers and signifieds of other signs within the system itself, there is, in Saussure's scheme of thing, no other value or meaning of the sign to be accounted for.

However, if it is part of the *general* principle governing the value of an "item" that it may be related to something dissimilar existing independently of and *outside* the system to which the "item" belongs, what then would be more obvious than to relate the value or meaning of linguistic signs to *things* dissimilar to and existing independently of and outside the language system -

i.e. to the thing to which they *refer*? Obvious, not least in view of the fact that when giving concrete examples to demonstrate the differential value of words (e.g. 'green', 'river', 'sheep', 'mutton', etc.), references are invariably made to things in the world and their features (e.g. colours of things, water flows, animals vs. their meat); indeed, no such demonstration can be carried out without making such references.

For Saussure, however, such references to things outside the language system would have the adverse effect that language "in the final analysis would be reduced to a mere nomenclature", i.e. a list of terms corresponding to just as many pre-existing concepts and things (op.cit). And it would have the effect, not least, that the language system could not, after all, be considered a self-contained system, being definable independently of everything outside itself. This, it seems, is captured in the following passage:

> "the process which selects one particular sound-sequence to correspond to one particular idea is entirely arbitrary. If this were not so, the notion of value would lose something. For it would involve a certain imposition from the outside world. But in fact values remain entirely a matter of internal relations, and that is why the link between idea and sound is intrinsically arbitrary." (*Course*, p. 111, [57].)

But how does the process whereby one particular sound-sequence is selected to correspond to one particular idea come about? And if the language is a self-contained system of signs, the values of which remain entirely a matter of internal relations, then how do signs acquire the property - which obviously they have - of "meaning" and "referring" to things outside themselves? The nearest we come to an answer to the first question in *Cours* are the following suggestions.

Philosophers and linguists, says Saussure, have always agreed that were it not for signs, we would be incapable of differentiating any two ideas in a clear and constant way. For,

> "In itself, thought is like a swirling cloud, where no shape is intrinsically determinate. No ideas pre-exist, and nothing is distinct, prior to the introduction of linguistic structure." (*Cours*, p. 155).

Sound, which lies outside this "nebulous world of thought", is just as shapeless and amorphous. However, it is a "malleable" material which can be fashioned into separate parts in order to supply the signals which thought needs. Hence,

> "we can envisage the linguistic phenomenon in its entirety - the language, that is - as a series of adjoining subdivisions simultaneously imprinted both on the plane of vague, amorphous thought (A), and on the equally featureless plane of

sound (B)." (Represented graphically by Saussure as in the figure below).[5] (*Course*, p. 110, [155].)

IV. Linguistic Value

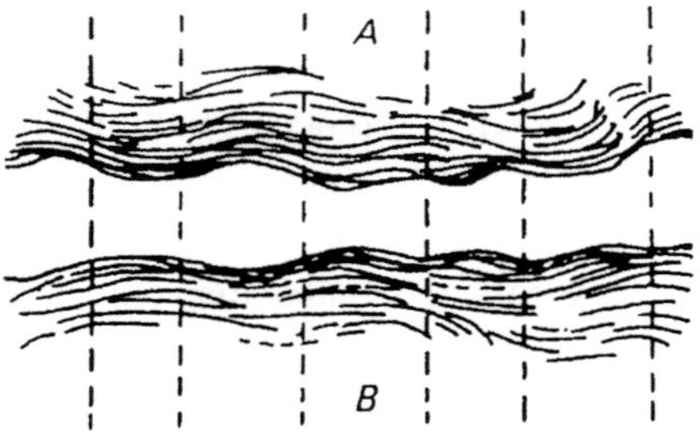

Thus conceived, language acts as an "intermediary" between thought and sound with the result that "both necessarily produces a mutually complementary delimitation of units". Thought, says Saussure, "which is chaotic in nature, is made precise by this process of segmentation". What takes place in this process is neither a transformation of thought into matter, nor a transformation of sounds into ideas, but, says Saussure,

> "a somehow mysterious process by which 'thought-sound' evolves division, and a language takes shape with its linguistic units in between those two amorphous masses." (*Course*, p. 111, [156].)

In this "mysterious" process by which concepts and their meaning are formed and the language system itself is constituted, nothing existing outside concepts and language plays a part. On the contrary. For, to answer the second question, according to the principle of the arbitrary nature of the sign, as interpreted by Saussure's structuralist followers, signs not only "organise", "subdivide" and "segment" our thought into individual concepts, but they also "organise", "subdivide" and "segment" the world itself into individual things and properties. Without this linguistic organisation and segmentation, the world would be just as "shapeless" and "amorphous" a mass" as sound and thought. This would seem to be the interpretation of this principle by Roy Harris, who in the introduction to his translation of *Cours* attributes the following view to Saussure:

> "Words are not vocal labels which have come to be attached to things and qualities already given in advance by Nature, or to ideas already grasped independently by the human mind. On the contrary languages themselves, collective products of social interaction, supply the essential conceptual frameworks for

[5] By permission of Gerald Duckwoth & Co. Ltd.

men's analysis of reality and, simultaneously, the verbal equipment for their description. The concepts we use are creations of the language we speak." (p. ix).

This interpretation is in keeping with how Saussure's views and principles are being presented by adherents of *Structuralism*, who claim that their -ism is based on exactly those views and principles. Thus, in commending the insights and ideas of Saussure's theory of language, A. J. Greimas, a founder-member of Structuralism, claims that "it is Saussure's fundamental view of the world that it is a gigantic network of relations, an architecture of "meaning-carrying" forms, which in themselves contain their own meaning - a view which he transforms into an epistemology and a linguistic methodology". (Greimas, 1956.)[6]

Several problems are posed by this structuralist interpretation. First, if it were true that the formation of ideas and concepts emerges out of a "mysterious" process, by which an arbitrary sound pattern and an equally arbitrary thought get united in a sign - with no "imposition" from the outside world - it would require not only reference to mysterious, but to *miraculous* processes to explain how it comes about that signs of different persons acquire the *same* conceptual counterparts or signifieds. Furthermore, if signs were thus formed in the minds of persons, no comparison would be possible by which to establish the identity - or differences - of the signifieds or concepts of the signs of different persons. Indeed, if the processes by which language produces ideas and concepts, and defines the value and meaning of the signs of which they are part, were totally independent of anything outside the system itself, it would be difficult if not impossible to explain how this could ever have come about as a product of *social interaction* (cf. the above quotation from Harris). And just as impossible to explain how it would ever occur to anyone that those signs, subsequent to having been formed in their minds, might be used to refer to things which are different from the signs themselves - indeed, might be used to "analyse", "segment" and "organise" into existence things and properties of the outside world.

Secondly, and just as seriously, if it were true that no ideas nor any concepts exist prior to having been thus produced by language, and also true that the meaning of signs and their combinations into linguistic utterances relied entirely on the rules and structures of the language system, then, as far as this system is concerned, *any* choice of utterances to pick out any particular thing or event in the world would do - quite arbitrarily. For this

[6] A similar interpretation of Saussure's views and assumptions has been argued by Jonathan Culler in his introduction to Saussure's theory of language (1986), to which I shall return later.

system, being independent of anything outside itself, does not, indeed cannot, prescribe what choices of signs and utterances may be made in any particular situation or about any particular thing or event. If the utterance only complied with the structure of the language system, then, according to this view of the language system, it would make no difference what the utterance were used to "significate". If so, a linguistic utterance, say, "the person standing in the doorway is my uncle", and another, say, "the thing lying on the table is an old hat" could just as well be used to "significate" one and the same thing or event.

One of the main problems with the structuralist view, as pointed out above, is that the assumption that the value and meaning of signs and concepts is entirely determined by internal structures of the language system with no influence from the outside world, is incompatible with the assumption that language and its concept is a product of the social interaction and conventions of people. Another problem, related to the structuralist interpretation of the second assumption, is to explain how people could begin to interact and make conventions about the use and meaning of signs to pick out and refer to properties of things in the world, if no things or properties exist or are "given in advance by nature", i.e. prior to the existence a collectively produced linguistic, conceptual framework by which the world may be "organized" and "analysed". That is, to explain how this social interaction and making of conventions by people could ever have got off the ground, *prior to* or in *advance of* people having identified the world they live in and the things and properties that their interaction and conventions may concern, and which, therefore, exist *independently* of this interaction and these conventions.

Well, for any further discussion of the problems and consequences of the principles of the arbitrary and relational nature of the sign, it would be appropriate to remind ourselves, firstly, that an *idea* or a *concept* - whether or not it be expressed in a word or appears in the dressing of some other kind of sign - is not an idea or a concept unless it has a meaning, and that it does not have a meaning unless it *refers* to something. This, I take it, must be uncontroversial. What *has* been vigorously debated within classical philosophy, and not at all agreed (*pace* Saussure), is the status and nature of the relation between our *concepts* and *language*, and of the relation of both concepts and language to that which they are about, i.e. *reality* and things and events in reality. The formulation of this "tripartite" problem dates back to *Locke*, and his search for an answer to the question as to how the ideas of the mind originate. After having dismissed the view that they could be inherent in our minds (i.e. innate), he suggested instead that the mind had some psychological faculties and mechanisms by which concepts and ideas were processed from raw sensory data originating in the physical world, and that language - via other psychological processes - is adapted to or derived from an experiential world. However, by this suggestion he did not solve the problem which was taken up by later philosophers, namely whether our description of the world is in fact about the world, or whether it is merely a description of our *ideas* or *concepts* of the world.

It appears that Saussure's assumption (or those of his structuralist followers) of a self-contained language system in which the meaning of words - and, as a consequence of concepts - are thought to be inherent, 'solves' all those problems in one stroke. They simply vanish in view of the assumption that language and concepts are indistinguishably one and the same thing, and that *things* in the world and their features come about by the very same "mysterious process" by which "thought-sound evolves divisions" of signs and words, and thus become *the products of our concepts*. If we may trust Greimas and other Structuralists, it is the world-view or epistemology of Saussure that in virtue of being segmented by the "network" of concepts which language "throws" upon the world, what exists in the world is itself *indistinguishably* of a linguistic, conceptual origin. This solution, however, not only leaves unsolved the series of problems pointed out above, but also entails unacceptable epistemological consequences.

The tripartition of traditional philosophy of *language*, *concepts* and *reality* into independent "entities" is of course just as unacceptable and nonsensical - precisely because it assumes that those entities are *independent* in the sense of: determinable, independently of each other. That this assumption is wrong does not mean that we cannot *distinguish* between language, concepts and reality. We can indeed distinguish between e.g. a description of a river, a concept of a river, and some actual flow of water of a particular quantity existing somewhere in reality, which the description or concept is about. A *description* or a *concept* of a river is not a *flow of water*, nor is a flow of water a description or a concept of a flow of water. However (and here traditional philosophy is wrong), that such distinctions may be made does *not* mean, that a linguistic description or a concept of a river may exist independently of or without reference being made to actual rivers existing in the world. But nor may the meaning and value of the signs for rivers, as wrongly assumed by Saussure, be determined independently of referring to actual flows of water.

Indeed, any linguistic analysis and categorisation of the systematic properties of linguistic occurrences of any language, necessarily relies on the fact that the linguist carrying out the analysis and categorisation, just as any other users of the language being analysed, may identify and determine *that* to which the occurrence in non-arbitrary ways may be used to refer to. Hence, such an analysis relies on the fact that he and they have knowledge and notions of what is the case or true, and not the case or false, about the thing or event in question. That for example in a comparative analysis of English and French, a linguist may determine the differential values of the terms 'river' and 'stream', and 'rivière' and 'fleuve', and define the sets or categories to which they belong within each of its language systems, presupposes and depends on his knowledge of the various properties of flowing of water to which those terms refer (e.g. their size, swiftness of flow, straightness or sinuosity, direction of flow, depth, navigability, etc.). Without presupposing such knowledge and determination of events and their properties, and presupposing, furthermore, that the terms belonging to both languages have determinate implications and correct uses *in virtue* of being about those events and their properties, no further comparison and categorisation of the meaning or value of terms would be possible. Let me give another example. A linguist may determine that the "fixed syntagmatic rules", which define the constituents of the sequence, 'The man frightened', only allows it to be followed by constituents belonging within a particular paradigmatic set of contrasting or opposing elements (e.g. 'the boy', 'the dog', 'his listener', but not 'stone', 'purple' or 'sincerity' (Culler, 1986, p. 46)). This relies on the knowledge he shares with other language users that the term 'frightened' is used to refer to a particular psychological state, and his knowledge about what sort of things may be frightening, and to whom. Without this knowledge on the part of the linguist, he could not begin to work out any paradigmatic sets of opposing elements that may replace each other in a syntactic sequence of elements.

The fact that a theory about the language system cannot be developed independently of how language is being used, nor independently of knowledge and conceptions about that to which language and its signs refer, means that both the view of the language system as a self-contained system and as an independent *methodological* unit has to be rejected. In what follows next, I shall argue that the existence of an interdependency or necessary relation between language, concepts and reality is fully compatible with a refutation of the classical linguistic "nomenclature-view" of language.

6.2.4 The "nomenclature-view" of language reconsidered and revised

Saussure's refutation of the classical linguistic nomenclature-view was based on the observation that the same reality in different languages may be described in many different ways. Any attempt to justify a natural, one-to-one relation between linguistic expressions and things in reality would have to fail, since it would require that the very different semantic structuring of vocabulary (of colours, of gender relationships, feelings, etc.) could be shown to derive directly from the "real" things and their "natural" or "fixed" properties. But this is not possible, nor would it be possible to pronounce any one of the different categorisations or semantic structures as more "natural" than any other.

Another way of refuting the assumption of a natural relation between language and reality has been to argue that if reality intrinsically had fixed or essential properties or qualities, and it were true that a natural relation existed between these qualities and the words and concepts of language, then a language user, when presented with an object and given the task of telling what it was, would come up with one and only one description of it, i.e. the "natural" one which the object, so to speak, induces or causes in him. But, it is a triviality to say that such a task is not well defined. Either it must be implicitly understood or explicitly stated (e.g. by instruction) from what perspective, purpose, motive or interest the object is supposed to be described. Otherwise one might equally well describe the object as "a man made thing", "merchandise", "a mechanical object", "a kitchen tool", or "a gift from one's aunt".

Well, a reasonable conclusion to draw from these observations would be that how we describe any particular thing may *vary* from one situation to another. Thus, we have many notions and "names" for the same things, depending on the situations in which we observe and act with or upon them, and on the intentions and purposes, or points of view adopted in those situations - a fact which is seen within the same language as well as in comparative studies between different languages. Conversely, the same name or notion may be applied to different features or properties of the same things, or to quite different things, which again may be observed in the same language as well as in comparative studies of different languages.

We may agree with Saussure and the Structuralists, therefore, that the "nomenclature view" of language and use of language *in its classical and most banal version* is wrong. This view cannot possibly explain the diversity of categorisations and descriptions of things and events in reality encountered in different languages, nor explain the diversity of descriptions and categorisations existing within the same language about the same things. But this is a shortcoming that this view shares with the theory of the language system, which Saussure proposes in its stead. Furthermore, for the reasons already argued in the previous section, this alternative view of a self-contained or immanent language system raises problems which are at least as serious as the view it is meant to replace.

In his presentation of the theory about this system and the principles applying to it, Culler attributes the view to Saussure that "language does not simply assign arbitrary names to a set of independently existing concepts or things" (Culler, 1986, p. 23). Language, as an object of any concern to linguistic studies, is a system, as Culler puts it, "which sets up an arbitrary relation between signifiers of its own choosing, on the one hand, and signifieds of its own choosing on the other" (ibid. p. 23). That different languages produce different sets of signifiers and signifieds means, according to Culler,

> "that each have a distinct and thus arbitrary way of organising the world into concepts and categories. [...] Not only can a language arbitrarily choose its signifiers; it can divide up a spectrum of conceptual possibilities in any way it likes." (ibid. p. 23 - 24).[7]

Culler then goes on to argue that the division of language made by Saussure between a self-contained, inherent language system and speech is not just a postulate which must be accepted "on faith", but *a logical and necessary consequence of the arbitrary nature of the sign*. For, if the essential feature of the linguistic sign is that its meaning and identity may only exist and be determined by contrast with coexisting signs of the same nature, we must, says Culler, look to the system of relations and distinctions which provide it with a meaning and an identity. Indeed, without a language system "where the elements keep one another in a state of equilibrium in accordance with fixed rules" (*Course*, p. 109, [154]), their meaning and identity would be so arbitrary as to vanish into thin air.

However, what to Culler seems a necessary and logical consequence of the arbitrary nature of the sign, seems to me rather to be a way of 'rationalizing' the unacceptable consequences of methodologically isolating language and its words from their relation to what they "significate" or are about. However, the postulated arbitrariness of words and signs *only* manifests itself when language is viewed independently of its use by speakers to refer to and communicate their knowledge about things and events in the world. When language and linguistic descriptions and signs are seen within the context of the situations in which they are being used, there is nothing arbitrary at all about them. Indeed, as argued in the previous section, *any well defined linguistic analysis and categorisation of the systematic properties of linguistic occurrences relies on this fact*.

In the section which follows, I shall discuss the consequences of this point. I shall start by presenting what according to Saussure characterises *the language system* as opposed to *speech*.

[7] We notice in passing how *language* in the passage just quoted is being equipped with capabilities of "choosing" and "making divisions and categories". This is nonsense; language *users* may choose and make divisions. We see how quickly sloppy use of language gets us on to the slippery road of Structuralist assumptions to the effect that not only does the language system *choose* what we say, but it also *talks* through us. In this case language, through Culler, has apparently chosen to talk rubbish about itself.

6.3 The logical relation between a systematic and a speech act description of linguistic occurrences

In view of his insistence on the necessity for linguistics to isolate the language system from its use, it is not surprising that Saussure has a lot more to say about what *distinguishes* the language system from speech, than what *unites* language with speech, and a lot more to say about how the language system *determines* speech than how this system *relies* on speech.

One of the important distinctions between language and speech, according to Saussure, is that whereas language is *social*, speech is *individual*. His argument for this distinction goes as follows. First, language conceived as a system of signs and rules for their combinations, is social in the sense that it is that which speakers, who share a language, have in common, and which ensures that they may make themselves understood and understand others. Secondly, language as it exists at any particular time, is stable and homogeneous, and it "eludes the control by the will whether of the individual or of society: that is its essential nature". (*Course*, p. 16, [34].)

Speech, on the other hand, although "executed" according to fixed rules, is unique in the sense that the utterances of individual speakers are put forward in particular situations at particular times, which are never repeated; they are the expressions of "the will" of individual speakers to communicate their thoughts and intentions - which could have been expressed differently, e.g. by choosing other utterances which convey the same intentions or meaning. So, while the language system is collective, there is nothing collective in speech - apart from the fact that speech is the sum total of what people say.

An individual, says Saussure, cannot develop signs on his own, and thus cannot develop thoughts, that only he understands. Concepts - and the signs in which they are expressed - are the product of language and only exist in virtue of language. As a collective phenomenon and a social institution, language

> "takes the form of a totality of imprints in everyone's brain, rather like a dictionary of which each individual has an identical copy. Thus it is something which is in each individual, but none the less common to all" (Course, p. 19, [38]).

However, language and its signs are social in yet another sense, i.e. they rely on agreement and conventions among speakers. The signs of language which make up our concepts only exist in virtue of a kind of contract agreed between members of a community. (*Course*, p. 8 - 15, [23 - 32].). The arbitrary nature of the sign, says Saussure, "enables us to understand more easily why it needs social activity to create a linguistic system. A community is necessary in order to establish values. Values have no other rationale than usage and general agreement. An individual, acting alone, is incapable of establishing a value." (*Course*, p. 111-112, [157].)

This arbitrariness of language and its signs, as we have been told, is no more clearly observed than when comparisons are made between the different ways in which communities of speakers belonging within different cultures and nations agree to create linguistic systems and thus different ways of conceptualizing the world they live in. However, although the way of life of a nation or a culture, i.e. its material conditions, its rituals and social institutions etc., may have "some" effect upon its language, says Saussure, "it is to a great part the language which makes the nation and its culture". (*Course*, p. 21, [40 - 41].)

Well, this notion of language as a social system or structure, the function and existence of which depends on agreement among members of the speech community, would seem to be difficult to reconcile with the notion that this same system is a self-contained system of "pure value" which is independent of all other manifestations of language or of things existing outside language. For, how could language and its system, being itself a social institution and geared to communicate the needs of people and the functions of their social institutions, be self-contained and independent of the material and cultural conditions and of the social practices of the members of the society in which the language is used and of which it is about?

Now, although Saussure does not deny that the way of life of a nation or culture does have "some" effect on its language, and that the study of speech and (other) external linguistic phenomena may teach linguists a great deal, "it is not true to say that without taking such phenomena into account, we cannot come to terms with the internal structure of the language itself". (*Course*, p. 22, [42].) More interestingly, though, he does not deny the possibility of a "linguistics of speech", nor the importance or relevance of such a linguistics. In a short chapter in *Cours* which presents Saussure's attempts to clarify the distinction and relation between a *linguistics of language structure* and a *linguistics of speech*, he contends that,

"In allocating to a science of linguistic structure its essential role within the study of language in general, we have at the same time mapped out linguistics in its entirety. The other elements of language, which go to make up speech, are automatically subordinated to this first science. ... The distinction drawn above and the priority it implies make it possible to clarify everything. That is the first parting of the ways that we come to when endeavouring to construct a theory of language. It is necessary to choose between two routes which cannot both be taken simultaneously. Each must be followed separately." (*Course*, p. 18, [36], p. 20, [38].)

However, he also contends, that the "two objects of study" are closely linked and that each *presupposes* the others. Thus, speech presupposes a language in order to be intelligible and to produce all its effects, and conversely, speech is necessary in order to establish a language. Indeed *historically*, speech always takes precedence, and it is speech which causes a language to evolve. "Thus there is an interdependence between language itself and speech". (*Course*, p. 19, [37].) But this, in Saussure's opinion, does not affect the fact that a sharp distinction must and *can* be made between the study of the structure of language as a closed system with its own order (linguistics proper), and of speech (which cannot be studied from the same linguistic point of view); nor does it affect the fact that in a *theory of language*, the description of the language structure must have priority, and that the study of speech is subordinate to and relies on the linguistic description of the language structure.

However, to take a step further the argument already made in previous sections: contrary to what Saussure seems to suggest, the interdependence which exists between the language system and speech is not merely of a historical nature, having to do with whether a language could exist before or without speech, or speech before or without a language. Rather, it is an interdependency which makes it impossible to develop any consistent theory of the language system *independently* of a theory of speech acts - and vice versa. These two theories and their descriptions of linguistic occurrences, as first pointed out by Mortensen (1972), are *logically* related, in that the conceptual systems of either are not well defined independently of the other. The conceptual apparatus used in the systematic description of a linguistic occurrence *presupposes* the conceptual apparatus used in a speech act description of the same occurrence - and vice versa. Thus, what characterizes the elements or components of a systematic paradigm is that they are mutually distinct and incompatible; however, what determines that they are thus distinct and incompatible, is that a speaker cannot simultaneously choose two different elements of a paradigm for the same function in the same speech act. Conversely, to the extent that a speech act may be conceived of as a linguistic choice, a speech act description presupposes a linguistic system that defines the linguistic possibilities of choice.

This means that a description of the language system, i.e. of its taxonomies, categories, syntagmatic and paradigmatic relations and rules for combining linguistic elements in sentences, relies on analyses of linguistic occurrences which has first been determined and understood by the linguist as *utterances*, i.e. as speech acts, and thus on his knowledge of what the utterance is being used for in a communicative situation. It is in virtue of being a speaker of the language that the linguist knows that to carry out a particular speech acts, e.g. to 'describe' or 'propose' something, is to assert that something is the case about something (e.g. the apple is red), or that something is being done to something (e.g. the boy eats the apple). Similarly, as a language user of this language he knows that particular words and terms are being used to express and refer to such states of affairs and the relationship between them (apple, boy, red, eat, and is). Knowledge of such states of affairs in the world *and* the words and terms being used to refer to and communicate about them, form the necessary point of departure for the definition and determination of terms within a systematic theory of the language of e.g. nouns, adjectives and verbs, as well as ways in which they combine in e.g. affirmative and indicative sentences.

Now, after having completed a systematic description of the language, any linguistic elements of an utterance, as well as its grammatical structure, may be conceived of as choices among the possibilities which the linguistic system provides - and allows. For example, from a systematic point of view, the choice of a *noun*, say, 'apple' in the *affirmative sentence*, 'this is an apple', may be seen as a choice among other possibilities which might have been chosen, e.g. 'house', 'leaf', or 'nightingale'. However, a purely formal description of the language system and its grammar does not include procedures for *how* to make choices between such possibilities. Nor are such choices, say, the choice of a speaker of the word 'apple' appearing in his *utterance*, say, 'this is an apple', merely "linguistic" in the sense of choices made within a linguistic system, which of itself prescribes what terms among its taxonomical possibilities would be *relevant* or *correct* in a given situation; but neither are they "arbitrary" or "accidental". One does not pick out the word 'apple' as opposed to 'house', 'leaf' or 'nightingale' at random. Rather, to talk about the word 'apple' as a *linguistic* choices, only makes sense when seen as a stance taken by a language user, who in an actual communicative situation reckons that he is willing to commit himself to the implication of the choice 'apple' to refer to and describe some particular thing in that situation.

Hence, the knowledge required by *both* the ordinary language user when making choices of utterances or speech acts in particular situations, *and* by the linguist attempting to determine the grammatical structure and paradigmatic oppositions and meanings of terms of linguistic occurrences, is knowledge which goes far beyond mere knowledge of "linguistic" possibilities and constraints inherent in the language system. However, this knowledge, being implicit and presupposed in any linguistic analysis of the structure of language and its elements, is often ignored by linguists - and *is* most certainly ignored and left out in any theory of language which suggests that language, or any part of language, may be conceived of as a self-contained and inherent system, from which speech may be "generated" or "executed" independently of the communicative function served by language. However, *no* linguistic system exists, nor any operations in a linguistic system, independently of their presupposed relation to acts being carried out by language users and to things existing outside the system. Therefore, the relation which exists between a systematic and a speech act description of linguistic occurrences is ***a necessary** relation in virtue of the presupposed relation between these descriptions of the occurrences and what they are used for and are about in communicative situations*. This is just another way of saying that in an integrated comprehensive theory of language and use of language - if such a theory could ever be made - neither the use of language describable in terms of speech acts, nor of the structures of the linguistic system which underlie them, may be conceived of as isolated or independent methodological units - in the sense: definable *independently* of reference to how language is used, and about what. Or put even shorter, any such comprehensive theory would have to be just as much a theory about reality as about language.

* * * * * *

It has to be a truism to say that individuals cannot together develop or be taught a language and its use, if they do not have any notions - if only rudimentary - of things and persons in the situations in which they find themselves, nor without being able with others to identify those things and persons as things about which something is the case or true, and something else is not the case or false. Likewise, it has to be a truism to say that nobody could learn a language and its use, who does not understand that linguistic terms and their combinations may be used in various ways to refer to and be about things which exist *independently* of and are *different* from such terms and references. In this matter we have not only logic and common sense on our side, but even empirical fact. Thus, contrary to what Saussure suggests about the precedence and priority of the language system, investigations of pre-linguistic children seem to show that, "historically", i.e. long before they acquire the use of language, its signs and grammatical structure, children are able to engage in a wide range of communicative actions, which have features in common with speech acts. Pre-linguistic children may both understand and distinguish between the communicative acts and intentions involved in describing or asserting something about a thing or event, of giving an order to someone, of making a request etc., and they may themselves - by using only a few words - indicate such communicative intentions in their cooperation with others. However, such communicative action necessarily requires the ability of the child to identify both persons and things, and thus requires concepts and notions - if only rudimentary - of persons and things, i.e. of what is the case about persons and things, and what acts one may carry out with persons and things. In other words, concepts and notions about things may indeed "pre-exist" language and linguistic signs.

This is well documented in Jerome Bruner's excellent observations of the communication and cooperative actions of pre-linguistic children, published in his book, *Child's Talk. Learning to Use Language,* (Bruner, 1983). Language acquisition, says Bruner, "begins" before the child utters his first lexico-grammatical speech. "It begins when mother and infant create a predictable format of interaction that can serve as a microcosm for communicating and for constituting a shared reality" (ibid., p. 18). The transactions that occur in such formats, according to Bruner, constitute the "input" from which the child later masters grammar, how to refer, mean, and request, and to realize these intentions linguistically. Thus, it seems that right from the beginning the linguistic acts performed by speakers, and the meaning and understanding of the functions they serve, are firmly rooted in acts they perform with others in a shared world of material things and persons.

6.4 Conditions for determining differences between different languages: examples of consequences for theories of language

In *Cours* Saussure several times illustrates the nature and properties of a language system by comparing language with a game of chess. A game of chess, says Saussure, is like an artificial form of what a language presents in natural form (*Course*, p. 87, [125]); the "likeness" of language and a game of chess is that everything which concerns the value of the pieces and rules of the system is purely internal to the game (*Course*, p. 23, [43]). Now, if we accept this analogy, two languages, each with its own distinctly different system of inherent signs and rules of combination, would be comparable, so it would seem, to *two* distinctly different games, say *chess* and *checkers*. However, granted that such analogies did indeed capture the nature of the differences between two different languages and the essential features of the value and identity of their elements, serious problems would arise. For, it would be just as impossible *in principle* for users of either languages to give an account of the differences between the two languages from "within" their own language, as it would be to give an account of the differences between chess and checkers from "within" either of those games. And just as neither game can "talk" about anything existing outside of itself and, furthermore, since the value of the pieces of the two games are incompatible, a speaker of one of the two languages cannot talk about and determine the meaning of the signs of the other - for the signs of the other language would have no well defined meaning within his own language.

However, we can in fact determine that languages are different, and therefore, necessarily, determine *how* they are different. As is well known, Saussure himself spent most of his time as a linguist making highly original analyses of differences - and similarities - between different languages, and of the same language at different times in history. Indeed, the principle of the intrinsically arbitrary nature of the sign is based on his observation of the different lexico-grammatical structuring of different languages. Ironically, though, such activity and determination of differences between languages and their terms are obviously inconsistent with his notion of the self-contained nature of natural languages - be it the language he is using when carrying out his analyses or the languages being analysed. In order to overcome this inconsistency, one would have to assume, for example, that the language in which the comparative analyses of different languages are carried out is not the same sort of language, i.e. a natural language, but a different one, say a "super-meta-language", which has all the features and properties that natural languages are being denied. That is, it would have to be a language in which non-arbitrary and correct descriptions can be made of the terms and systems of both of the languages being compared. This would necessitate that the "super-meta-language" somehow contained the signs of *both* of the languages being analysed, in such a way, however, that, in spite of being in the super-meta-language related to *other* and different signs, their values remain unchangeably the same as in the systems being compared.

At the risk of over-killing the issue, it would have to be required, furthermore, that in this "super-meta-language" non-arbitrary and *correct* determinations could be made of the things and properties in reality that the terms of the two languages are supposed to "segment" or "significate". Otherwise it could not be claimed that the observed differences of terms concerned different ways of "segmenting" and "organising" the *same* things - indeed, both *things* and *terms* as well as their differences would cease to be well defined.

To the problems above may be added that if the signs of a language system were really *intrinsically* arbitrary, and the value and meaning of its signs were *inherently* given in the system they make up, then *translation* of terms and texts from one language into another and different language would also be impossible. For how could a translation of texts and terms in one language ever be equivalent to texts and terms of another and different language? Indeed how could such equivalence, or lack of equivalence, ever be determined - except, once more, by assuming some kind of "super-meta-language" which could incorporate the texts and terms of both languages?

But, there is no need for assuming the existence of a special super-meta-language thus equipped, if we can agree that *natural languages* themselves are necessarily similarly equipped, i.e. they allow their users *in their own language* to talk about their own language and determine the differences between terms in this and those of a foreign language, and thus allow them to determine and describe in their own language the meaning of terms and texts of a foreign language. That is, if we can agree that *it is a condition of any natural language that it may be used by its speakers to talk about language and its use* - and thus that natural languages already inherit this "super-meta-function".

6.4.1 Cognitive Relativism: an example from linguistics

The fact that remarkably different taxonomical and grammatical structures may be found in the languages of different cultures has led to the popular view among anthropologists and linguists that not only do different languages "structure" reality in different ways - but the language of the community in which we grow up determines how we conceive of ourselves and the world around us. This view, first proposed by the pre-romantic philosopher Wilhelm von Humbolt in the eighteen century, has in more recent times been further developed by the American amateur linguist Benjamin Lee Whorf (1956). According to Whorf the cognitive structure that different languages produce in their users, causes them to be aware of and describe phenomena and properties of the world in ways so different as to be inconceivable to members of different language communities. Whorf was led to this view by his studies of the cultures and languages of North American Indian tribes, which are not only different from one another, but also differ drastically semantically as well as grammatically from his own "Standard Average European" culture and language. Since Whorf shared with Saussure and the Structuralists the idea of a holistic structure of languages - and thus of the holistic nature of *cognitive structures*, which determine how we view the world - the consequence of his theory is both that the world-views of the members of one speech community would be inconceivable to those belonging to a different one, and also that only a poor translation of one language into a different language is possible.

Well, it is arguable that knowledge of reality acquired by members of different cultures and the way this knowledge is expressed linguistically may differ greatly. And so may the "theories" of reality and things developed within different cultures and societies. Moreover, both such "theories" and the material conditions and social institutions of a culture or society may significantly affect the possibilities of action open to its members, as well as the conditions for observing things and reality developed within the society. Indeed, such differences of knowledge, description and possibilities of action and observation available to different cultures may even be determined - *and be determined precisely as such*. But if such determinations of differences can be made, we cannot at the same time maintain that our cognitive structure and that of members of foreign cultures are *fundamentally* different, nor that the forms and terms of our own language and that of a foreign language are incompatible. On the contrary, both cognitive structures and forms and terms must necessarily be - or may be made - perfectly available to us.

This may be immediately obvious when the differences in concepts and terms of different languages reflect the differences in the material conditions of different cultures. Thus, according to the anthropologist Elenore Smith Bowen (1964), plants play a major role in the life of an African society she has studied, indeed, they are as important as persons. Even small children of this society know a large number of different plants, and the names and concepts by which they are described in their language. However, they do not have many names for or concepts of snow - as do Eskimos - who, conversely, do not have many names for or concepts of plants - especially not of the kind of plants which only grow in Africa. Not surprisingly, the African people studied by Bowen and Eskimos may not share the same concepts of the world, in particular not of plants and snow; nor are their linguistic terms of such matters equivalent. However, a member of, say, the African society may on personal inspection, i.e. by going to Greenland, come to know and understand that every one of the terms and concepts of snow in the Eskimo language have determinate implications and correct uses, i.e. they refer to very real properties of snow, and to contexts and possibilities of action and observation, which are essential for an Eskimo to know about in order to survive. The African may come to understand, for example, that one of the terms for snow refers to the type of snow suitable for building igloos, while another refers to a different type of snow, which falls when a particular change in weather is imminent. Thus, he may learn that there is nothing *arbitrary* or *relative* in how any of those terms are used, or between the implications of the terms and what they refer to. And he may do so because *both* the structure of his cognition *and* the conditions for description in his language are not *fundamentally* different from but, on the contrary, very similar to that of an Eskimo.

Because we may uncover and determine differences in the knowledge of the world, as well as in the language and concepts used to express this knowledge in different cultures, such differences do not and cannot reflect differences of a fundamental nature in the cognitive structure of people belonging within different cultures. If we did maintain this, we would be contradicting ourselves. Much more likely, the differences and variations we may find in different cultures reflect the differences in possibilities of action and observation which have been developed - either due to the difference in the interests or purpose which govern the social institutions of different cultures - or due to sheer necessity - that is, in order to cope with prevailing environmental conditions.

It may be extremely difficult to acquire the language, concepts and outlook of the world of a foreign culture, indeed, it may take most of a lifetime to do so. And it may be extremely difficult to make satisfactory translations of texts and terms of a foreign language into one's own. But the conditions for us to *begin* to acquire a foreign language and its concepts - and thus to be able to tell the difference between our own and those of the foreign language - must be that we can cooperate and communicate with its users about the objects and events being described in their language, and that *somehow* their descriptions and assertions about them may be translated into our own. It must be a condition that we and foreigners can together identify the things in the situations in which we find ourselves, as well as ourselves as persons with bodies who may observe and act with these things in particular ways. Thus, no matter how different our own, or their, description of things and events may be *in addition* to those necessary for identifying ourselves and the things being talked about, we could not even begin to discuss or determine these differences unless we presuppose that

1. we may indeed come to share with the foreigners the knowledge and concepts they have of these things and events

and

2. in the foreign language, as in our own and any other natural language, descriptions and assertions have meaning and implications in virtue of referring to and being about particular, determinable things

and

3. it is a fundamental property of statements in both their and our language that they have logical implications and correct uses, and thus that both languages may be used to talk in non-arbitrary ways about things which we may together determine and identify.

6.4.2 Cognitive Constructivism: an example from psychology

The works of Whorf and anthropologists have inspired some researchers within psychology to assume that language "structures" our cognition of reality and the way we think about ourselves and the world. Conversely, conceptual and linguistic relativity have inspired others to assume that the concepts and categories of language serve the function of "structuring" an otherwise unstructured reality. Thus in 1956, the same year that Whorf's thoughts about the influence of language on our cognition were published, Jerome Bruner proposed the view that linguistic concepts and categories are *created* or *invented* by us in order to make up groups of things and qualities in the world; such concepts and categories, says Bruner, do *not* refer to categories or groups existing in the world:

> "Our intellectual history is marked by a heritage of naive realism. For Newton, science was a voyage of discovery on an uncharted sea. The objective of the voyage was to discover the islands of the truth. The truth existed in nature. Contemporary science has been hard put to shake the yoke of this dogma. Science and commonsense inquiry alike do not discover ways in which events are grouped in words; they invent ways of grouping. The test of the invention is the predictive benefits that result from the use of invented categories. The revolution of modern physics is as much as anything a revolution against naturalistic realism in the name of new nominalism. Do such categories as tomatoes, lions, snobs, atoms, and mammalia exist? In so far as they have been invented and found applicable to instances of nature, they do. They exist as inventions, not as discoveries." (Bruner et al., 1956, p. 7.)

This quotation is taken from the introductory chapter of *A Study of Thinking* - and the following from the conclusion.

> "The categories in terms of which we group the events of the world around us are constructions or inventions. The class of prime numbers, animal species, the huge range of colors dumped into the category "blue", squares and circles: all of these are inventions and not discoveries. They do not exist in the environment," (ibid. p. 232).

Bruner's suggestion that our categories and concepts of things in the world, such as lions, tomatoes, and, presumably, human beings, are inventions or constructions by which we group "events" in the world - which may be said to exist "in so far as" these categories may be found "applicable" to "instances of nature" - immediately prompts this question: if our linguistic categories and concepts were mere arbitrary inventions, by what means, then, do we determine that they may be found applicable to events or instances of nature? It would seem that such tests of applicability of concepts and categories to instances and events in the world require *both* that instances or events of nature do actually exist, *and* that we have at least some knowledge of what is the case or true about them. And it would require that we are able to distinguish the instances and events in question from *other* instances and events about which different things may be known to be the case or true - and for which different terms would be applicable. And it would require, furthermore, that the descriptions and concepts we use to determine such events or instances of nature were not themselves mere products of our inventions or constructions. Otherwise, such tests of the applicability of categories and concepts could only be tests in which *constructions* were tested against other *constructions* - and not against the events and instances of nature themselves. So, if there is to be any sense in saying that our concepts and categories serve the end of "structuring" and "ordering" instances of nature and events in the world, it must be assumed that such events and instances do in fact exist, indeed exist as things about which something is the case or true, and that we are able to determine what is the case or true about them. These assumptions are, necessarily, a condition for *any further development* of categories and concepts with which to organise and group these events and instances.

So let it be said once more, what Eleanor Rosch and many others researchers of concept formation have already said (cf.,e.g., Rosch et al, 1976), namely that the basis for our different categorical systems are *things* and *objects* existing in reality having the properties referred to in these different conceptual systems.

The view expressed by Bruner in the quotation above is a good example of how the fact that we cannot talk about things in reality and their properties without describing them, is mistakenly seen to imply the assumption that, therefore, reality and the properties of things only exist *in virtue* of being talked about and described; indeed, that reality and things do not exist independently of our description of them. The epistemological consequences of this *Anti-realistic* view, which has been carried to its extreme by the American philosopher Nelson Goodman, to whom Bruner ardently adheres (see Bruner, 1986), will be discussed thoroughly in Chapter 17.

I cannot conclude this very summary rejection of the views of Bruner just quoted without adding that his commitment to *Irrealist* or *Anti-realism* has not reduced the value of his experimental work on concept formation. This work is classic in its field, for good reasons, as are so many other of Bruner's contributions to psychological inquiry. Like Gibson's work within perception psychology, Bruner's original research within cognitive psychology will undoubtedly have a lasting influence not only for cognitive psychology, but for psychology in general.

Chapter 7

Situations, action and knowledge

In the previous chapters I have argued that the meaning and truth of linguistic occurrences cannot be determined independently of the communicative function that language serves for people finding themselves in concrete situations in reality, nor without referring to the knowledge they have and the acts they carry out in those situation. In this chapter I shall attempt to clarify some of the formal implications of this necessary relation between knowledge, situations and acts on which communication and use of language relies.

7.1 Linguistic and non-linguistic knowledge and concepts

It seem necessary to assume that the knowledge about things in reality on which is based the co-action and acquisition and use of language of persons is knowledge of a *propositional* nature, i.e. knowledge that something is the case and true about the things - and that something else is not the case or false. In this sense the knowledge and concepts of persons involved in their non-linguistic identification and determination of things and events in the world is, formally speaking, just as "linguistic" as the knowledge and concepts involved in linguistic descriptions and assertions about things. And in this sense there is no fundamental difference between the logical structure of the knowledge and concepts of linguistic descriptions and assertions, and that of the knowledge and concepts involved in our non-linguistic determination and identification of things and events. We do not have two different sets and sorts of concepts and categories - one for our non-linguistic identification and determination of that which we are talking about, and another which we use when making linguistic assertions or descriptions of them. Let me give an example. My visual identification of the object lying at my desk as being a cigarette lighter implies that it is cylinder shaped and hollow, although I cannot see it all the way round or look through it. And it implies knowing that if pressed at a given point, a flame will be emitted from another part of it. To have perceptually identified the object as this particular object implies having identified a whole range of properties and possibilities of action on or with the object - some of the properties being visible and others being non-visible at the moment of observation - all of which form part of the meaning and implication of the perception of the object as being my cigarette lighter. In this respect to perceive a thing is equivalent to that of describing a thing, and my perceptual cognition and identification of the object is not conceptually different from a linguistic description of it. In fact, it would be difficult if not impossible to point out any differences between the categorical and semantic structuring involved in the visual cognition of the object and a verbal description of the same object.

Now, there are of course ways of knowing things, which are not propositional in nature; moreover, a lot of the knowledge we have of things and events in reality is not easily put into words - if we have words for it at all. For example, the kinaesthetic knowledge we acquire when learning to ride a bicycle and the proprioceptive coordination it involves, certainly do not come in propositional form. Even skills of a highly intellectual kind, as for example the type of skills developed by expert operators of complex industrial process plants, seems to a large extent to rely on a "feel" for the processes in the plants, which is not available to the operator in propositional form - and of which linguistic expressions can only be made with difficulties. (The same has been found to be the case about the so called *tacit* knowledge of experts within other areas, cf. the studies reported in Dreyfus and Dreyfus, 1986.) And although a lot more can probably be put into words than what is in fact being put into words (see Praetorius and Duncan, 1988), it is debatable whether the knowledge, subsequent to having been put into words, still has the same cognitive structure as before. Rather, research seems to indicate that previously unverbalized knowledge, after having been formulated verbally, in many cases leaves the person with greater opportunities for action as well as problem solving skills within a given knowledge domain, than before attempting to make the knowledge verbally available (Duncan, 1987).

However, this does not invalidate the claim of equivalence of the conceptual, categorical structuring entailed in our linguistic descriptions of things, and the non-linguistic identification and determinations of the same things, on which are based our linguistic description of them. This claim does not imply that there is no difference between perceiving a thing and describing a thing. On the contrary, there are countless differences between perceiving an object (e.g. seeing, touching, smelling it etc.) and putting forward a verbal description of an object. Besides, in an actual situation in which we perceive an object, we may walk around the thing, lift it up, give it a closer look etc., and thereby acquire more knowledge about the object, which we cannot do by linguistic means. But with respect to perceptually identifying an object as a particular object, having these particular properties and towards which we have these particular possibilities of action, observations, etc., i.e. in an *epistemological* respect, our perceptual identification and knowledge of an object does not differ conceptually from a corresponding linguistic description of it. Rather, it seems reasonable to assume, as already noted, that we do not have one set of concepts and categories for our non-linguistic cognition of the things in the world, and a different set of linguistic concepts and categories of the same things, which we so to speak make use of when we talk about these things. If we had, it would be difficult to see how we would go about determining that the two sets of concepts and categories in fact concern the same things.

To say that our linguistic speech acts are part of or imbedded in the nonverbal acts of identification and observation that we carry out in concrete situations with other language users, only makes sense on the condition that an equivalence exists between linguistically expressible knowledge of things and the cognition involved in perceptually observing those things. Hence, it must be assumed that to the non-linguistic cognition on which our linguistic descriptions rely, principles and conditions apply which correspond to those assumed about language and linguistic statements. Among those are that the knowledge and conceptions involved in our non-linguistic determination of things and acts in reality are as propositional as the knowledge expressed in linguistic assertions and description of the same things and acts.

7.2 Situations

That the notion of 'situations' is as fundamental as 'reality', 'language', knowledge, 'true', and 'persons' becomes clear when we consider that we cannot talk about persons and users of language, who do not find themselves in concrete situations in reality, about which they have knowledge. Conversely, we cannot talk about reality without referring to different concrete situations that persons may be in, and about which they have knowledge and a language in which they may to put forward propositions about these situations, which are correct.

In Chapter 5 the term *logical space* was introduced for the limited descriptive system within which a description or a statement about reality is put forward, and within which the implications and correctness of the description or statement may be analysed. This term was introduced to emphasise that the range of implications of a statement about a thing or event to which a speaker has committed himself by putting it forward depends on the circumstances under which it is put forward. To those circumstances belong the intentions, purposes and points of view from which the thing or event is being described, as well as the knowledge and possibilities of observation and action available to the speaker in the situation.[1] Now, this notion of a 'logical space' to capture the range of implications of statements put forward within a descriptive system would seem to be a rather "elastic" notion. However, this weakness of the notion is only apparent. In a discussion of the implications of a description or statement it will always be possible *in practice* to make explicit the range of implications of the descriptive system within which the discussion is carried out.

The same applies to "situations". A situation is something concrete, which we cannot talk about without referring to physical space and time, yet a situation cannot be determined only by referring to space and time. As is the case for logical space the determination of a situation and what belongs to it depends on particular interests, purposes and intentions - and it may refer to the past as well as to the future; indeed, it may refer to a decade or to most of a person's life. This is the case when, for example, we talk about how our present situation is determined by our upbringing or education - or how *Kant's* situation was critically dependent on the development of classical physics. However, as in the case of logical space, it will always be possible *in practice* to explicate arbitrarily exhaustively and unambiguously the purposes and relevant circumstances that define an actual situation.

Now, although we cannot talk about reality and its objects independently of talking about concrete situations in which persons may find themselves, and not talk about situations without referring to objects which exist in these situations, the existence of objects does not dependent on the existence of situations. Objects are not part of situations in the sense of *identical* with situations, nor do objects exist in virtue of being part of the situations in which persons find themselves. Just as objects and events in reality do not cease to exist when or if they are not described or cognized by persons, these things do not cease to exist when the situations in which persons describe and cognize them no longer exist.

[1] I refer the reader to the previous chapter for examples of how the meaning of notions may differ in different circumstances. That the implication and meaning of statements and descriptions depend on the circumstances in which they are put forward becomes perhaps even clearer in the following different, but quite ordinary usage of a statement, 'this is red'. In one situation the meaning of the statement is simply 'it is not yellow, green, nor brown', whilst in another, e.g. a situation in which I am selecting apples in an orchard, its implication may be 'this is ripe'; said by one bricklayer to another while building a yellow house it may be 'this is the wrong kind of brick'.

But neither are persons identical with the situations in which they may find themselves - in the sense that a person is identical with one or more concrete situations. Nor are situations part of or identical with persons. If we said so, we would be saying that a situation is something which may have consciousness or knowledge, or the like. But nor can we ask what belongs to a person and what belongs to the situation in which he find himself, for example ask whether his body belongs to the situation or to his person.

So, where does this lead us? Hopefully, it leads *and* forces us to acknowledge both the *inter-dependency* and *asymmetry* which exist between notions of persons, situations, and objects. The inter-dependencies consist in the fact that it is impossible to talk about persons independently of situations in which persons may find themselves in reality, and of which they have knowledge; and impossible, moreover, to talk about objects of reality independently of *describing* objects, and independently of presupposing that they exist as things which we may talk correctly about. The inter-dependencies consist in the fact, furthermore, that although objects cannot be defined in terms of persons and situations, and persons cannot be defined in terms of situations or objects, and situations cannot be defined in terms of persons or objects, neither concepts are well defined without or independently of referring to the other. The *asymmetry* consists in the fact that it is a defining property of objects that they exist independently of persons and the situations in which persons find themselves, and independently of the knowledge they have and descriptions they may put forward about the objects in those situations.

7.3 Actions

Let me start by giving an account - from the point of view of a person - of the formal implications of being a person in a concrete situation in reality of which one has knowledge and of which one is part.

First, being a person in concrete situations in reality implies knowing that one has a physical body, which exists in a world made up of physical things and other persons, from which one may distinguish oneself and be different. Indeed, to be a person is to be someone who may identify and *designate* oneself as someone, an "I", who exists at particular places at particular times *relative* to other things or persons existing around or outside oneself in reality at those times. Hence, part of the knowledge of being able to identify and designate oneself as a person, an "I", is knowing that one can thus identify and designate oneself only *relative* to other things and persons from which one differs.

Whilst objects and other persons exist independently of me, I do not exist as a person, of whom I have knowledge, without or independently of other things or persons in the world, of which I also have knowledge. I cannot be a person who knows only about myself, but of nothing else. However, although the knowledge of a person about himself and other things cannot be *reduced* to the knowledge he has about himself and other things in any of the situations in which he finds himself in reality, his knowledge about himself as a person or an "I" who exists, does not exist in *isolation from* or *independently* of the knowledge he has of himself and of other things existing in those situations. Indeed, a person or an "I" who knows that he exists, cannot be someone existing beyond the "here and now" of concrete situations in which he has this knowledge, i.e. he cannot know of himself that he exists (or has existed) independently of being at the same time a person who knows that he exists (or has existed) in some concrete situation.[2]

An essential part of the knowledge of a person of things or other persons is knowledge about the possibilities of observation and action one has with regard to those things or persons - e.g. what one may *do* to or with them, and what may be observed to happen to them - or to oneself as the case may be - as a consequence. Indeed, to be a person is to be someone who may find oneself in *different* situations, i.e. situations in which one may have other or different possibilities of observation and action with regard to the same things - or to different things - existing in those situations. Thus, to be a person is to know that more and different things may be known about things and oneself than are already known, in the present situation as well as in different situations. Now, if one's knowledge of what things *are*, implies knowing what one can do to them, then to be a person having this knowledge implies being someone who may *initiate* such acts, and thus implies being someone who knows that one is the *agent* of the actions thus initiated. If, furthermore, knowledge of what things *are*, implies knowing what happens - to the things or to oneself - as a consequence of the acts one carries out with them, then to act with things must imply *anticipating* such consequences of one's acts, i.e. consequences which change one's future possibilities of action with regard to things. Indeed, I think we shall have to agree with Zinkernagel (1962) that *to act must be understood logically as changing one's possibilities of action*.

An act may be characterised as an event - an event which produces other events, which are intended by the agent who initiates the act. Thus, crucial to an event characterized as an act is that what happens is *made* to happen by someone who may designate himself an agent of the act and, thus, someone who *intends* to initiate some change or other in his possibilities of action. Put

[2] So, in passing, farewell thou troublesome transcendental "I" or "ego". And farewell to the equally troublesome Wittgensteinian version of this notion of an "Ego" as a singular perspective centre being the limit of the world, but none of its contents. (Wittgenstein, 1958)

differently, we cannot talk about *action* without or independently of a notion of *intention*, and cannot talk about intentions without or independently of a notion of persons capable of designating themselves as *agents* of action.

An action of a person, i.e. some activity or other carried out with the intention to change the states or events of reality and, thus, to change one's possibilities of action, may be opposed to that of a *reaction* or a *response* being causally determined by some state or event in reality. Moreover, if to act implies having possibilities of action - and of knowing what they are - then *to carry out an act may be understood logically as having made a choice among possibilities of actions* - as opposed to reacting or responding in causally determined ways.

I contend that action thus defined is something which may only be ascribed to persons, having knowledge about states and events in reality, and to whom intentions may be ascribed to changes those states and events - but not to physical systems or organisms, the processes and behaviour of which may be characterised solely in terms of causally determined reactions or responses to states and events in reality. Conversely, a person is someone who, in contrast to such physical systems and organisms, may carry out actions.

Now, it may be objected that not all behaviour of persons may be characterised as actions, and that at least some of the behaviour of persons are reactions to events or responses to states in reality. Indeed, it has been argued that the actions of persons, because the *reasons for* their initiation are some prevailing states in the world, may themselves be described purely as events causally determined by such states.[3] Thus, when describing the action of persons we may make do with terms which refer to acts as events, and describe the occurrence of acts, their cause and effect, in the same causal and deterministic terms by which events in physical reality are described and characterised. The purpose of this argument has invariably been to narrow down the difference between human beings and physical systems or biological organisms, and to argue that the difference - if there is any - is not fundamental, nor one of principle. If so, the possibility exists of somehow *reducing* the action of persons to events describable in terms of the conditions and principles that govern events of physical systems or biological organisms. And it has been argued that the possibility exists, conversely, of somehow *deriving* the conditions and principles governing the action of people from those governing the processes and behaviour of physical systems and biological organisms. According to the assumptions of the latter view, at least *some* of the conditions applying to the action of persons may

[3] Arguments for this view may be found within *Behaviourism* in psychology and in *Logical Behaviourism* in philosophy, as well as in the various versions of eliminative materialism. A similar view has been proposed by the American philosopher Donald Davidson in his defence of "Anomalous Monism", which will be discussed in Chapter 9.

be found - at least in a primitive or rudimentary form - in the activities and responses of physical systems and biological organisms, e.g. organisms which made their appearance on the evolutionary ladder prior to that of human beings, or physical systems (such as computers) that have been designed to carry out tasks which, if carried out by human beings, would be described in terms of actions.

However, I think we shall have to admit that it is not at all clear what is meant by saying that the processes of physical systems or the behaviour of biological organisms bear in them the rudiments of actions of persons, or that features of such processes or behaviour are comparable to those of action - whilst in embryo. What does it mean to say, for example, that the processes of a physical system or the behaviour of a biological organism may be characterised as that of making *choices* between possibilities - albeit in a primitive or rudimentary form - or that such choices rely on *knowledge* about the world and the possibilities of action in the world available to the system or organism - if only in a primitive form? Or, what does it mean to say that physical systems or biological organisms may designate themselves as agents of *intentional* acts - if only in a rudimentary way?

And if only *some* of these primitive forerunners comparable to features of action are present, it is not at all clear in what way they may be said to be comparable. Does it mean, for example, that a physical system or biological organism, say an amoeba, may carry out activities according to *intentions* (or that its activity is prompted by something comparable to intentions), but without having knowledge about possibilities of action available to it, among which it may choose? Indeed, what sense would it make to talk about the existence of intentions (or something comparable to intentions) of a physical system or an amoeba, which does not find itself in concrete situations in physical reality where it has knowledge about such possibilities of action, and relative to which such intentions may be determined? Or, does it mean that a physical system or an amoeba may act, and thus initiate changes in its possibilities of action - but without being an agent, who may designate itself as the initiator of these changes, or without being an agent who may anticipate the consequences of its action in terms of changes in its (future) possibilities of action? Indeed, if taken one by one, *what* features and *how many* entailed in the action of persons may be missing or left out in the processes of a physical system or behaviour of an amoeba, and *still* be said to be comparable to features constituting action of persons?

It seems to me that the discussion above has shown that it does not make sense to talk about the presence of any one of the features constituting an act without or independently of talking about the presence of all the others. Nor do terms for any one of the features of an act have well defined meanings independently of or without reference to the others. That is, it seems that the relation between these notions is a *logical* relation. For this reason I do not find it particularly interesting to discuss whether it may be said of an amoeba or a physical system (say, a computer) that it may designate itself as an initiator of its action, *or* of making choices, *or* of having intentions, *or* of having the knowledge of reality which is required in order to form intentions about what acts to carry out in reality - that is, I find it uninteresting to discuss whether such features, *taken one by one*, may be said to be part of the processes or states of physical systems or biological organisms. For, again, if none of the features of the action of a person may be said to exist independently of the others, then none of the terms referring to such features have well-defined meanings and applications for instances of activity or behaviour lacking any one of the features of action.[4]

The foregoing arguments seem to be just as much arguments why the action of persons cannot be *reduced* to or explained in terms of the processes and functioning of physical systems or biological organisms. Nevertheless, let us for a moment assume that the action and functioning of persons are describable in purely physical, causal and deterministic terms. So to assume, however, would be to assume *both* that persons and their acts may be described in terms of physical systems and biological organisms, *and* that this may come to be known by persons, who may carry out investigations and, thus, carry out intentional acts upon and correctly describe the processes and functioning of physical systems and biological organisms in these terms. That is, persons about whom it is presupposed, therefore, that they cannot *at the same time* be described in those terms, but who may realize that the conditions and principles for their own functioning differ fundamentally from those governing physical systems and biological organisms.

* * * * * *

In previous sections I have talked about cognition and identification of objects and events as being part of actions carried out by the cognizing person, indeed, that cognition or identification of something may itself be considered as an act, which depends on the intentions and purposes of persons in particular situations. It may seem a bit odd, to say the least, to talk

[4] It has to be admitted that most of our highly developed behavioural skills are also highly *automated*, that is, they are executed with little or no conscious awareness of the individual behavioural components and their sequence which constitute such skills. The point is well taken, however, that faults ("slips" and "lapses"), which occur during the execution of such skills are best characterized - indeed, they are experienced by us - as *actions not as planned* (Reason, 1979). Hence, such faults in automated skills of persons, and by implication such skills, cannot be accounted for without or independently of reference to the intentions, knowledge and anticipated outcome of persons who execute them.

about our everyday cognition and identification of cups, furniture, trees, and houses, or cats and dogs, as something which relies on *choices* depending on purposes, interests, conditions of observations and knowledge. Such talk only makes sense within an analysis of the formal and logical conditions involved in our cognition, perception and description of things.

Please note: when I say that identifying a thing as a particular thing is an act which relies on a choice it is *not* to be understood in the sense that it is a matter of choice that I see the dog outside my window as a dog – that, for example, I could have chosen to see it as something else. Nor does it mean that it is a matter of choice that when I describe it, I describe it as a dog. In all the situations I can think of, it will be part of my perceptual identification of the dog that I see that it is a dog, and part of my description of it that I describe it as a dog. However, it may, formally speaking, be a matter of choice, knowledge, purpose and circumstances in the situation, that I see and describe it as 'the bigger of the two dogs playing together on the field outside my window'; or that I see it as 'the rather scruffy dog with the twisted tail belonging to my neighbour'; or as 'a friendly and harmless animal belonging to a particular species - as opposed to the bull grazing on the same field, which belongs to some other species'. However, to maintain that my seeing or describing the dog in any of the above capacities as *causally determined* by events or 'stimuli' in the world, as opposed to the stance of a person in a concrete situation which depends on the interests, knowledge and possibilities of observation and action in that situation, would be tantamount to missing those crucially important formal aspects of perception and description which make both the description and perception of persons an act.

Normally, we directly experience objects, trees, houses, chairs and tables as something concrete and tangible. However, we are seldom - if ever - aware of the general conditions and presuppositions implied in talking about cognizing trees or houses, etc. It is not difficult to imagine, then, how hard it must be to accept that we cannot talk about our knowledge and description of the most concrete, tangible and everyday matters like trees and houses without relying on a whole series of rather intangible inter-dependencies, logical relations and asymmetries between such matters and such knowledge.

Chapter 8

Scientific and other descriptions of reality

8.1 The dependence of scientific descriptions on non-scientific descriptions of reality

In everyday as well as in scientific situations we describe reality from different points of view, involving different possibilities of observation and action, and according to different conceptual schemes. However, it is the intuition of many philosophers that there is a way in which reality *really* or *essentially* is independent of the variety of different descriptions of it, and that only descriptions which capture or are in accordance with this essential nature of reality are true and objective descriptions of reality. Hence, the fact that reality exists independently of our descriptions and conceptions of it is taken to mean that there must be something which reality is "in itself" *beyond* points of view, and descriptions put forward from within different points of view.

It is the intuition of other philosophers that since the way we describe and conceive of reality, in everyday as well as in scientific situations, is determined by our central nervous systems and restricted by the capabilities of our minds to construct conceptual schemes and languages,[1] there is no certainty that any of our descriptions of reality are actually in accordance with the non-linguistic reality existing beyond our descriptions of it. Indeed, the conceptions and experiences we have of reality may have very little - if anything - in common with reality itself. As already argued in previous chapters, this view must be rejected since it implies a polar opposition

[1] See for example Maturana and Varela, 1980

between *reality* and *descriptions* and *conceptions* of reality according to which neither notions of reality, nor of descriptions of reality have well defined meanings. Worse still, since both *reality-in-itself* and our current descriptions of reality can only be conceptual-linguistic constructions of our minds, no meaningful distinction - and thus no polar opposition - can be made between reality-in-itself and descriptions of reality.

An alternative view which has gained almost universal acceptance, not only within science and philosophy but also in common sense, is that descriptions actually do exist of reality and its essential properties and structures, namely the descriptions developed within the natural sciences - and in particular by physics. Such scientific descriptions are said to be *objective*, in that they do not rely on, but are totally independent of the subjective, species specific points of view of the human mind. This alternative view shares the assumption with the view presented above that it does indeed make sense to talk about reality having *essential* properties, and that only descriptions and notions of such properties represent what objectively and truly exists in reality. The views only differ on whether such notions and descriptions are or may become available to us.

Now, the assumption that the essential properties of reality are in fact available to observation and description by us seems to make much better news than the assumption of the alternative view. This more optimistic view is not unrelated, of course, to the progress made within the natural sciences during the last 300 years. However, the very notion of reality and its things having *essential* properties, and that only descriptions of those properties are objective or true, has consequences which are no less contentious than those of the first view. For if it were the case that only descriptions belonging within particular conceptual systems (say, those of the most advanced disciplines of the natural sciences) were true descriptions of what really or objectively exists in reality, it would automatically follow that all other descriptions of reality were either false - or not about reality at all. Indeed, by thus making concepts such as 'true', 'reality', and 'properties of reality' as well as 'true assertions about reality', dependent on particular conditions of observation, purposes and conceptual, descriptive systems, these concepts would be deprived of any well-defined meanings in all other situations in which we act, observe and describe reality. What is seldom appreciated, however, is that this inevitably has consequences not only for non-scientific descriptions, but also for scientific descriptions.

Let me begin my argument for why this must be so by stressing that it is uncontroversial that scientific descriptions provide more *precise*, *systematic* and *extensive* descriptions of particular features and events of material reality than ordinary everyday descriptions of the same things or events. Indeed, descriptions of features and events may be given within the natural sciences which are not observable in an everyday context, and for which we have no concepts in our everyday language. And it is uncontroversial that such more precise, systematic and extensive descriptions and observations have given rise to formulations of impressive theories of universal generality about the constituents and behaviour of physical matter, which are far superior to any "folk-theory" or notion about physical matter based on everyday observation and description of things. However, if it is right, as pointed out by Niels Bohr, that descriptions of the conditions for carrying out experiments, say, in particle physics (including descriptions of the working of measuring instruments) rely on and are expressed in terms of classical physics - which in their turn rely on ordinary everyday descriptions and observations of macroscopic objects (including descriptions of measuring instruments and other objects belonging to the experimental setting), then it would of course be untenable to maintain that descriptions in terms of particle physics are more true than (or true as opposed to) descriptions of the conditions for carrying out investigations in particle physics. To do so would be to maintain that it is possible to develop true and correct descriptions on the basis of false and incorrect descriptions. The example already given in Chapter 1 illustrates the contradiction involved in this view. If we say that the essential and true properties of reality are those being studied in, say, particle physics, and that only descriptions of those properties are correct and true descriptions of what objectively exists in reality, then the consequence would be that a physicist, after having discovered the structure and laws of the particles of an atom, would no longer be able to give a correct or true description of the laboratory, in which he carried out his experiments, nor of the apparatus and measuring instrument, which he used in these experiments. In general terms, he would be unable to give a correct or true account of the conditions under which an experiment in particle physics is carried out.

Some, if not all, versions of *language-reality materialism* involve the further assumption that if only *that* truly exists in reality which is describable in terms of the current most advanced part of physics, then things described in everyday terms as tables and chairs, rocks and trees - as well as experience and beliefs about such things - *quite simply do not exist*. Granted this assumption, we may well ask what *are* things like tables and chairs, rocks and trees? And if what is being described in physics in terms of atoms, protons, molecular structures, etc., are not descriptions of the kinds of thing of which *objects* like tables and chairs, trees and rocks, are made up, then what *are* they descriptions of? If, moreover, *beliefs* or *conscious experiences* of objects - scientific and non-scientific alike - are not the sort of thing which is describable in physical terms, and thus do not come under what exists objectively and truly in reality, what then *are* beliefs and conscious experiences, and *where* do they exist?

Russell, who was the first in modern times to promote this view, which in various guises has dominated science and philosophy throughout this century, proposes the following answers (Russell, 1948). The objects we experience as being "out there" are nothing but private events of our minds. Such things as the sun, tables and chairs, trees and rock, do not exist in reality. What exists in reality, i.e. external to our perception of it, are particles of different kinds, structured in different ways - which may affect our sensory systems and central nervous systems so as to produce the experience of objects in our minds. Indeed, if there were no human beings with sensory systems and central nervous systems, such things as tables and chairs would not exist. To this one may presumable add - not even the chair I am now sitting on would exist, nor for that matter my body sitting on the chair, or its sense organs, or any other things I may perceive in non-scientific situations. For if it were true, as maintained by Russell, that what I experience and observe in non-scientific situations does not exist in reality, but only exists as "psychological events" of my private mind, then my body, my eyes and ears, and everything else I may experience about myself and my body in such situations, do not exist in reality. Here is Russell's view as he himself presents it.

> "What I know without inference when I have the experience called "seeing the sun" is not the sun, but a mental event in me. I am not immediately aware of tables and chairs, but only of certain effects they have on me. The objects of perception which I take to be "external" to me, such as coloured surfaces that I see, are only "external" in my private space, which cease to exist when I die - indeed my private visual spaced ceases to exist whenever I am in the dark or shut my eyes. [...] They are only external to certain other percepts of mine, namely those which common sense regards as percepts of my body; and even to these they are "external" only for psychology, not for physics, since the space in which they are located is the private space of psychology (ibid., p. 241). ... [A] serious error, committed not only by common sense but many philosophers, consists in supposing that the space in which perceptual experiences are located can be identified with the inferred space of physics, which is inhabited mainly by things which cannot be perceived. The coloured surface that I see when I look at a table has a spatial position in the space of my visual field, it exists only where eyes and nerves and brains exist to cause the energy of photons to undergo certain transformations. The table as a physical objects, consisting of electrons, positrons, and neutrons, lies outside my experience, and if there is a space which contains both it and my perceptual space, then, in that space the physical table must be wholly external to my perceptual space (ibid., p. 236). [...] The common sense world results from a [...] correlation combined with an illegitimate identi-

fication. There is a correlation between the spatial relations of unperceived physical objects and the spatial relations of visual or other sensational data, and there is an identification of such data with certain physical objects." (ibid., p. 237).

Well, the first problem which springs to mind is this: If what is described within physics does not concern the reality we observe, i.e. the sun, tables and chairs, and bodies having eyes and ears and sitting on chairs, we would be hard put to say what those physical descriptions concern. Granted Russell's view, we could not even say, as he does, that "the table described as a physical object consists of electrons, positrons and neutrons", etc. - for it would be illegitimate to identify "such sense data" as *experienced tables* which only reside in our minds with the *physical events*, which exist in reality external to the perceptual field of our minds. And if it is illegitimate to say, moreover, that the sun or a stone exists - as anything but mental events - it would be pure nonsense to claim that a description of the sun and a stone *in the terms of physics* is a description of something which exists in reality *external* to our minds.[2]

More recently Sperry has expressed virtually the same view, however in a slightly different way, which makes obvious other problems to which it gives rise. Thus, Sperry contends,

"As we look around the room at different objects in various shapes, shades, and colours, the colours and shapes we experience, along with any associated smells and sounds, are not really out where they seem to be. They are not part of the physical qualities of the outside objects, but instead, like hallucinations of the sensations from an amputated phantom limb, they are entirely within the brain itself. The perceived colors and sounds etc., exist within the brain not as epiphenomena, but as real properties of the brain processes. When the brain adjusts to these perceived colors and sounds, the adjustment is made not merely to an array

[2] In a rather obscure passage, Russell seems to have doubts about the *absolute* truth of scientific descriptions. He writes: "In all that has been said hitherto about the world of science, everything has been taken at its face value. I am not saying merely that we have taken the attitude of believing what men of science tell us, for this attitude, up to a point, is the only rational one for any man who is not a specialist on the matter in question. What I do mean is that the best scientific opinion of the present time has a better chance of truth, of approximate truth, than any differing hypothesis suggested by a layman. The case is analogous to that of firing at a target. If you are a bad shot you are not likely to hit the bull's eye, but you are nevertheless more likely to hit the bull's eye than to hit any other area. So the scientist's hypothesis, thought not likely to be quite right, is more likely to be right than any variant suggested by an unscientific person" (ibid., p. 251). To this it may well be objected that without well defined knowledge and descriptions of the target of our shooting, no one could decide whether it had been hit, let alone how far from the bull's eye our shots had hit.

of neural excitations correlated with the colors and sounds but rather to the colors and sounds themselves." (Sperry, 1976).

According to this view of Sperry, which is shared by an alarming number of neuroscientists and psychologists, the shapes and shades we perceive are not part of the physical properties of objects existing "out there" in reality. Rather, they are qualities of our brains - and as such have the status of "mental phenomena" on a par with hallucinations. But if e.g. the shape of objects were not really a physical property of outside objects, i.e. of things existing outside our mind, it would inevitably imply that *none* of the objects "we look at in the room" exist in reality. Or, would it make sense to say that these objects exist - but that they do not have a shape? This view, as far as I can see, implies a total annihilation of our familiar world of objects, and of the existence of such objects in reality, as something we can talk sensibly about - and therefore the annihilation of *perception* and *knowledge* of such objects as something which we can talk sensibly about. We cannot even (or consistently) say, as Sperry does: "as we look around the room at different objects in various shapes, shades ...etc". For our cognition and knowledge of these properties of tables, chairs and rooms, etc., and thus of tables, chairs and rooms themselves, are nothing but hallucinations, something which only occurs in our brain - which amounts to saying that such objects and rooms do not exist at all in physical reality.

The implications of this version of "physicalism", if taken to its conclusion, must be that our consciousness, cognition and descriptions of physical and neurophysiological events and phenomena necessarily fall victim to the same fate, i.e. it will not be possible either to talk consistently and sensibly about the *physical* properties of the objects we perceive in reality nor about the *neurophysiological* properties of the brain. For, the descriptions of the physical qualities of reality, as well as the descriptions of the neurophysiological properties and processes of brains, would have to be founded on descriptions of *hallucinations* of reality and brains. It would be a situation in which we were merely *hallucinating* that we are persons with bodies of particular shapes, finding ourselves in particular rooms or laboratories studying physical properties of reality, and neurophysiological qualities of brains. In the final analysis, it would be a situation in which there would be no way of distinguishing descriptions of such studies from descriptions of *hallucinations* of such studies, etc. ad infinitum.

As pointed out by Nagel (1994), the problems of *physicalism* is undoubtedly the unfortunate result of *misinterpreting* the conditions which lead to the overwhelming success of the methods and principles of the physical sciences. I agree entirely. However, it seems to me that rather than clarifying the misinterpretation of these conditions, Nagel at the end of the following passage exemplifies this misinterpretation.

"What has made modern physical science possible is the method of investigating the observable physical world not with respect to the way it appears to our senses - to the species-specific view of human perceivers - but rather as an objective realm existing independently of our minds. In order to do this it was necessary to find ways of detecting and measuring and describing features of the physical world which were not inextricably tied to the ways things looked, sounded, and felt to us; and this resulted in the discovery of objective, essentially spatiotemporal properties of the physical world which could be mathematically described and related by general laws of extraordinary power and universality, thus enabling us to transcend the rough and particular associations available at the level of merely human appearances. The result is an understanding of objective physical reality almost unrecognizably different from the familiar world of our theoretically unaided experience. But it was a condition of this remarkable advance that the subjective appearances of things be excluded from what had to be explained and described by our physical theories. [...] They are excluded *because they are tied to species-specific points of view that the objectivity of physical science requires us to leave behind.*" (Nagel, 1994. p. 65-66, italics added.)

I agree with Nagel that the condition for the success of the physical sciences - starting with classical mechanics where it all began - was that ways were found to determine the qualities and features of objects belonging to our familiar world which were relevant for a precise mathematical description of their movements in space - and in later developments of physics of their internal atomic and sub-atomic structure. And it was a condition that these features and qualities could be delimited from other features of the objects, such as their taste, smell, colour, functions, aesthetic appearances or looks, which were of no relevance for our descriptions and understanding of the "physical causation" involved in the movement of objects, or of their internal atomic and sub-atomic structure. What went wrong was to take the possibility and success of constituting such conditions to mean that *only* the features and qualities of physical objects and events as observed under these conditions, by the methods of physics and expressed in its (mathematical) terms, were part of *objective reality*. For this does not follow at all; it is precisely at this point that misinterpreting the conditions for the success of physical science sets in - and with it the problems and absurdities of physicalism already encountered. If it did follow, it would also have to follow that the features of objects of our familiar world, which play no role in scientific accounts of the structure and cause of physical events, but which *do* play a role in our identification of objects in this world, do not exist independently of the human mind and its subjective, species-specific points of view - and thus in effect, have no objective existence in reality. However, so to maintain would be to maintain that the familiar world of objects of which we ourselves, our bodies, etc., are part, and in which we carry out investigations of the physical properties of those objects, is a product of our minds which, literally, exists - and *only* exists - in the "realm" of our minds.

More seriously, granted that our knowledge and determination of this familiar world and its objects necessarily forms the point of departure for our further scientific investigation in and of this world and these objects, how then can we sensibly say that when carrying out scientific investigations we have left behind what in Nagel's terms are subjective, species-specific points of view of the human mind? And how can we be sure that such scientific investigations do not themselves rely on some *other* species-specific points of views of the human mind, and thus rely on views which are just as subjective as those on which rely our ordinary everyday perception, knowledge and description of the world?[3]

If on the other hand we assume, as Nagel does, that the laws of physics and its descriptions concern features and structures existing objectively in reality and, moreover, that the universality and generality of these descriptions and laws means that *they apply generally to the objects of our familiar world*, then we shall necessarily also have to assume that *that* world and its properties exist just as objectively in reality as do the properties and structures of them described by physics. That is, we shall have to assume that objects and properties of our familiar world - in whatever way they appear to us - exist in reality just as independently of our cognition and descriptions of them, and just as objective as do the physical features and structures of those same objects described by physics. The only other alternatives to these assumptions would clearly be nonsensical:

Either,

1. The universality of the laws of physics and the generality of its descriptions of features and structures existing objectively in reality (such as molecules, atoms, elementary particles and their composition) apply to and are the features of which are composed the objects in our familiar world, known and described by us as tables and chairs, rocks and trees. However, the objects and properties to which these laws and descriptions of physics apply do not themselves exist objectively in reality; nor, therefore, are our descriptions and knowledge of them objective or true.

Or,

[3] In favour of the latter view, it could well be argued that no non-human species has yet been known to have developed accounts of physical reality in terms of classical mechanics, let alone in terms of particle physics.

2. The physical features investigated by the natural sciences and described in terms of molecules, atoms, elementary particles, etc., are features of things existing in some different world or reality - i.e. a reality which is different from the "species-specific" familiar one of which we ourselves are part and in which we carry out these investigations. Therefore, physics and its general laws do not apply to objects of our familiar world.

Now, the only way to avoid the absurdities arising from the views that things in our familiar world - or indeed those observed by the methods of the natural sciences - do not exist independently of the species-specific points of view of our minds, is to accept, first, that just as our everyday observations and descriptions of the world rely on "points of view", so do our scientific observations and descriptions. Indeed, it is to accept that *any* observation and description of physical, material reality in any of the situations in which we find ourselves - be they everyday or scientific - invariably rely on points of view, in the sense: *rely on the possibilities of action and observation as well as on descriptive-conceptual systems available to us in those situations*. And it is to accept, secondly, that this does *not* imply that things thus observed and described are things which do not exist in reality *independently* of those points of view, possibilities of observation and descriptive-conceptual systems, and all the rest. On the contrary; as argued in the previous chapters, we would not have any consistent or well defined notions of *descriptions* and *knowledge* of reality, nor therefore of *reality*, without presupposing that reality exists *independently* of our knowledge and descriptions of it, and as something about which we may have knowledge and put forward true statements - *no matter the nature of this knowledge and these descriptions, and what part or properties of reality they concern.*

And it is to accept, thirdly, that the descriptions of features and properties of things as observed by the methods of physics and expressed in the terms and expressions of mathematics, necessarily rely on other *non-scientific* descriptions and determinations of things existing objectively in reality, and thus rely on descriptions and concepts which are not themselves part of the language and concepts of the natural sciences. If we say otherwise, for example say that the non-scientific observations and descriptions of things on which rely our scientific observations and descriptions, are not about things which exist objectively in reality, then we would be saying that it is possible to develop descriptions of what exists objectively in reality on the basis of determinations and descriptions of things, which do not exist objectively in reality, and which are not themselves objective or true. In other words, we would be contradicting ourselves and talking nonsense.

As rightly pointed out by Nagel in another passage of the text quoted above, the condition for the success of the investigations of natural sciences was that

> "[...] stripping familiar objets of all but their spatiotemporal, or primary, qualities and relations [made possible] their reduction to more basic and law-governed objective phenomena from which the familiar properties can be seen to arise by necessity, as the mass and hardness of a diamond can be seen to result from the combined mass and structural arrangement of its carbon atoms, or the liquidity of water can be seen to result from the relations of its molecules to one another." (Nagel, ibid., p. 67).

Thus, far from explaining or reducing "away" objects in our familiar world and their non-scientific properties, physics explains why they exist *objectively* as precisely the kinds of object, having precisely the properties by which they are determined and correctly described in our ordinary every-day language as, say, diamonds and their hardness, or water and its liquidity.

We may contend therefore that there needs be nothing odious, epistemologically, ontologically, or otherwise in the fact that things in the world - be they tables and chairs, diamonds and water, or atoms and the structure of their elementary particles - inevitably are described and observed by us according to "points of views". Nor need this fact necessarily make descriptions and observations of things *subjective* as opposed to *objective*, nor make the thing described into things which do not exist perfectly independently of such views, and descriptions put forward from within such points of view. However, what *may* be said to be subjective in the sense: left to persons who observe and describe, are the *choices of points of view* made in particular situations to describe and observe particular feature and properties of things existing in reality, as well as the purposes which determine these choices. Indeed, it *characterizes* beings with human minds that things may be observed and described from different points of view, and that descriptions and observations from different points of view may concern the *same* things. And it characterizes *the same thing* that it is that which in different situations may be correctly described in different ways. Without such a notion of 'same thing', and thus of 'same' and 'thing', there could be no notion of further investigations of things by which we may come to know more and different things about them. (This point and its consequences will be discussed more thoroughly in Chapter 16 of Part III).

Furthermore, what *is* dependent on, or a feature of our human minds, is that humans may *perceive* and have *experiences*, *beliefs* and *knowledge* about things and their properties, which exist in reality *independently* of this knowledge and these beliefs and experiences. Indeed, this feature of our mind must be a presupposition for any further investigations – scientific or otherwise – of things existing in reality. And what *is* dependent on, or is a feature of our human mind, is the capability to develop languages and descriptive-conceptual systems in which to put forward statements and propositions about these things and properties, as well as about our knowledge, beliefs and experiences of them, which may be *true* or *correct*. If we so prefer, we may say that such true beliefs, experiences and knowledge and propositions - be they of a scientific or of an ordinary everyday kind - are "features" of the "realm" of our minds, which exist just as objectively as do the things and features they concern, and to which they are necessarily related (cf. the arguments in Chapter 5).

However, as pointed out by Nagel (and here I wholeheartedly agree with him), such beliefs, descriptions, experiences and knowledge about physical things, are *not* themselves "physical things", and *therefore* are not the kinds of things which physics may account for. Nor are the notions and terms in which are described their content and experiential qualities, even less their referential and logical properties of being true or false, part of the vocabulary of physics to describe properties and states of physical matter. Hence, the reason why physics has nothing to say about our experience, knowledge, and beliefs of objects in our everyday world and their features, is *not* that such objects or features do not exist. Rather, the reason is that physics is not concerned with features of objects in our familiar world *as they appear to our senses*, but with the general and universal features of these *same* objects, as they may be observed by means of methods developed by physics, and accounted for in the language of physics. And the reason why physics has nothing to say about the referential, intentional and other logical features of beliefs, knowledge and descriptions about *anything*, is that these properties are not among the features of things that physics may account for, but rather, are features which physics has denounced as irrelevant in descriptions and explanations of phenomena falling within its domain. Conversely, to maintain that the body of theories, observations and knowledge of physics does not depend on persons and the points of view of persons, is to maintain wrongly that the knowledge, observation and theories of physics may exist independently in the sense of "outside" the minds and cognition of person, who have developed these theories and this knowledge, and who carry out physical observations.

Lastly, the world in which we act and make sense of and understand the acts we carry out, is still (science or no science) the familiar world of macroscopic things. It is on the basis of our knowledge and descriptions of *this* world, and our understanding of the actions performed on objects in this world, that any of our further scientific investigations are carried out - and of which, therefore, we necessarily have to assume that they are as true and objective as the knowledge and descriptions provided by such further scientific investigations.

* * * * * *

The aim of the foregoing discussion and argument has been to show that to the extent that scientific descriptions presuppose and rely on non-scientific descriptions and knowledge of reality and ourselves, we shall have to agree that what we say about the possibility of non-scientific descriptions of reality to be objective and true, necessarily has consequences for the possibility of scientific descriptions of reality to be objective and true as well. More specifically, we shall have to assume that the non-scientific descriptions and determinations of reality and its things, on which scientific descriptions rely, are descriptions of things which *exist objectively*, and that we may put forward description and propositions about them, which may be true and correct.

Conversely, no descriptions of reality put forward in any situation in which true descriptions may be made about reality, can be said to be more true or objective than (or true or objective as opposed to) other descriptions of the same reality, put forward in any other situation. Thus, we shall have to assume that any descriptions put forward within any concrete situation, the implications and application of which are correct, are equally true descriptions of reality. Neither the truth of descriptions, nor the objective existence of what is described, then, can be made dependent on particular possibilities of observation and description - be they scientific or non-scientific.

These points may be put in more general terms. First, any description, scientific and non-scientific, *rests upon an abstraction* in the sense that it is put forward in a particular situation with regard to particular descriptive systems or conceptual apparatuses relevant to the actual conditions of observation and opportunities of action in that situation. Secondly, a scientific description *depends on an abstraction* to the extent that it depends on an ordinary everyday determination and description of reality, its objects and properties - that is, to the extent it is possible to maintain that it is this *same* reality, these same objects, existing in particular places at particular times etc., having been determined in an ordinary everyday way in these and those ways, which are the objects to which scientific descriptions apply. Without this dependency on or *logical* relation between scientific and ordinary descriptions of reality, no consistent claims can be made about *what* scientific studies concern. (For example, that it is the molecular structure of things correctly described as tables and chairs, trees and rocks and so forth, with which they are concerned.)

Because of the dependency of any description on other descriptions put forward in other situations, no description, not even a physical, scientific description, is in itself *sufficient* to determine the objects or events being described. Hence, no description can be *identical* with what it describes, nor may things be reduced to any particular description of their properties put forward in any particular situation. However, this is precisely what the supporters of the assumption of the existence of essential, objective properties and descriptions of reality maintain. In the case that such essential properties are assumed to be the physical properties as described by natural science, the assumption implies that that to which we can give *both* an everyday *and* a physical description is the same as that which can only be *correctly* described in the terms of physics. So to say would be tantamount to saying that a physical description of a thing is *identical* with the thing - in which case well defined meanings of "same thing", and thus of "same" and "thing", would no longer exist, nor therefore of scientific as opposed to non-scientific description of things .

Such a reduction will not do. On the contrary, it seems a logical necessity to assume that no assertion about anything made in any concrete situation can be identical with that which it is about. And this in two respects: first, no assertion about anything can be exhaustive; we shall always be able to say *more* about it - in the concrete situation as well as in other situations. Reality is in this respect inexhaustible. We shall always be able to ask new or additional questions and to develop new possibilities of observing and investigating reality, which will give us new information and knowledge about reality. Or, could we possibly imagine that, one day, there would be no more questions to ask about reality, ourselves and the situations in which we find ourselves? If we cannot so imagine, we shall have to agree that our knowledge and description of reality will always be *incomplete*.

Secondly, the assertions and descriptions forming part of the very determination of objects and events being scientifically investigated, and about which new and additional knowledge is being developed, cannot be identical with the content of the descriptions by which the objects are being determined (e.g. descriptions of objects in ordinary everyday language). Otherwise, it would be impossible to say that more detailed investigations of these objects and events may be carried out, in the same as well as in other situations, and by means of which we may come to know *more* about the same object and events. Neither can assertions *resulting* from such further investigations (e.g. assertions about the physical properties of reality) be identical with what is asserted in those assertions. They cannot be so because they presuppose and are based upon the assertions by means of which the objects of the investigation have been determined: That it is *these* objects or *these* events, the *same* objects or events which these (physical) descriptions or assertions concern. (The consequence of this for Computational Functional models of language and cognition together with the consequences for these models of the *incompleteness of knowledge and descriptions*, will be further dealt with in Chapters 12 and 14.)

In other words, *truth*, what is true about reality and its things, cannot be dependent on particular conditions of observation and description; nor can the determination of what objects and properties of things *exist* in reality. In view of the dependency of any description put forward in any situation on other descriptions put forward in other situations about the same thing or phenomenon, we cannot question or doubt the truth of any of those descriptions without questioning or doubting the truth of all the others. In particular, we cannot question the truth and objectivity of everyday descriptions of things in our everyday world on which, logically, depend scientific investigations and descriptions of the physical properties of those things, without questioning or doubting the truth of such scientific descriptions. In effect, we cannot do so without rendering the notion of truth meaningless.

According to eliminative materialism and other versions of physicalism, mental states, beliefs, consciousness and other psychological states do not exist - or, more precisely, they do not exist as the kinds of phenomenon conceived of and described in our "folk psychological" notions and theories about them. What exists are physical or physiological states and processes of the brain. In so far as such phenomena as consciousness, belief and other psychological states may be said to exist, this view continues, they only exist as something which is describable in terms of physics and physiology. In the next section I shall argue that although physicalism evolved at a time when it became possible to begin to formulate the conditions for scientific research, physicalism is incompatible with these very conditions.

8.2 The limits of scientific theories and descriptions

Whether an area of inquiry becomes establish as an area of scientific research depends on the possibility of determining precisely what constitutes the object (thing, event, phenomenon, etc.) being investigated within the area, as well as the particular aspects and properties of the object that the investigations concern. Secondly, it requires that the concepts developed to describe and determine the particular part or aspects of the object being studied, be so precise as to delimit unambiguously these parts and aspects from other parts or aspects of the object, and to delimit them from other objects of a different constitution or "nature". That a particular area of knowledge and inquiry may attain the status of scientific research, then, depends on the possibility of indicating precisely which, limited, part or aspect of reality the science concerns. We may say not only what the investigations concern, but equally importantly, what it does not concern, and we can indicate with what part of language a science is particularly concerned, and with what other parts it is not, i.e. isolate the words and concepts of the science from other words or concepts of language.

Thirdly, a research area is determined by the methods developed to investigate the particular phenomena and aspects of the object being studied within the research area, as well as by the purpose of the research. These standards and requirements for scientific research date back some 350 years, and with them the Galilean method, widely being used by all sciences today, of *methodological reduction.*[4] This method aims, first, by successive *isolation* of the constituents and properties of the object to determine its *irreducible* entities and, secondly, to isolate and systematically control the external and internal influences affecting these entities and properties with the purpose of establishing lawful relations and causal reactions. In particle physics, for example, the *quantum* is regarded as an irreducible entity, whereas *the cell* is considered an irreducible entity in biology. However, the method of reduction is the same whether the object of investigation concerns inanimate matter or biological substance.

Following these methods of reduction and criteria for scientific research, reality has been divided into three broad categories of objects, i.e., a) inanimate physico-chemical matter, and b) living bio-organical substances and organisms, and c) psycho-social phenomena of human beings, constituting respectively the area of research of the physical, biological and psycho-social sciences and their many sub-disciplines. Physics, for example, is divided into several areas of research - classical mechanics, nuclear and particle physics, thermodynamics, and so forth. Within the science of living substances and organisms we find zoology, botany etc. - each studying phenomena which in well-defined ways delimit them from phenomena being studied in other scientific areas of research, and each using concepts which have well-defined implications and uses within the scientific area in question. In mechanical physics, concepts are used, such as *solid objects* with *measurable mass* and *extension in space*; objects which may be at particular determinable places at particular determinable points in times, i.e. concepts with implications not being part of the concepts used to describe or account for atoms or sub-atomic phenomena. In biology we find concepts like genes, proteins, hormones and cells which refer to phenomena completely alien to inanimate matter studied within physics. And within psychology we have concepts such as meaning, understanding, knowledge, truth, language and use of language, which are necessary for describing the cognition and knowledge of humans about everything they describe and have knowledge

[4] Cf. Chapter 2 of Part I.

about; concepts, however, which are not part of the vocabulary of the natural sciences or biology to describe physical matter or biological substances.

Among the variety of features which uniquely distinguishes the objects of study within these different sciences are *the principles of transformation,* or change of states, which apply to their objects and properties, as well as the particular *structure* and *degree of complexity of organization* of their entities.[5] The mechanisms of transformation of inanimate matter are physical and chemical interactions, whilst the mechanism of transformation of biological substance is (currently thought to be) that of evolution by *natural selection*. The molecular compounds of atoms by which inanimate matter of the universe is made up do not exceed half a dozen, whilst organic molecules of living substance contains tens or even hundreds of thousands of atoms, making them many times bigger than the molecules of inanimate matter. Furthermore, the degree of complexity of organization of organic substance is almost infinitely greater than that of the organization of the inanimate physico-chemical matter. Each single tiny cell of an organism has a highly complex organization of its own, with a nucleus, chromosomes, and genes, and other cell-organs, and is built out of a number of different kinds of proteins and other types of chemical units, mostly large and complex. However, what crucially distinguishes biological substance - ranging from organic molecules to cells and organisms - from inorganic physico-chemical matter is its *self-reproductive* capability and, as already mentioned, the mechanisms of transformation by natural selection, neither of which figure in the domain of inanimate physico-chemical matter.

In contrast to this self-copying facility and mechanism of transformation by natural selection of biological substance, we find in the domain of the psycho-social of human beings a principle of transformation by *cumulative experience* combined with *conscious purpose*. This is a principle of transformation which has provided humans with resources, not only to adapt genetically to the circumstances in which they find themselves, but to *control* and deliberately *transform* these circumstances, just as the acquisition of languages have allowed them to communicate and make conventions as to how to organize and conduct their lives, satisfy their basic needs, their spiritual and artistic desires, in the setting of a wide variety of societies and cultures. What came to make human life crucially different from the lives of all other organisms was the development of language; it provided humans - individually as well as collectively - with new ways of acquiring, expanding and transmitting their knowledge, crafts and skills, and a freedom to develop ways of thinking and acting which is potentially open - whereas even higher

[5] Cf. Huxley (1953), and Køppe (1990) on which is based the following account of the distinguishing features, structures and principles concerning physical matter as opposed to those of biological substance and psycho-social phenomena.

mammals in comparison only have few patterns of behaviour and cognition, the possibilities of which are both limited and easily exhausted.

All this has led to a rate of change in the collective domain of the psycho-social, which is totally unparalleled in the domain of physico-chemical matter and bio-organical substance. If we are to believe the theory of the Big Bang, the major transformation and organization of physico-chemical entities took place in a few seconds, and the principles by which they have since been organized, and the causal laws by which their interaction has been governed have stayed unchanged. Other millions of millions of years had to pass before circumstances, probably unique in our part of the universe, evolved to allow for self-copying organic molecules to come into existence - and still other millions of years have been taken up to develop - by the mechanism of natural selection - the genetic pool of the currently existing range of biological organisms, including those of higher mammals. In the long prologue of human evolution, each major change in the psycho-social domain demanded something of the order of a hundred thousand years, and immediately after the end of the Ice Age, something like a thousand years. In contrast, during much of recorded history of human life, the time-unit of major change in this domain has been around a century, while recently it has been reduced to a decade or two.

The division of reality and its objects into three distinct areas of scientific research coincide with how these objects evolved in time. Thus, the existence of physico-chemical matter preceded and was a condition for the evolution of the biological substance, the existence of which in its turn preceded and was a condition for the evolution of psycho-social phenomena. But, although the evolution of these radically different objects and phenomena may be described as a continuous process in time, the *transition* from physical matter to living biological substance, and from biological substance to psycho-social phenomena, *cannot* be described in terms of a continuous process. Each transition marks the emergence of objects with properties and irreducible entities which are *qualitatively* different from and which cannot be described in the terms of those which existed *prior* to the transition. Thus, neither may the organization and lawful relations and interactions between entities, nor the mechanisms of transformation and change of states, which constitute the *constraining conditions*[6] for the existence of objects belonging within any of these areas be *deduced* from or *reduced* to those existing prior to them. This means that although particular physical and chemical conditions must necessarily have existed for biological substance to occur, the constraining conditions for the existence of biological substance can only be described at the level of description of biological objects and properties, and thus are only determinable *after* the emergence of biological

[6] Polanyi, 1968, Køppe, 1990.

entities and properties. Hence, attempts to deduce living bio-organical substance from or reduce it to that of inanimate physico-chemical matter and the organizational structures and mechanisms of transformation which apply to such matter, would be attempts which ignored that the constraining conditions for the existence of biological substance necessarily are determined at and, thus, *presupposes* the level of description of biological substance. Similarly, such attempts would ignore that the very method by which these constraining conditions have been *scientifically* established - i.e. by systematically isolating and methodologically reducing the constituents of bio-organical substance into their irreducible biological entities and lawful relations and interaction - is incompatible with physical reduction.

The same holds *mutatis mutandis* for psycho-social phenomena and attempts to reduce such phenomena and their constraining condition to or deduce them from those applying to biological substance or physico-chemical matter. Indeed, it is perhaps even easier to see why such a reduction or deduction is impossible if we consider how psycho-social phenomena and fact differ ontologically from material matter and biological substance. Thus, whilst physical matter and biological substance exist independently of human beings and their action, cognition and description,[7] phenomena and things in the psycho-social domain do not exist independently of human beings and their action, cognition and description. This is to be understood in the sense that most of the things in the everyday world of humans, who live in so called civilised cultures and societies, are created by humans - ranging from manufactured material artefacts to the institutions of these cultures and societies themselves, their political, socio-economical, judicial, and educational systems, as well as the conventions, norms and values by which these societies and systems are constituted. This does not of course mean that psycho-social phenomena and things do not exist *objectively* in reality, nor that they do not exist just as objectively as do physical matter and biological substance. However, unlike physical matter and biological substance, these phenomena and things and the significance, value, function and meaning by which they are constituted as precisely the sorts of thing they are, i.e. serving the purposes they do in the domain of the psycho-social, only exist in virtue of being *recognized, known and sustained in action* by human beings. This means that we here have a unique instance of an interdependency between, on the one hand, *knowledge* and *description* of facts and properties of things existing in reality and, on the other, the facts and properties of things that this knowledge and description concern *without* the asymmetry or independent existence of facts and properties of things, which I have argued for so

[7] - indeed, both physical matter and biological substance existed *prior* to the arrival on the scene of human beings, except maybe for the bits of biological substance necessary for the development of human cognition and language.

vehemently in previous chapters. It is an instance, one might say, in which - by incorrigible necessity - the ontological and epistemic existence of properties of things in reality coincide. (For detailed analyses and arguments of this point, I refer the reader to Searle's book: *The Construction of Social Reality*, 1995.)

A few examples may illustrate this point. A five dollar bill is a thing which may be used in exchange for goods and services. Once it has been produced and is used to serve this purpose, it exists just as objectively as a monetary token as do the paper and the molecules of the paper on which its value is printed. However, the five dollar bill and its property to serve as a monetary token only exists because people have *agreed* that it may serve this purpose, and it only exists in so far and for as long as people so *believe* and actually *use* it to serve this purpose. The same holds for other manufactured thing such as tables and chairs, houses and washing machines, or for exhibits of art such as paintings and sculptures. However, as material things, both a five dollar bill and the table it lies on and the chair I sit on do certainly exist *objectively* in reality; and despite the fact that the functions, meaning and uses of five dollar bills, chairs and tables do not exist independently of persons having notions about these functions and uses, five dollar bill notes, tables and chairs and their functions do not exist merely as "constructions" in the minds of persons, nor does their existence depend on the perception of persons in the sense that they cease to exist when they close their eyes, turn their backs to them, or the like. These things exist objectively as things which may be used as monetary tokens, things to sit on, and things on which one's dinner may be served. But, again, neither five dollar bills, nor tables and chairs exist as things *having these particular functions, meaning and uses* independently of persons who have made them to serve these functions, and who recognize and believe that they may have these functions, meaning and uses.

It is not difficult to show that the same holds for educational and judicial systems, marriages and other societal institutions, as well as for the conventions, norms and rules agreed among persons which constitute and define these institutions, and which guide the conduct of people living in societies having such institutions. However, since Searle has already presented and analyzed an abundance of examples showing this (ibid.), I shall again refer the reader to his book. The point argued for here (and, I take it, in Searle's account) is that the objective existence of facts and properties of things in the psycho-social domain requires persons, who cognitively as well as in action provide the conditions for the objective existence of these facts and their psycho-social properties. Thus, things and facts which have been created and exist in the psycho-social economy, are not "natural kinds" or things - although a lot of the physical stuff which goes into producing them (e.g. five dollar bills, banks, or schools, universities and libraries of educational systems) *are* natural things, which exist independently of persons and their knowledge and use of them.

Hence, in the domain of psycho-social phenomena and properties of things, neither the knowledge and recognition of these phenomena and properties exists without or independently of persons, nor do the phenomena or properties of this domain exist objectively independently of person having made them to exist. The very existence of psycho-social facts and their properties requires human beings or persons who intentionally create, know of and recognize them as *psycho-social* facts and properties, and who by purposeful, intentional action sustain and provide the existence for these facts and their properties. Persons, furthermore, who know of and are able to distinguish between "natural kinds" and "psycho-social kinds", and who may come to cognize the possibilities and uses of "natural kinds" in the creation and production of "psycho-social kinds", and who by way of language are able to convey or transmit this knowledge and these uses.

Well, given that

a) psycho-social facts and their properties, including the cognitive and linguistic resources of persons on which their existence rely, differ so radically *ontologically* from both (mere) physical matter and biological substance;

and that

b) the constraining conditions for the existence of psycho-social facts and their properties, including these cognitive and linguistic resources, cannot be given *prior* to the evolution and existence of such cognition, language and facts:

and given, *therefore*, that

c) the determination and description of these constraining conditions presupposes and *only* have well defined meanings at the level of description of psycho-social facts,

then the impossibility of deriving psycho-social facts and their properties from mere physical matter or biological substance would seem obvious. Conversely, it would seem obvious why it is impossible to reduce psycho-social facts and their properties to mere physical matter or the biological substance of organisms. And since, not least, it requires persons with a reflexive mind and language, who have knowledge about and may distinguish between the different ontological status of things and fact existing objectively in reality, it ought to be obvious why neither the minds of persons, nor their knowledge and descriptions of those things and facts, may be reduced to any of the ontologically different things and facts that they may have knowledge about and may use language to put forward assertions about.

What has been said above does not conflict with the fact that the existence in reality of material matter and biological substance is a condition for the existence of human beings or persons, nor does it deny that the bodies and brains of human beings, which are conditions for humans to function as cognizing, language using and acting persons, are made of physical matter and biological substance. Furthermore, what physics is saying about physical matter and what biology is saying about biological substance is applicable to the brains and bodies of persons, and in this sense the processes and event of the brains and bodies of persons are amenable to physical and biological description. For a start, human bodies, just as the bodies of birds and plants, do obey the physical laws of gravity; and to the extent that everything is built up of atoms and molecules, what is said about atoms and molecules of material things being studied within physics applies to everything, including the material matter of human, birds and flowers. Likewise, to the extent that living cells, genes and proteins in birds are similar to those found in the human body and in plants, what is said about cells, genes and proteins within zoology applies equally well to the cells, genes and proteins described in others areas of biology.

However, in view of the arguments summarized above, it would seem obvious that what e.g. classical mechanics says about the movements of material objects, or what particle physics says about atoms and elementary particles etc. *only* applies to those features and properties of humans, birds, plants etc., which they have in common with all other material objects. By definition, physics, as a well-defined, delimited research area with a limited field of application, has nothing to say about all other properties, which humans, bird or plants may have *apart* from and *in addition* to their physical properties. Likewise, what modern biology says about cells, proteins, genes and other properties of organisms only applies to those properties of human beings which they have in common with other living organisms. By definition, biology as a well-defined, delimited research area, has nothing to say about all the other properties which human beings may have *apart* from or *in addition* to other living organism. Hence, neither biology, nor physics have anything to say about the properties and phenomena of a psycho-social nature which has been *discarded as irrelevant* to, nor part of, the matters and substances studied within physics and biology. Neither do the concepts used to characterise and determine psycho-social properties and phenomena figure in the vocabularies of physics or biology to describe physical matter or biological substance, i.e. *in vocabularies with concepts which have acquired well defined meanings precisely because it has been possible to isolate them from, among others, the concepts and terms used to describe psycho-social properties and phenomena of persons*.

Nevertheless, it is the view of the growing number of scientists and philosophers who adhere to *Naturalism* that everything which exists in reality is *matter*, indeed only that exists *objectively* in reality which is describable in the mathematical terms of physical matter. Thus, to the extent that biological organisms and psycho-social objects and phenomena exist objectively, they are in principle (or will eventually be) describable in the terms of physics. This claim of Naturalism is based on the following reasoning: (1) because everything which exists in reality has physically describable properties, then (2) everything which exists is describable in purely physical terms *only*. What is said in (2) does not, of course, follow from (1) - and to make matters worse, it is claimed (3) that because all objectively existing properties and attributes of things are either physical, having physical causes and effects, or are causally related to things which are physical, then all properties and attributes of all objectively existing things may be described in purely physical terms. Non-physical things and properties, on the other hand, which are not related in a causal way to objectively existing physical things, and which cannot be described in terms of, or be reduced to something physical, simply do not exist.

This view and the reasoning behind it may probably be traced back to Galileo and his doctrine about the mathematical nature of material reality. Like other mathematicians, Galileo was convinced that God had created the world from simple mathematical principles and formulas, and that the behaviour of physical matter was governed by and could be described in terms of mechanical laws. He did concede, though, that even physical things are ascribed non-mathematical properties and attributes to which no quantitative measures may be assigned; however, such properties and attributes played no role in "physical causation", and, therefore, they were not part of physics. However, he then made a mistake for which, presumably, the belief in the doctrine about the mathematical nature of the world is wholly responsible. He drew the conclusion, that if what *exists* in the world is accountable in mathematical terms, then attributes, properties or phenomena which could not be given an exact mathematical expression, simply did not exist in reality. Such properties or phenomena were merely "names residing solely in the sensitive body" (Galileo, 1953).

It is a mistake, though, which has been repeated over and over in every generation of scientists and philosophers since Galileo - although in different guises. The root of the mistake is the same fundamental claim, namely that because all physical matter, its behaviour and properties, may be precisely described in the mathematical terms of physics and explained by its immensely powerful universal laws, then *only* what is describable in those terms of physics exists objectively in reality; everything else does not. Or, if it is granted some kind of existence (for example in our minds), it only *is* because it is thought to be caused by, or is a *property* of something physical (i.e. the brain), and thus *in principle* is reducible to something describable in purely physical terms. Indeed, everything which is so describable, it is claimed, will eventually be accountable in terms of the most advanced branch of physics, namely quantum mechanics. According to this view, mental or psychological phenomena - human cognition, language and use of language included - are either *caused by* or *properties* of physical processes and states of the brain and the physically describable interaction of such states with the outside physical world. Hence, both such interaction, processes and states of the brain, and the psychological phenomena and events which are caused by, or properties of, such interaction, processes, and states of the brain, will eventually be describable in terms of quantum theory.

Well, within the natural sciences themselves, a "unity" or total reduction of any of its autonomous, formalized theories (e.g. of chemistry, electronics, classical mechanics) with more basic and comprehensive theories of physics (or even to the most basic of them all, i.e. quantum theory) has (as yet) proven to be impossible; that is, it has proven impossible to equate or identify extensions of predicates and descriptions concerning phenomena of any of these autonomous theories with these more basic theories of physics (Oppenheimer and Putnam, 1958),[8] and thus to replace predicates and descriptions of these theories with those of the basic theories of physics. In view of this fact, it would seem incredibly farfetched to assume that phenomena and states within the domain of the psycho-social (or for that matter within biology) could be reduced to and be described in terms, say, of quantum physics. This is not to deny that, in scientific *practice*, cases of "reduction" within - or between - areas of the natural sciences may be encountered; however, such cases of reduction are examples of a kind of reduction in a much more loose and piece-meal sense than spelled out in the classical doctrine of reduction by Oppenheimer and Putnam. Examples of such more loose and imprecise senses of reduction are those in which a particular phenomenon is explained in terms of *identification*; i.e. lightening is identified as electrical discharge; chemical bonds as various modes of electron transfer

[8] Perhaps with the exception of the reduction of thermodynamics to mechanics via the hypothesis that gases are composed by many tiny independent particles which interact mechanically according to the laws and principles of classical physics. This and other hypotheses have allowed the derivation of equations in mechanical terms which are similar to the equations of thermodynamics, as for example Boyle's law.

and sharing amongst atoms, and genetic information is identified as sequences of amino acids in the DNA double helix, etc. Others are those in which a phenomenon is explained in terms of what *causes* it (albeit not in the sense of: may be *predicted* from), as in the case in which the hardness of a diamond is said to be caused by its atomic structure, or the liquidity of water is caused by the bond between hydrogen and oxygen ions, or - to quote a much used example in the philosophical literature - the heat which develops in the bearing of a wheel rolling down a slope may be said to be caused by the movement of the molecules of the wheel.

These practices notwithstanding, it would be both wrong and nonsensical to say that "hardness" is a property of an atomic structure, or that "liquidity" is a property of bonds between hydrogen and oxygen, or that "heat" is a property of moving atoms or sub-atomic particles. The terms "hardness", "liquidity" and "heat", and the properties they refer to, are terms and properties which occur at the level of description of macroscopic things such as solid things and liquids and the measurement of their properties - but *not* at the level of description of molecules, atoms or sub-atomic particles, and the measurement of their properties. Nor may these properties of solid things or liquids be *derived* from (let alone be predicted from) the properties of their atomic structure, the velocity of molecules, or from the bonds of atoms or sub-atomic particles from which molecules are constituted. If this is agreed, how much easier would it not then be to agree that the claim that mind - and all that goes with or is a "product" of it - is *matter*, in virtue of being caused by, or being a property of physical processes of the brain and its interaction with its physical environment, is a claim which is just as unfounded as it is void. Even if it were possible to account entirely for the physical processes and states of the brain in quantum physical terms, such an account would not amount to an explanation of how mental phenomena are caused by, are identical with, or "emerge" out of purely quantum physical phenomena. Nor would such an account in itself offer sufficiently extensive links between psychologically specifiable and explainable phenomena and properties of the mind, and the physically specifiable and explainable states and properties of the brain, which is a condition for an ontological "unification" or reduction of mental states and psycho-social phenomena and properties to physical phenomena and properties.

The proposition that everything existing objectively will eventually be accountable in terms of the most advanced theories of physics - which to-day happens to be quantum theory - is a dogma which only may be brought to bear either by denying the existence of everything which proves non-describable in strictly physical terms; *or* by claiming that non-physical properties of things may be identified as or are properties of something, which *is* describable in physical terms. The insurmountable problems and difficulties of this claim already becomes obvious when we attempt to make clear how it is to be understood. To take a simple example. What does it mean that, say, *a house* may be identified as, or is a property of phenomena describable in terms of quantum physics? Well, the atoms and property of the sub-atomic particles of all the different physical material of which a house is constructed, i.e. its bricks, concrete, cement and the like, may of course in principle be accountable in quantum physical terms. But needless to say, a quantum physical description of these materials and physical properties could not be a description of *a house* unless it formed part of, or could be derived from the implications and meaning of a quantum physical description that a house is the sort of construction, which is built from solid bricks, and used by people to live in, and which may occupy a particular area in space at some particular time, etc. However, it is a triviality to say that such concepts are not part of particle physics, nor is that to which they refer properties of particles. Neither is there any reason to believe that these concepts ever will be part of the vocabulary of any branch of physics.

The same problem applies to an account in terms of quantum physics of the content and logical properties of the proposition, "this is a house". It may well be that the physical properties of the proposition, i.e. the sound waves produced when uttered, may be accountable in terms of quantum physics. But the fact that the proposition is *referential*, and that it may be *true* or *false*, is certainly neither reducible to, nor derivable from any quantum physical phenomena, nor from any of the expressions used to describe such phenomena. That would require *not only* that the intentionality or "aboutness" of propositions and their truth value, and thus the very notion of 'truth', could be derived from quantum physics; but it would *also* require that these key notions of language and propositions could be derived from something *which is more elementary or basic than the notions of 'intentionality' and 'truth'*. However, to account for such more elementary (quantum physical) phenomena, whatever the language being used, requires the use of propositions or expression by which to refer to these phenomena, as well as the notions of 'true' and 'false', and therefore, neither propositions or expressions, nor notions of 'true' and 'false' can possibly be explained in terms of or be derived from such more elementary phenomena, which do not imply the existence of such notions. Alternatively, it would require that it were a property of quantum physical phenomena that they could refer and be true or false *in the same sense* as propositions about them may be; however this alternative would be just as much at variance with any well defined notion of quantum physics phenomena as it is at variance with logic.

A variation on the same theme: Let it be granted that quantum theory - or any future even more basic physical theory - were "the measure of all things", of what *is*, *that* it is, and *what* it is; and let it be granted, furthermore, that such a basic theory *par excellence* is a theory couched in mathematical expressions and formalisms - indeed that it is the nature of the phenomena it concerns that they are only thus expressible. All this granted, we would still have to assume that the mathematical expressions and formalisms of the theory are *about* phenomena existing in reality *independently* of the mathematical descriptions and expressions used by the theory to account for them. To say otherwise, to say, for example, that these expressions and descriptions are part of the very ontology of the phenomena that the theory concerns, would leave us (i.e. physics) with intolerable epistemologically problems.[9]

Now, if the physicalist claim of a possible reduction and explanation of the properties and capabilities of our minds to brain-states, being themselves describable in terms of basic quantum physical phenomena, were to have *general* validity, and thus applied in general to the capabilities of our mind to develop cognition, language and use of language, it would of course also have to apply to the capability of our minds to develop the mathematical systems and expressions required to account for and formulate the theory concerning these basic physical phenomena. If so, we would come full circle. For, it would be a claim to the effect that the capability of our minds to develop these mathematical systems and expressions, and thus these mathematical systems and expressions themselves, could be reduced to and explained in terms of these basic physical phenomena - and the mathematical expressions and formalisms in which they are couched. Or put even shorter: mathematical systems, their expressions and formalisms, may be explained by and in terms of mathematical expressions and formalisms being part of and originating from within the systems themselves. However, if Gödel's *theorems of undecidability*[10] are right, this is not a feasible proposition.

[9] To spell it out, it would be to assume that mathematical formalisms and expressions are not just *means* for describing quantum physical phenomena, but rather are *identical* with quantum physical phenomena. For example, expressions and formalisms concerning electrons and their properties *are* electrons having these properties or, electrons and their properties *are* mathematical formalisms and expressions. Now, if things were that simple, there would be no need, nor any justification for carrying out physical experiments in order to discover quantum physical phenomena and their properties, but we could make do with investigating mathematical systems from which the relevant expressions and formulae could be derived. However, not even the most ardent adherents of the doctrine that God has created the world from simple mathematical principles and formulae, I think, would adhere to this kind of mathematical constructivism.

[10] - the first of which is that if set theory is consistent, there exist theorems which can neither be proved or disproved; that is, no matter what are the set of axioms being used there would be questions concerning the theory which could not be answered.

From the account above of the implications and consequences of the assumption of a psycho-physical reduction of "mind" to "matter", it seems that *reasons of principle* exist for the impossibility of such a reduction.

8.3 Conclusion

Whereas the method of reduction to explain the existence properties and phenomena on one level of description by their causal relation to those accounted for on others may be defensible with regard to physical matters, no similar *psycho-physical* reduction would seem possible. Despite our growing knowledge of *correlation* between some particular neurophysiological and biochemical states of our central nervous system and some conscious psychological states and phenomena, as well as our growing understanding of some of the physical operations of our brains, which are necessary for the existence of conscious psychological states, neither this knowledge of correlates nor this understanding of operations comes anywhere near *explanations* of how conscious psychological states and phenomena arise out of neurophysiological and biochemical states or physical processes of our brains.

Rather, despite the claims of adherents of psycho-physical reductionism, it seem that *reasons of principle* exist why conscious psychological states and their properties cannot be explained in terms of, nor be reduced to neurophysiological or physical states and processes of our brain. The possibility of such reduction would require *either* that conscious psychological states, including knowledge, beliefs and their logical properties of referentiality and truth, could be accounted for in terms of the current types of theories within neurophysiology and physics - which patently they cannot. *Or*, it would require that that which patently cannot be described in terms of the current types of theories within neurophysiology and physics somehow be "eliminated" out of existence.

The latter alternative is precisely what followers of so called *eliminative materialism* have suggested - and maintain to be a necessity in order for accounts of psychological states to attain the same kind of scientific rigour and rationality as those of the natural sciences. (cf. Churchland, 1981). According to this "-ism", the possibility remains of some future developments of physics and neurophysiology in which conscious psychological states, including knowledge, beliefs and "other propositional attitudes", as well as their logical properties to *refer* and be *true* and *false*, could be explained in terms of something more basic and elementary, which does not imply the existence of conscious psychological states, nor of the logical properties that such states share with the language in which propositions may be put forward about such states – as well as about all other things. Indeed, it is envisaged to be possible for persons to develop knowledge and scientific theories about such more basic or elementary phenomena, as well as a language in which to describe and account for these phenomena, which at the same time would abolish the notions of both knowledge and language, and the logical properties of knowledge and propositions. To spell it out, it is the self-defeating assumption of eliminative materialism, as it is of all other kinds of physicalist reductionism, that we could have knowledge and formulate theories, and thus put forward *true* and *false* descriptions *about* such elementary or basic phenomena, which at the same time explained away and made void the very notions of knowledge, descriptions, reference, and truth.

Well, it is almost too easy to show the absurdity of the assumption of such future development of a science in which accounts could be given of "this" more elementary something to which e.g. 'truth' and 'reference', and with them all other notions of cognition and language, could be reduced in such a way as completely to eliminate and render these notions superfluous. For one thing, granted this assumption, the "language" employed by the science in which accounts be given of "this" more elementary or basic to which truth and reference may be reduced, or rather by which these notions may be eliminated, would have to be a "language" in which the notions of 'truth' and 'reference' did not exist and were not implied – and, as a consequence, nor any other notions and key concepts that language shares with logic. Indeed, it would have to be a science and a "language" which, unlike our current sciences and their languages, did not rely on nor necessitate the employment of formal logic and its principles. If this is not immediately obvious, one only has to consider the contradiction involved in assuming that notions such as 'reference' and 'truth' – and with them all other notions of language and logic – could be eliminated or explained "away" by a science, the accounts and theories of which themselves relied on the employment of these notions and the principles of logic, and thus on their existence. Or consider, the contradiction *and* impossibility of assuming that a science, capable of eliminating or reducing these notions of language and principles of logic to something more elementary or basic, is a science in which both true and consistent accounts and theories could be formulated *about* "this" more elementary or basic, albeit in a language which did not to rely on these notions and principles.[11] This, I think, should suffice to make

[11] A similar argument is made by Hilary Putnam in his book, *Representations and Reality* (1991). He writes: "The idea that there are properties of reference and truth (or falsity) possessed by words and sentences in anything that deserves to be called a language would appear to be as much of a myth [to eliminative materialists] as the idea that there are "propositional attitudes". But reference and truth are the fundamental notions of *the* fundamental exact science: the science of logic. Why don't the eliminationists speak of "folk logic" as well as of "folk psychology"? I once put just this question to Paul

clear why no theory which build on scientific rigour and rationality, can ever eliminate or reduce psychological states of persons, nor the cognition and notions being part of such states and their logical properties, to something more elementary or basic which does not imply the existence of such states and properties.

To the positions which holds that only those psychological phenomena are scientifically "kosher" which are strictly causally and deterministically related to the basic phenomena observed and described within the natural sciences, present or future, and thus are themselves phenomena the existence of which may be described and explained in terms of the causal, deterministic laws of the natural sciences, the following objection obtains. The fact that human beings under particular conditions of observations may come to know about and describe the constituent parts and features of things, e.g. atoms and sub-atomic particles, is certainly *determined* by the fact that atoms and sub-atomic particles exist and may be observed by human beings under those conditions. Indeed, we shall have to presuppose so. However, that human beings may develop conditions of observations - or points of view - from which they may come to know about and describe (say, in appropriate mathematical terms) those constituents of things, is *not caused* by the existence of atoms and other elementary particles of which things are constituted, nor by their mathematical descriptions. So to say or assume would clearly be nonsensical. But this is precisely what is assumed in attempts to account for and explain psychological states and phenomena, and hence knowledge, language and use of language, by their causal relations to things and events in physical reality, and in terms of the deterministic laws of natural science which apply to such things and events. Such attempts not only ignore the reasons of principle which exist against the possibility of such explanations, but grossly simplify issues and conditions for cognition and use of language of persons, which we have not even begun to understand, and of which we still lack precise analyses and adequate descriptions. Indeed, such attempts would mean the end of a science of psychology even before it had got started. But they would certainly also be attempts which violate the very condition on which the development of *any* science necessarily rests – all of which should be clear from the discussions of examples in the next three chapters, namely Chapter 9 (of Davidson's arguments for *Anomalous Monism*), Chapter 10 (of Stich's argument for his version of a

---

Churchland, and he replied, "I don't know what the successor concept [to the notion of truth - H.P.] will be". This is honest enough! Churchland is aware that the notion of *truth* is in as "bad shape" as the notion of belief from his point of view, and accepts the consequence; we must replace the "folk" notion of truth by a more scientific notion. But the innocent reader of Churchland's writings is hardly aware that he is also being asked to reject the classical notion of truth!" (Putnam, 1991, p. 60).

Syntactic-causal Theory of Mind, and 12 (of Fodor's arguments for a *Causal Theory of Content*).

Chapter 9

Physicalism and Psychology

9.1 Introduction

In its most general form, physicalism in psychology involves the view that since psychological events have physical causes and effects, then psychological events have physical descriptions. For example, since mental states, such as conscious experience, perception, thinking, and beliefs are causally linked with physical processes of the brain, then mental states are accountable in terms of physical brain processes, and hence are reducible to, or may be identified as physical processes in the brain.

To adherents of psychological physicalism, this view fixes the limits to what may and what may not be acceptable psychological theorizing; a psychological theory which is not compatible with the assumption of the purely physical nature of the human mind-brain system, or a theory which claims non-physical psychological entities, is not scientific and can be ruled out *a priori*.[1] Indeed, so it is claimed, neither theories nor data from research on psychological phenomena and events can attain scientific status unless they can be made consistent with the rest of the natural sciences.

The problem with this view, as pointed out by Seager (ibid. p. 21), is that in order to establish a causal relation between two phenomena, not only is it required that the *explanans* (brain states) and the *explanandum* (conscious experience) be independently determined, but also that they both have an equal claim to *truth* and *existence*. However, this is clearly incompatible

[1] For a presentation and extensive discussion of a variety of different forms of psychological physicalism, see Seager, 1991.

with the claim of an identity of mind and what goes on in mind with physical processes in the brain. Thus, for the sake of argument, let it be assumed that it makes sense to say that a description of some brain processes, to which some particular conscious experiences are causally related, are physical descriptions of those conscious experiences; and let it be assume that it makes sense to say, therefore, that the conscious experiences in question are identical to the brain processes being described. Granted these assumptions, the problem remains of maintaining that what goes on in the mind and consciousness is causally related to particular processes in the brain, *unless* what goes on in the mind may be determined and described independently of these processes and *in terms of* conscious experiences and thinking. Hence, physicalism will always be in need of two different and independent descriptions, both having an equal claim to truth and objective existence - which is incompatible with the claim that conscious experiences are identical with and, thus, accountable purely in terms of physical brain states.

As already pointed out earlier in Chapter 1, a related problem with the assumption of the possibility of reducing conscious experiences to physical brain processes is that, in spite of the widely accepted claim to the contrary, we would quite simply not know what this assumption means. What could possibly be meant by saying, and how would we go about determining what amounts to a correct description, say, of my seeing the letter box and its red colour - in purely physical terms? And even if it is true that my visual experience of letter boxes and their colours are causally related and correspond to some particular physically describable brain processes, accounts of this relation and correspondence does not amount to an explanation of how physically describable brain state may gives rise to conscious experiences of letter boxes and their colours. As rightly pointed out by Nagel (1986, 1994), McGinn (1991), and many others, psycho-physical reductionist theories are not - in any serious sense of this term - *theories* which explain or yield anything to our understanding of how conscious experiences and other psychologically describable phenomena come to exist out of physical states of the brain.

Now, there are probably several reasons why physicalism has had a strong appeal to psychology in its quest to gain respectability as an area of scientific research on a par with the natural sciences. For one thing, experience from other areas of research seems to prove that the method of reducing things to their more elementary parts has led to far more accurate accounts of the properties of things and far more precise predictions of their behaviour and the effects they produce - in our familiar world as well as in the not so familiar part of reality described by the natural sciences. So, if the reduction involved in psycho-physical identification were possible, the advantages would be considerable. We would then be in a position in which orderly, deterministic and causal accounts could be given of the seemingly unpredictable and unorderly thinking and action of persons, *that is*, if only such thinking and action be reduced to ever more precise accounts of the causally predictable physical processes of the body and brain of persons. In the following section I shall discuss arguments for a proposal to this effect.

9.2 Anomalous monism, or Psychology as Physics

In a paper entitled *Psychology as Philosophy* the American philosopher Donald Davidson round up a series of condition which must obtain for identifying psychological phenomena as physical events (Davidson, 1974). He starts his discussion of the possibility of psychophysical reduction by pointing out that in describing behaviour, i.e. things we do or action we carry out - whether intentionally or not - the concept of intention is central; and so are other concepts, such as beliefs, reasons, desires, and hopes. These concepts have "traits", says Davidson, which "segregate them conceptually from other families of concepts - particularly physical concepts". Conversely, he defines an "event" as an action, "if and only if it can be described in a way that makes it intentional" (ibid. p. 41-42) - which e.g. random movements or reflexes are not.

He then asks the question of whether intentional human behaviour can be explained and predicted in the same way that other phenomena can (and by this he obviously means physical phenomena). On the one hand, "human acts are clearly part of the order of nature, causing and being caused by events outside ourselves". But on the other hand, "there are good arguments against the view that thought, desire and voluntary action can be brought under deterministic laws, as physical phenomena can". The fact that it would be impossible to account for the necessary and sufficient conditions for an intentional act to occur, let alone to determine in advance of the act whether these conditions were satisfied, precludes any serious laws connecting such conditions and action. It is an error, says Davidson, to compare a truism like, 'if a man wants to eat an acorn omelette, then he generally will if the opportunity exists, and no other desire overrides', with a law that says how fast a body will fall in a vacuum. It is an error, according to Davidson, because in the latter case, but not in the former, we can tell in advance whether the conditions hold, and we know what allowance to make if they do not (ibid., p. 45).

Davidson presents the following arguments against the possibility of connecting the psychological in a deterministic law-like way to the physical:

> "What lies behind our inability to discover deterministic psycho-physical laws is this. When we attribute a belief, a desire, a goal, an intention or a meaning to an agent, we necessarily operate within a system of concepts in part determined by

> the structure of beliefs and desires of the agent himself. Short of changing the subject, we cannot escape this feature of the psychological; but this feature has no counterpart in the world of physics.
>
> The nomological irreducibility of the psychological means, if I am right, that the social sciences cannot be expected to develop in ways exactly parallel to the physical sciences, nor can we expect ever to be able to explain and predict human behaviour with the kind of precision that is possible in principle for physical phenomena. (ibid., p. 42).
>
> The constitutive force in the realm of behaviour derives from the need to view others, nearly enough, as like ourselves. As long as it is behaviour and not something else we want to explain and describe, we must warp the evidence to fit this frame. Physical concepts have different constitutive elements. Standing ready, as we must, to adjust psychological terms to one set of standards and physical terms to another, we know that we cannot insist on a sharp law-like connection between them. Since psychological phenomena do not constitute a closed system, this amounts to saying they are not, even in theory, amenable to precise prediction or subsumption under deterministic laws." (ibid., p. 52).

This would seem to be compelling arguments not only against deterministic psychophysical laws, but also against the possibility of identifying psychological events as physical events, and thus of explaining psychological events in terms of the same kind of deterministic laws applying for events in the world of physics. For if "the family or sample of terms" for psychological events and phenomena and that to which they refer (such as knowledge, thoughts, beliefs, desires, intentional action initiated and carried out by persons), have no counterpart in "the family or sample of terms" for physical events and phenomena and that to which they refer, then what psychological terms refer to cannot possibly be reduced to or be identified as that to which physical terms refer. If we said so, we would be guilty of committing a category mistake - or quite plainly of talking nonsense.

Nevertheless, Davidson also thinks that an adequate theory of behaviour must do justice to both the insight that "action is part of nature being causally related to physical events", and the insight that "voluntary action cannot be brought under deterministic, causal laws". Indeed, an adequate theory of behaviour, according to Davidson, is a theory which shows how, *contrary to appearances*, these two insights may be reconciled. Short of indicating what such a theory of behaviour might look like, he gives the following reasons why: Although behaviour may not be predictable in a serious law-like way, "this does not mean there are any events that are in themselves undetermined or unpredictable; it is only events as described in the vocabulary of thought and action that resist incorporation into a closed deterministic system. *These same events*, described in appropriate physical terms, may be as amenable to prediction and explanation as any. I shall not argue here for this version of monism, but it may be worth indicating how the parts of the thesis support one another." (ibid., p. 42 - 43, italics added).

Before venturing further into Davidson's attempt to indicate how the "parts of this thesis support one another", we would do well to ask - in view of the arguments just presented - how psychological events could possibly be *the same events* as physical events described in purely physical terms - if psychological events and the terms by which they are described, *have no counterpart in the world of physics*. And ask, furthermore, how psychological events, belonging within an open system, and which *defy being subsumed under the closed system of physics and its laws*, may nevertheless become just as perfectly causally lawful and predictable as physical events belonging within a closed physical system - if only psychological events be described in appropriate physical terms.

Well, the possibility of incorporating events belonging within an open system into that of a closed one, would probably obtain - provided it were possible to secure and determine that it is *the same events* which are described in both systems. But this condition, it would seem, is not met according to Davidson, who in so many words has stressed that if it is psychological events and their features we want to explain and describe, e.g. behaviour, action, intentions, beliefs and desires, "and not something else", such features have no counterpart in the world of physics - and thus cannot be the same as physical events, "which have different constitutive elements" (op.cit.).

Now, *if* in spite of Davidson's arguments to the contrary, the possibility still remains of identifying psychological events as physical events, and thus to describe psychological events in purely physical terms, the determination of what would be an appropriate physical descriptions of a psychological event and its physical constituents could not be made *independently* of or without first having determined and described the event and its constituents in *appropriate psychological terms*; in the case of an intentional act, for example, independently of having described it in terms of a choice made by an agent among the possibilities of action open to him in the situation, with the intention of bringing about some specific future changes in his possibilities of action, and thus independently of the knowledge of the agent about the situations and things that his action concern, and so forth. For, after all, a physical description of the display of *behaviour* of the agent cannot in itself specify whether it constitutes, say, an intentional act as opposed to non-intentional random movements, or reflexes. Therefore, no determination of the appropriate physical terms by which to describe the act as a physical event, required for a deterministic, law-like account of it, may be made *independently* of a determination and description of the event in psychological terms. Nor, of course, can such a physical account of an action be made *in advance* of a description of the outcome of it, specified in psychological terms, i.e. in terms of the changes in possibilities of action open to the agent subsequent to the act.

If all this means that a physical account of the act as a physical event necessarily relies on a psychological description of the act, and also that the psychological description on which the physical account relies, concerns phenomena which, for the reasons just given, are not "amenable to precise prediction or subsumption under deterministic laws" (op.cit.), how then could any physical description be so "appropriate" as to *make* them amenable to such predictions and laws? Here, in Davidson's own words, are the premises on which rest his arguments to this effect.

> "Take as a first premise that psychological events such as perceiving, remembering, the acquisition and loss of knowledge, and intentional actions are directly or indirectly caused by, and the cause of, physical events.
>
> The second premise is that when events are related as cause and effect, then there exists a closed and deterministic system of laws into which these events, when appropriately described, fit.
>
> The third premise is that there are no precise psycho-physical laws.
>
> The three premises, taken together, imply monism. For psychological events clearly cannot constitute a closed system; much happens that is not psychological, and affects the psychological. But if psychological events are causally related to physical events, there must, by premise two, be laws that cover them. By premise three, the laws are not psycho-physical, so they must be purely physical laws. This means that the psychological events are describable, taken one by one, in physical terms, that is, they are physical events. Perhaps it will be agreed that this position deserves to be called *anomalous monism*: monism, because it holds that psychological events are physical events; anomalous, because it insists that events do not fall under strict laws when described in psychological terms,."(ibid., p. 43).

However, the conclusion drawn from the premises is not valid - for the reasons given above, as well as those which now follow.

First, it may well be that psychological events are causally linked with physical events; in the case of an act, for example, to carry out an act on things necessarily involve an agent causing physical changes on them. In this sense, *part of* the constituents of an act involve physical events causing physical events, both of which may be described in purely physical terms and explained according to strict, deterministic causal physical laws. However, with respect to this part of the act, there is nothing *psychological* whatsoever, nor therefore has anything psychological been accounted for or explained in accounts of that part in those physical terms and according to those physical laws. This is trivial - and trivial that the parts of the act which constitute it as a *psychological event*, i.e. an act involving intentional choices between possibilities of action, and all the rest, have yet to be accounted for in physical terms and explained according to physical laws. But how could this be achieved if, as Davidson has himself pointed out, there quite simply is no possibility of describing that which is *psychological* about a psychological event in purely physical terms, and therefore none of explaining in strict deterministic physical laws the role played by the psychological in the causation between psychological and physical event? If the problem at hand is not immediately obvious, we just have to consider, once more, the impossibility of describing the psychological constituents of an act carried out by an agent, i.e. knowledge of the agent of possibilities of action, and choices made by the agent between possibilities of action with the intentions to change future possibilities of action, etc., in any of the terms available within the vocabulary of physics to describe physical events and their constituents. And consider, furthermore, the impossibility of accounting for the nature of the act, and thus of *any* of its constituents, in advance of its intended outcome, and consider, therefore, the impossibility of accounting for the act in terms of strict deterministic causal laws. Indeed, if to act logically implies that one intentionally changes one's future possibilities of action, and thus it is part of the *statu nascendi* of and act to effect subsequent changes of events, then the initial state of an act must be caused by events not yet present - a notion which is not part of the physics Davidson refers to.

As far as I can see, it is only if we disregard or "shave away" those parts of the act which make it a *psychological event* that it would make sense to talk of the act as a physical event. However, when those parts are disregarded and shaved away, there is nothing left of the act as a psychological event, nor therefore any psychological part of the event playing a role in the causation being explained - and hence no psychological event being accounted for or explained as a physical event. In effect, there is no psychological event being a physical event. So, if subsequent to having accounted for that of an act which may be identified as a physical event, and for which a causal, deterministic law could be expressed, this law could certainly not be applied to a psychological event, say an intentional act. For, in identifying the event as something purely physical, we may well have succeeded in getting rid of the act as an event described in psychological terms. But we would also at the same time have got rid of and reduced away that to which the law was meant to apply and explain, namely the act as a psychological event.

It is not difficult to show that the arguments against Davidson's position apply, *mutatis mutandis*, to the other psychological events he mentions, such as knowledge, believing, perceiving, and thinking. The fact that these other psychological events may have physical causes and effects, just as have acts, does not make any of these psychological events into purely physical events. And as applies to the example of an act, to the extent it is possible to reduce such psychological events, "taken one by one", to purely physical events, to which applies physical laws and causal explanations, such reduction inevitably entails reducing away that which constitutes these events as psychological event, and thus of reducing that away, to which these causal laws and explanations were meant to apply.

It is in more than one sense the underlying view of Davidson's paper that, "By evaluating the arguments against the possibility of deterministic laws of behaviour, we can test the claims of psychology to be a science like others" (ibid., p. 42). In other words, only in so far as it is possible to account for human behaviour and other psychological events in terms of deterministic, causal laws, can psychology be a science like others. This condition may obtain, according to Davidson, if we realize that the limits imposed on psychology and the social sciences in how they describe the subject area of their research "is not set by nature, but by us when we decide to view men as rational agents with goals and purposes, as subject to moral evaluation" (ibid. p. 52). So, psychology can indeed become a science on a par with others - if only we give up the decision to describe human beings in those terms, and instead describe them and their behaviour as physical events, governed by the causal, deterministic laws of physics. This, Davidson contends, is perfectly compatible with his argument that a "serious" deterministic law, and serious uses of the terms "cause" and "effect", have no place in the science of psychology - for all the other good reasons, he provides.

Well, let us agree with Niels Bohr and other physicists (cf. for example Bohr, 1939/1972) that causal explanation and deterministic laws only have strict definitions and applications within a very limited area of scientific research, namely mechanical physics. If, therefore, it is the nature of the phenomena studied within the natural sciences at large that they defy descriptions and explanations in causal, deterministic terms, I see no reason to expect that the phenomena studied within psychology could be accounted for along causal, deterministic lines. Nor do I see a glimmer of a reason why psychology, in order to be considered a proper science, would be well advised to describe psychological events along lines, which probably only apply within mechanical physics. However, I do see a host of reasons of principle and logic against it; indeed, reasons which would render psychology and its approach to its subject area most un-scientific, should it elect so to do.

What is needed in a scientific psychology in its pursuit of a more precise understanding of the phenomena which falls within its subject area, is the same as is needed for any other area of inquiry in order to gain scientific respectability. That is, careful analyses of the features and properties of the events and phenomena being studied, and development of appropriate terms and methods by which precisely to describe and delimit them from other events and phenomena, which are not of concern to psychology. Such analyses would reveal that psychological events do indeed have features and properties without which they would cease to be psychological events; and they would reveal that the determination and description of psychological events, expressed in *psychological* terms, are just as necessary for psychology's understanding of the nature and occurrence of psychological events, as are the description of physical events described in physical terms, for physics' understanding and explanation of the nature and cause of physical events. This is how other sciences since Galileo have proceeded - and progressed. The attempts in this book to determine some of the principles and *logical* relations as opposed to *causal* relation between concepts used to describe the cognition, language and action of persons, may be seen as a modest beginning to such an approach within psychology. What psychology does *not* need, however, is to resort to describing and explaining psychological phenomena by way of models and terms developed by other sciences, to account for events which are totally alien to psychological events - in an attempt to understand what is yet not understood, and which, therefore, has not yet been analysed and described in appropriate psychological terms.

Davidson's position on the physicalist lines along which psychology should proceed in order to make progress as a science, is by no means unique. Nor is his assumption that to the extent psychological events are causally linked with physical events, then they must somehow be physical events. It is an assumption which is shared by adherents of physicalism in general, and which underlie every physicalist theory within psychology - whatever the form and vocabulary in which it is coached. In particular, it is the same assumption of the effect of a causal relation between psychological and physical events, which underlie the theories of mind and consciousness - to which I shall turn in the next two chapters.

Chapter 10

Context, content and reference - the case for beliefs and intentionality

> "It is a curiosity of the philosophical temperament, this passion for radical solutions. Do you feel a little twinge in your epistemology? Absolute scepticism is the thing to try. Has the logic of confirmation got you down? Probably physics is a fiction. [...] Nobody has yet suggested that the way out of the Liar paradox is to give up talking, but I expect it's only a matter of time. Apparently the rule is: if aspirin doesn't work, try cutting off your head" (Fodor, 1986).

10.1 Introduction

During the last couple of decades the views and assumptions of so called Folk Psychology about our cognition and action have come under attack in North American philosophy of mind. These are the common sense views and assumptions we hold about ourselves and our fellow human beings to the effect that we are rational agents whose actions are motivated by and, thus, may be explained in terms of, our knowledge, beliefs, desires and related propositional attitudes. For example, according to Folk Psychology the fact that on my way to work this morning I carried an umbrella, may be explained by referring to my knowledge or belief that it was raining, my desire not to become wet, and my belief that an umbrella would protect me from becoming wet. And it is part of the assumptions of Folk Psychology that our every day descriptions and cognition of ourselves and the ways in which our beliefs and desires motivate our action are generally correct, and that we are very often correct when ascribing such mental states and motives to other people.

According to the critiques of Folk Psychology (henceforth FP) these views and assumption are not only wrong in practice, but in principle. Indeed, it is the very notions of people having beliefs, desires and intentions which are wrong. The central tenet of the various arguments against these notions is that what they refer to is not scientifically specifiable - from which, it is claimed, it may be safely concluded that the notions of beliefs and of people having beliefs, desires and all the rest are void, and what they refer to simply does not exist. According to Quine (1960, 1970), for example, we are only committed to entities quantified over in our best scientific theories, which to Quine are the theories of the natural sciences and behaviourism within psychology; and because notions such as beliefs and desires, and of people having beliefs and desires occur nowhere in these theories, then the entities denoted by FP do not exist. In another version it is argued that if, somehow, FP can be shown to be radically wrong about the organisation of our cognitive system, then "the right thing to say is that there are no such things as beliefs" (Stich, 1979, p. 231). A third version (Paul Churchland's, 1981) combining the arguments of the first two, has it that since FP "suffers explanatory failures on an epic scale" (ibid. p. 76) as to how our beliefs and other propositional attitudes affects our behaviour, and since, furthermore, the entities denoted by these terms have no prospect of reduction to neuroscience (ibid. p. 75), then no such things exist as beliefs and desires, or of people having beliefs and desires.

The aim of this chapter is not so much to show that a rejection of FP - and thus of our conception of ourselves as persons, having beliefs and knowledge about ourselves, other persons and the situations in which we and they act, etc. - is self-defeating. This would seem immediately obvious. Rather, it is the aim of the various sections critically to discuss some of the central arguments leading up to this rejection, among them the arguments against the possibility of developing scientific theories about the cognition and behaviour of persons, which invoke notions of beliefs and of persons having beliefs. In the first section, "*Against Stich's case against belief*", I shall discuss Stephen Stich's arguments to this effect and argue that they rely on mistaken assumptions, not only about the conditions for developing scientific theories in general, but in particular for developing rigorous scientific theories of cognition.

After this follows a discussion in which I shall attempt to clarify the implications of ascribing beliefs to people, which may be true or false, as opposed to ascribing properties to things, which may or may not exist, and analyse the conditions for ascribing properties to things as opposed to ascribing beliefs to people. This discussion and clarification is motivated by a longstanding debate among philosophers and logicians according to which the language and statements of science – i.e. of physics and biology – is largely extensional and thus is expressible approximately in familiar predicate calculus. In contrast, belief ascription statements suffer from some "peculiarities" and "anomalies", which prevent them from being thus expressible, and hence prevent statements about beliefs and other psychological states from being scientific. It is the aim of my analysis of the conditions for ascribing properties to things as opposed to beliefs to people to show that the differences in these conditions notwithstanding, they have no consequences whatsoever for our possibility of ascribing beliefs and knowledge to people just as scientifically as we ascribe properties to things.

10.2 Against Stich's case against beliefs

To-day, nobody except Quine and his followers seriously thinks that behaviourism exerts a threat to the existence of entities and events denoted by the terms of FP. Behaviourism has long since ceased to be viewed as the best, let alone a credible scientific theory about the behaviour and cognition of humans. Behaviourism is as dead as the rats in Skinners experiments; not, however, because its adherents finally agreed with their critics that the assumptions of behaviourism were radically wrong, but because it was generally agreed that the cognitive science which took over offered a better alternative. As to the arguments by Quine and Churchland against the existence of the entities denoted by FP, I think we shall have to agree with Stich that

> "it is plainly mistaken to infer from the fact that a term occurs nowhere in science to the conclusion that the entities putatively denoted by the term do not exist. Consider, for example, such terms as 'favored by Elizabeth I', 'slept in by George Washington', or 'look like Winston Churchill'. Surely none of these terms occurs in any currently received scientific theory. Nor is it likely that they will find a place in scientific canon of the future. But it would be simply perverse to deny, on these grounds, that there are any beds slept in by George Washington or any men (or statues) that look like Winston Churchill" (ibid. p. 222).

I think we shall also have to agree with Stich that

> "it would be comparably perverse to deny the existence of people who believe that p on the grounds that 'believes that p' is not invoked in a cognitive science, or to deny the existence of beliefs that p because '(is a) belief that p' is ill suited to cognitive theory building" (ibid., p. 222).

For, how could a cognitive science, which did not invoke 'believes that p', and which did not account for the fact that people do actually have beliefs ever claim to be about the cognition of persons?

However, according to Stich, to whose idea of cognitive theory building I shall now turn, such a theory is not only a possibility, but a necessity. His arguments for this view follows three main lines. First, a theory of cognition, in order to be scientific, must rid itself of all considerations entailed in FP and its notions of belief and other propositional attitudes of persons. Such notions are notoriously "contextual", "historical" and "ideological" in that accounts for the content of beliefs and propositional attitudes, as well as accounts of their effect on behaviour, invariably necessitates that reference be made (explicitly or implicitly) to the contexts or situations in which they occur. Hence such notions have no place in scientific theory building. Secondly, he argues that a cognitive theory which "adverts" to content is likely to lose important generalizations in that it cannot account for cognitive development of children, nor characterize the anomalous cognitive states of people suffering from mental diseases, let alone for the radically different cognition of "exotic" people. Thirdly, Stich argues that if it can be shown both that FP is generally wrong about our beliefs and how they affect our behaviour, and that the best cognitive theory about "the general organization and gross functional architecture of our cognitive system" is significantly different from the functional architecture suggested by FP, then FP and its notions about belief and of people having beliefs must be false.

In its stead Stich proposes, or rather stipulates, a so called Syntactic Theory of Mind which has as its basic idea that

> "the cognitive states whose interaction is (in part) responsible for behaviour can be systematically mapped to abstract syntactic objects in such a way that causal interactions among cognitive states, as well as causal links with stimuli and behavioral events, can be described in terms of the syntactic properties and relations of the abstract objects to which the cognitive states are mapped. More briefly, the idea is that causal relations among cognitive states mirror formal relations among syntactic objects. If this is right, then it will be natural to view cognitive state tokens as tokens of abstract syntactic objects." (ibid. p. 149).

I shall not go into how Stich envisions that this proposed Syntactic Theory of Mind may be worked out, nor whether or how it may work and provide a better paradigm for cognitive theorizing than cognitive theories "whose generalisations appeal to the notion of content" (ibid. p. 149). These problems have already been thoroughly discussed - and seriously questioned - by others. (See e.g. Pylyshin (1980), Fodor (1981, 1986), and Patricia Churchland quoted in Stich (ibid.)). What I shall go into and critically discuss is, firstly, Stich's argument as to why it might turn out that FP "is radically wrong", and the consequences he thinks one may draw from it. Secondly, I shall discuss the claim that a cognitive psychology built from folk notions is likely to lose important generalizations, because it cannot account for cognitive states and processes that differ radically from our own. Thirdly, I shall argue that his contention that a scientific cognitive theory, just like any other scientific theory, needs no reference to content, context or interests, misses important points about the conditions for generalizing scientific theories. Fourthly, and not least, I shall argue that no attribution may be made of cognitive states, or, for that matter, of the "abstract syntactic objects" onto which they map, nor may any experimental test of their behavioural effect be carried out, without appeal to both content, beliefs, desire and other notions of FP.

Here, first, are Stich's arguments why FP may turn out to be wrong. He writes,

> "The first argument, the less serious of the pair, is an inductive generalization from the sorry history of folk theories in general. It starts with the observation that folk psychology really is a folk theory, a cultural inheritance whose origin and evolution are largely lost in pre-history. The very fact that it is a folk theory should make us suspicious. For in just about every other domain one can think of, the ancient shepherds and camel drivers whose speculations were woven into folk theory have a notoriously bad track record. Folk astronomy was false astronomy and not just in detail. The general conception of the cosmos embedded in the folk wisdom of the West was utterly and thoroughly mistaken. Much the same could be said for folk biology, folk chemistry, and folk physics. However wonderful and imaginative folk theorizing and speculations has been, it has turned out to be screamingly false in every domain where we now have a reasonably sophisticate science. Nor is there any reason to think that ancient camel drivers would have greater insight or better luck when the subject at hand was the structure of their own minds rather than the structure of matter or of the cosmos. [...] And I think the general failure of folk theories is reason enough to think that folk psychology might suffer the same fate." (ibid. 229-230).

Well, Stich "inductive generalizations" about folk theory and psychology have not remained unchallenged.[1] Among the objections[2] is that neither folk theory, nor folk psychology are stagnant theories; on the contrary, they have been quick to adopt the knowledge and notions provided by the sciences - whenever they were seen to contribute to a better understanding and to more accurate predictions and explanations of matters of concern for people in managing their every day lives. This would seem to suggest that neither the structure of reasoning or assumptions of the sciences, nor their procedures of

[1]- nor have quite similar arguments against FP by Patricia and Paul Churchland and Dennet.
[2] E.g. by Horgan and Woodward (1985), and Fodor (1986).

observing and interpreting data, are fundamentally different from or at odds with the structure of reasoning and procedures of observation entailed in folk theories. But apart from this point, to which I shall return later, Stich's bleak view on folk theories invites other critical comments.

Firstly, it seems to have been so much easier for Stich and others to spot what according to our present knowledge can only be conceived of as mistakes and "screamingly false" beliefs of pre-scientific theories, than to see the things which they, somehow, must have got dead right. If, for example, the focus is shifted from exotic theories about the cosmos of camel drivers and shepherds to something a bit closer to both their and our homes, we find that even people living in pre-scientific eras had some pretty good ideas and theories about physical matter and ways of calculating its properties, as well as highly developed capabilities of inventing technologies to handle such matter. These were the ideas and technologies with which they were able to construct pure "Wonders" - just think of the ancient Egypt pyramids, the Greek temples, the Roman aqueducts, the great wall of China, the medieval cathedrals - indeed they knew how to make them so durable that even to-day, hundreds if not thousands of centuries later, we may still enjoy their greatness; a fact which should not fail to impress most people living in our current western scientific civilizations with its very poor track-record in constructing anything which lasts for more than a generation. - It may well be that prescientific physical theories were wrong about what makes physical things fall to the ground when not supported. But the beliefs and conceptions of prescientific people that they do, and their observations and theories about the conditions in which they are adequately supported, were not. And although it may be true that the speculations and mythologies of prescientific people about the cosmos were false, they were not so badly woven into their general conceptions and observation of physical matter as irredeemably to corrupt them; they knew presumably as well as we do how to distinguish speculations and mythologies from observations about matters of fact. If they had not - indeed, had they allowed themselves the sort of "imaginary" reasoning and "inductive generalizations" when dealing with the problems of how to raise camels and sheep and keep them healthy that Stich does when discussing FP - camel drivers and shepherds would most probably not have survived for so long.

Another thing, just as obvious that it is easily overlooked, is that pre-scientific people were able, much as we are to-day, to develop and live together in highly complex societies which - for all their failings - still work. The reason why they work is that people (then and now) are able to acquire knowledge and beliefs about themselves, others and the situations in which they find themselves, and to learn to behave according to extremely complicated sets of rules, beliefs and theories in these situations. These theories are none other than the good old folk psychological theories, which generally enable people fairly accurately to account for their own knowledge, beliefs, and other propositional attitudes, and to explain how these attitudes affect their behaviour; theories, moreover, which also enable them to account for the beliefs and predict the behaviour of others - because they can rely on the fact that others entertain beliefs and propositional attitudes which are comparable to their own, and thus may be ascribed to others as well. As noted by Fodor (1986), these theories and systems of beliefs work so well and have so compelling predictive powers that "if we could do that well with predicting the weather, no one would ever get his feet wet. Yet the aetiology of the weather is surely child's play compared with the cause of behaviour". In contrast, nothing useful about the behaviour and cognition of persons follows from present scientific theories about the physical or biochemical composition of their bodies, nor from neurophysiological theories about the states underlying this behaviour and cognition - and even less from the stipulated formal relations between "abstract syntactic objects" onto which cognitive states may be mapped. Folk Psychological theories about our cognition, beliefs, other propositional attitudes and how they affect our behaviour not only work; they are also generally true. What more can one expect from a theory? Indeed, the very existence of societies and social institutions in which people may collaborate and communicate about this as well as other matters relies on the fact that the assumptions of FP *are* generally true. Just try to deny this and you will not only be denying the existence of societies, social institutions, collaboration and communication between people, but also the very possibility of communicating and discussing any concrete examples of how or when FP assumptions are right or wrong.

Now, it may well be that in, say, a computational theory about the machinery underlying our cognition and other mental processes in some sense or other are describable in "purely syntactical" terms, i.e. without recourse to intentional terms. But, as once more pointed out by Fodor, "the generalizations that such theories account for are intentional down to their boots. Accordingly, computational theories of mental process don't replace the common sense story about propositional attitudes and their effects" (ibid. p. 422). To this must be added that neither may such theories be developed nor tested without or independently of the concepts and notions of this common sense story (for reasons which should become clear from the discussions in the section, "The dependency of scientific beliefs and propositions on contexts and interests", to appear later in this chapter).

However, according to Stich the common sense generalisations and stories told by FP about beliefs and other propositional attitudes have no interest for a cognitive science. In his opinion they not only must, but can be replaced with some other story and other notions analogous to what we have, say, in chemistry about chemical compounds that things are made up of and their chemical reactions, which can be told independently of other stories told about the same things, and which do not invoke the notions or entities talked about in these other stories. Thus, according to his principle of autonomy, only such states and processes are of concern to the psychologist, and may play a role in the psychologist's theory of cognition, which supervene on the current, internal physical states of the organism, and hence are themselves states and processes which are describable without reference to historical, intentional or ideological factors (ibid. p. 164 - 171). Stich uses the analogy quoted below to illustrates the generalizations of FP vis à vis the generalizations of a scientific cognitive theory.

> "The generalizations of Folk Psychology may be thought of as analogous to rule-of-thumb generalizations in cooking. Consider, for example, the generalization that separated mayonnaise can usually be repaired by beating a bit of it into dry mustard, then gradually beating in the rest. This generalization, like the generalizations of folk psychology, suffers from a certain vagueness. It is not clear just what counts as separated mayonnaise, nor is it clear where mixing gives way to beating and beating to whipping. Another point of analogy between the generalizations of cooking and those of folk psychology is that neither sort of generalization will find a place in a serious explanatory science. But nonetheless many of the generalizations known to good cooks are true, and wonderfully useful. A final analogy between folk psychological and culinary generalizations is that in both cases it is plausible to suppose that serious science will be able to explain why the rough-and-ready folk generalizations are generally true, and perhaps even explain why they are sometimes false. In the case of mayonnaise, I would guess that the explanation will come from the physical chemistry of colloids along with some detailed investigations of what, from the physical chemist's point of view, a separated mayonnaise actually is. And in the case of belief, the explanation will come from an STM cognitive theory along with some detailed investigation of what syntactic state or states are generally describable as beliefs that p. [...] The prospect we are envisioning could be described as the reduction of folk psychology to STM cognitive theory." (Stich, ibid., p. 227)

However, in my view the envisioned reduction and replacement of folk psychology and its notions about beliefs to a scientific cognitive theory, would be just as impossible as a reduction and replacement of the folk theories and notions of cooking about mayonnaise and its separated version to chemical theory and its notions. Here are the most obvious reasons why, expressed in the terms of Stich own analogy:

To the extent that a chemical account of chemical compounds may be said to apply to mayonnaise - separated or otherwise - such an account necessarily requires an ordinary every day description of mayonnaise and notions of what mayonnaise is (cf. the arguments in Chapter 8). Likewise, for a chemical theory about the chemical reaction of compounds of separated mayonnaise and dried mustard to explain how separated mayonnaise may be repaired by being whipped into dried mustard, necessitates and relies on precisely this cookery description of the event. Therefore, it would be "screamingly false" to say that our ordinary notion of mayonnaise and the rules-of-thumb of cooking of how to repair separated mayonnaise may be reduced to or replaced by a chemical theory and its explanations of the reaction of some or other chemical compounds of mayonnaise. Likewise, it would be false to say that because a chemical account may be given of mayonnaise and its separate versions, which do not invoke notions of mayonnaise and the entities they refer to, then we may make do without the notions of mayonnaise and the entities they refer to. And just as obvious, it would be wrong to say that a chemical analysis and description on its own accord, i.e. without recourse to notions and descriptions of cooking, may explain what separated mayonnaise actually is. Indeed, it is precisely because a chemical theory does not invoke notions of mayonnaise, and because of the dependency of a chemical account of separated mayonnaise on everyday cooking descriptions and notions of mayonnaise and its separated versions, that no account from a chemical point of view may *alone* explain what mayonnaise - or it separated versions - actually is.

For the same reasons, neither could a chemical explanation of mayonnaise and its various forms invalidate cooking's description and observation of mayonnaise and these forms. Suppose that the reason why separated mayonnaise may be repaired by being beaten into dried mustard were explained by Folk Theory in terms of the "mustard spirit" uniting the "oil spirit" and the "egg yolk spirit". Even if this explanation is wrong - indeed, irrespective of whether or not it is wrong - the observation and descriptions by cooking of the effect of beating separated mayonnaise into dried mustard still hold. But nor could a chemical explanation be true which predicted an outcome which was at variance with the observation and description of cooking. This follows from the principle, generally accepted in all sciences, that if a (new) scientific law or explanation predicts an outcome of events which is in conflict with descriptions and observations necessary for determining the very event to which the prediction and laws are supposed to apply, then it is the scientific law or explanation which is wrong.

Now, it is part of Stich's arguments in favour of his Syntactic Theory of Mind that if it turns out that FP is wrong, then a reconciliation between cognitive science and folk psychology would collapse. If the points above are taken, it would indeed collapse - for the same reason as a reconciliation between chemistry and the observations and rules of cooking would collapse if the observations and rules of cooking were wrong. However, whether or not the rules derived from the observations and descriptions of cooking are wrong, is a matter which relies on and only may be determined on the level of discourse of cooking - and not from the point of view of chemical theory, which does not invoke the notions of cooking; nor for the same reason may chemical theory on its own suggest alternative cooking notions or true cooking rules. The same would seem to apply *mutatis mutandis* in the case of notions and generalizations of FP about our cognition, and those of a cognitive theory about the states underlying our beliefs and cognition.

However, not to Stich who thinks that evidence from cognitive science could exist which proves FP radically wrong about the overall organisation of our cognitive economy. From this it would follow that "the right thing to say is that there is no such things as beliefs" (ibid. p. 231), and "nothing to which the predicate 'is a belief that p' applies" (ibid. p. 229). Well, if again we turn to the proposed analogy, this, I think, is comparable to saying that if the rule of cooking as to how separated mayonnaise may be repaired turns out to be wrong, then the right thing to say is that there is no such thing as separated mayonnaise, and nothing to which the predicate, 'is separated mayonnaise', applies. So to say would clearly be both false and nonsensical. This analogy notwithstanding, even if most of the beliefs and generalizations of FP about our beliefs and their effect on our behaviour should turn out to be false, it could never follow that such things as beliefs do not exist, but merely that those beliefs are false - in which case, necessarily, true beliefs must exist by which those false beliefs could be proved to be false. What else?

Well, according to Stich it so happens that the notions and explanations of FP may be proven wrong, namely if "our best theory of cognition" is one whose general organization and gross functional architecture is significantly different from the functional architecture that FP suggests. Indeed, it may turn out that a reconciliation between FP and cognitive science does not hinge on the assumption that a correct cognitive theory is one "which cleaves reasonably closely to the pattern proposed by FP". This assumption may be wrong - for much the same reasons as it may turn out that a physio-chemical explanation of the reaction of colloids does not cleave closely to our ordinary everyday observations and descriptions of oil, egg yolks, dried mustard and their behaviour; or for the same reason as it has already turned out that, generally, physical descriptions of molecules from which all material things are composed, and physical explanations of the behaviour of things in terms of the characteristics of their molecule structures has proven not to "cleave" closely at all to the FP's descriptions and explanations of material things and their behaviour.[3] This granted, and granted furthermore

[3] This interpretation seems consistent with what Stich writes in the closing paragraph of his book (p. 246): "If the empirical presupposition of folk psychology turn out to be false, as

that our best cognitive theory is Stich's own Syntactic Theory of Mind, it then follows, says Stich, that

> "we could no longer say that belief sentences stand a good shot at being true. For in saying 'S believes that p' we are saying that S is in a belief-like state similar to the one that would underlie our own ordinary utterance of the content sentence. And if it turned out that the overall structure of the human cognitive system is significantly different from the structure postulated by folk psychology, then this claim will be false. There will be no belief-like states, and thus S will be in none. (Nor, of course will there be one underlying our own normal utterance of 'p'). If folk psychology turns our to be seriously mistaken about the overall organization of our cognitive economy, then there will be nothing to which the predicate 'is a belief that p' applies." (ibid., p. 229)

However, even if most of the generalisations of FP are wrong, and the best cognitive theory about the organisation and functional architecture is radically different from that of FP's, then, again, it would not follow that belief sentences must be false, let alone rule out the existance of beliefs, nor of belief-like states underlying beliefs that are false. This would not follow any more than it would follow from the best physical description and theory of houses that they are build up of structures of molecules that, then, houses do not exist, nor the bricks from which they are built, but only molecules and structures of molecules.

More seriously though, granted the best scientific theory of cognition is one which suggested that it would be false to say that our knowledge and theories about ourselves, others, and the situation in reality in which we carry out acts, is expressible in terms of the *beliefs* we hold about these things, situations and acts; and granted it followed therefore that the "normal utterances" and propositions we use to state our beliefs of 'p' are invariably false, then we would seem to have landed in something indistinguishably similar to the Liar's paradox. For, no true proposition could be communicated about the beliefs, knowledge or thoughts of anyone to anybody about anything - not even about the best theory of cognitive science. On all counts, not a comfortable position for a cognitive theory. Rather, it would seem necessary to concede that a cognitive science, like any other science, cannot make do without beliefs and people having beliefs. Indeed it (or they) cannot

---

well they might, then [...] our age-old conception of the universe within will crumble just as certainly as the venerable conception of the external universe crumbled during the Renaissance. But that analogy ultimately suggests an optimistic conclusion. The picture of the external universe that has emerged from the rubble of the geocentric view is more beautiful, more powerful, and more satisfying than anything Aristotle or Dante could imagine. Moreover, the path from Copernicus to the present has been full of high intellectual adventure. The thrust of my argument is that we may be poised to begin a similar adventure".

so make do any more than it (or they) can make do without our normal utterances 'that p', in which - in ordinary everyday and in scientific situations - we express our beliefs, knowledge and thoughts about ourselves and the reality in which we find ourselves.

10.3 The problem of generalizing across radically different cognitive states

However, Stich has yet another argument why psychologists should chose to describe cognitive states in purely causal, syntactic terms, rather than in terms of "folk notions". If, for example, the cognitive states of young children and "exotic" people (i.e. victims of strokes, injuries, retardation, and the like) differ radically from our own, and there is no content sentence theorists can use to describe their cognitive states, then

> "in opting for a cognitive psychology built from folk notions we are likely to lose important generalizations, since folk psychology cannot characterize the mental states of young children or "exotic" folk. So if there are generalizations which apply equally to their cognitive processes and to our own, a cognitive theory couched in the language of folk psychology will be unable to articulate them." (ibid., p. 216)

Thus, if we insist on constructing cognitive theories whose generalizations "advert" to content, says Stich, then we may well have to do without comprehensive theories of cognitive development and cognitive anonymity. However,

> "If, by contrast, we adopt a taxonomy of cognitive states which individuates them along narrow causal lines and which takes no account of similarities or differences between the subjects and ourselves, we will avoid the obstacles (...) that many cognitive generalization will be beyond our grasp and many of those which we can state will be plagued by the vagueness inherent in the language of content." (ibid. p. 148)

Well, it is a defining property of taxonomies and generalizations that distinctions are made, but differences are ignored. However, it would of course be a mistake to assume that any generalization or discrimination could start by taking no account of similarities and dissimilarities. It is a well known condition, recognized in any other field in which generalizations, discriminations and taxonomies are made, that an agreed cannon of similarity-dissimilarity is a necessary antecedent to any generalization or discrimination. So, to whatever purpose and use the generalizations and taxonomies of cognitive states of the kind proposed in Stich's theory could be put, they would have to rely on some *other* descriptive system than the one in which those generalizations and taxonomies are written, i.e. a descriptive system in which terms exist for similarities and differences between cognitive states by which *distinctions* could be made between radically different cognitive systems.

But what are the terms in which we may identify differences and similarities in our own cognition and that of young children or "exotic folk", and on what level of discourse is the nature of these differences determined in the first place? They are, as far as I can see, determined at the level of ordinary everyday observation and description of behaviour and action, and according to similarities and differences in behaviour and action being determinable on this level of discourse. Indeed, we have, as far as I can see, no *other* means by which to begin to distinguish between the radically different cognitive processes and states of people, even less to determine that or how they differ.

Now, at this level of discourse - no matter in what other terms the mechanisms underlying our cognition may be described - it is a feature of our own cognition and action that it is describable in terms of the states of *intentional* systems, and thus in terms of systems having beliefs, propositional attitudes, and other features denoted by the terms of FP. Hence, the descriptive system by which similarities and differences are determined between our own and the radically different cognitive states and processes of others, is one which includes terms such as 'intentional states', 'beliefs', 'content of beliefs', 'propositional attitudes', and all the other terms which are part of our ordinary everyday descriptions and determinations of intentional behaviour; indeed, terms of which it is necessarily presupposed that they have well defined meanings and correct uses.

For the sake of argument, let us now take the extreme case in which, on this level of discourse, it has been determined that the cognition of young children or "exotic" folk is so radically different from our own that their cognition may not *in principle* be describable in those terms. That is, for all we know and from the behavioural criteria we have, we are unable to describe them in terms of intentional systems, but only as systems the overall processes of which may be correctly described in non-intentional terms. Now, if the cognition that psychologists theorize about is so radically different from our own, it is difficult to see how any comprehensive generalizations could be lost, "which apply equally to their cognitive processes and to our own". Conversely, in the event this theorizing entailed taxonomies and generalizations "along narrow causal lines [...] which took no account of either the similarities or differences between the subjects and ourselves", then this theorizing could not possibly be so comprehensive as to account for any *difference* in the cognitive states of such "subjects" and ourselves. For no criteria within the theory itself would exist by which to discriminate between cognitive systems and their states being so radically different.

However, quite similar problems arise in cases in which the differences between ourselves and young children or "exotic" folk are assumed not to be radical in the above sense. Thus, for all we know and from the behavioural criteria we have, we are compelled to say that these children and folk are intentional systems, who like ourselves act according to desires and beliefs - however, we may not be able to say precisely *what* intentional states they are in or *what* the content of these states are. Indeed, the difference which exists between them and us may best - if not only - be expressed by saying that the content of their cognition cannot be described in terms of the content of the beliefs and knowledge of the sort that we ourselves may entertain and express. From this, however, it cannot be concluded that "there is nothing they believe", unless, of course, our notion of belief is such (which it is not) that people may only have beliefs if what they believe is what we ourselves may believe. Rather, if the right thing to say about others people is that they may be in intentional states, and thus are "intentional systems" having beliefs about themselves, others, their behaviour, and all the rest, then generalizations would exist which apply equally to their cognitive processes and to our own. However, such generalizations are of little practical use, since, in view of the nature of the differences supposed to exist in this case between our cognitive states and those of young children and "exotic" folk, we would be no wiser as to the similarities or dissimilarities of any *concrete* cognitive states of such children or folk and our own.

The situation in this case seems almost paradoxical. On the one hand, we may contend that other people are similar to us in that they are intentional systems, yet the content of their beliefs and the cognitive states they may be in are significantly different from those we may have or be in. On the other, what makes us determine that the content of their belief and cognitive states are thus different from ours, is our inability *in practice* to articulate the difference. In other words, granted the assumption that cognitive systems exist which are thus similar and different from our own, we have at the same time precluded ourselves from saying precisely what the differences are. Indeed, the very notion of 'difference' is left ill-defined. For the same reason, granted a theory *could* be developed that generalizes across intentional systems in which the content of cognitive states are inexpressibly different, such a theory cannot be so comprehensive as also to account for how or the ways in which the intentional states of such systems differ. That is, the result is the same as before.

However, the way out of what appears to be a paradox is not, as suggested by Stich, to attempt to describe the states of different cognitive systems in purely non-intentional terms, nor is it to describe both our own and those which radically differ from our own intentional cognitive system in the same non-intentional terms. Rather, I think, the paradox compels us to accept that we are up against the limits of what we can sensibly say about cognition which is either radically or inexpressibly different from our own. What we shall have to accept is that attempts to develop generalizations about cognition so differently both starts and leaves us with a paradox.

In contrast, no paradox occurs when we confine ourselves to talking about the similarities and differences encountered in cognition of a nature we ourselves know about and have access to. Thus, it has to be admitted that a lot of the beliefs and knowledge of a person about him or herself, others and things in situations in which the person finds him or herself may differ from the beliefs that others may have about the same things in the same situations. Indeed, we may even be able to talk in well defined ways about the existence of such differences. If this is admitted, however, such differences can only be determined, articulated and precisely accounted for on the condition that *enough* similarities in our cognition exist, and on the condition that we share a lot of other beliefs about these things and situations.

Well, to say that enough similarities of cognitive systems and their states have to exist in order for generalizations and discriminations between them to make any sense and be of any use, is probably merely to say what is generally accepted as a condition for making generalizations and discriminations in any other field of inquiry.

10.4 The dependency of scientific beliefs and propositions on contexts and interests

Stich's third line of arguments against FP's notions of beliefs and of people having beliefs is that a cognitive theory, which invoke notions such as 'believes that p', and '(is a) belief that p', are ill-suited to scientific cognitive theory building. For, beliefs and the content sentences by which they are expressed suffer from a vagueness which may only be sorted out by referring to contexts; likewise, no account of how beliefs may affect our behaviour can be given without referring to intentions or interests. Such notions are notoriously "ideological" and therefore, says Stich, have no place in constructing a scientific theory about the mechanisms underlying our behaviour. (We notice in passing that such a claim can hardly be made without implying that (even) scientific descriptions and theories entail reference both to particular contexts and particular interests).

But now that we have seen where Stich's arguments against the existence of beliefs and of people having beliefs have taken us - and seen in particular that no scientific theory about our cognition can make do without notions of beliefs and our "normal" utterances '(is a) belief that p', nor invalidate such notions and utterances without inflicting paradoxes, it would seem obvious to admit that a cognitive theory cannot start by reducing away these notions and utterances and what they denote, nor finish by so doing. Conversely, it would seem obvious that if cognitive theory building, which aims to account for human cognition and, thus, for the beliefs and intentional actions of persons which derive their significance from the contexts or situations in which they occur, cannot handle the issues and concepts which constitute the phenomena they aim to account for, and at the same time be scientific, then the right thing to say might well be that there can be no such thing as a cognitive science.

Alternatively, we would probably do worse than considering the possibility that if our "normal" cognition, beliefs, other propositional attitudes and their content are context specific and rely on interests, points of view, etc., this might well be the case also when our cognitive activities concern matters of scientific interests. Now, this surely goes against the received wisdom about what makes scientific theories scientific - as expressed by Stich in a footnote on page 139:

> "A word about context. [...] The vagueness of belief ascriptions is often markedly reduced by the context in which the attribution is made. But the vagueness-resolving capacity of context is of little use to the cognitive theorist. This for two reasons. First, the various pragmatic factors that help to reduce vagueness are generally irrelevant to the business of scientific explanation. The cognitive scientist seeks to explain or predict a subject's behaviour independently of any consideration of what the scientist's own previous conversational interest have been. To put the point in another way, the scientist's attributions of psychological states to subjects are, or ought to be, largely independent of context. Second, to the extent that the theorist's descriptions of the psychological state of subjects are not context independent, his theory itself is methodologically suspect. For in this case the question of whether or not a given theoretical generalization applies to a given subject may be answered both yes and no, depending on the setting in which the theorist asks the question. Terms with the acute context sensitivity of folk psychological content ascriptions make poor tools for the building of scientific theories." (Stich, 1983, p. 139)

Let me first comment on the claims made by Stich when applied to the conditions for attributing properties to physical objects appropriate for scientific theorizing.

Contrary to the view expressed by Stich, it seems that the achievements of the natural sciences are due exactly to their taking into account "pragmatic factors which may help reduce vagueness". By so doing they have been capable of delimiting precisely the area of events and properties of things in reality, which their observations concern. This delimitation involves two things: (1) establishing the context of observation appropriate for studying these events and properties, i.e. the context in which these events and properties occur, may be isolated, manipulated etc.; and (2) the development of theoretical terms or conceptual apparatus appropriate for describing and determining the individuating and constituent features of the event or property in question. Thus, the scientific study of, for example, *electrons* requires that conditions be established in which it is possible to determine the material occurrence and attributes of electrons. This requires experimental and technological constructions of various sorts, as well as instruments suitable for measuring the mass, charge etc. of electrons. These are attributes of electrons which are determined under idealized laboratory conditions, but which are assumed to characterize *that* reality in which the occurrence of given phenomena may be adequately described and accounted for as a consequence of the behaviour of electrons (for example, chemical bonds may be accounted for as various modes of electron transfer or sharing amongst atoms). But without the development of appropriate theoretical terms by which electrons are conceptualized (cf. 2), it would not be possible to determine what would constitute the technical constructions and experimental conditions appropriate for assessing the attributes of electrons, among them their mass, charge, etc.

We may contend then, that within the natural sciences there seems to be nothing suspect methodologically or otherwise in the fact that the business of attributing physical properties to physical matter is context dependent in the above sense - nor does the "acute context sensitivity" of physical terms make them poor tools for the building of scientific theories. On the contrary, these are the rules of the game of science, which ensure that the descriptions and explanations of the sciences may be rigorously determined, delimited and generalized. We may say to what they apply and under what conditions they apply; and we may say to what and under what condition they do not. In short, we may determine the correct implications of scientific descriptions and their correct use, i.e. determine under what conditions and in what contexts the descriptions hold true of a given phenomenon.

However, the fact that any description is context dependent in the above sense, does not mean that the *existence* of what is being described is context dependent, i.e. dependent on the situation, time and place etc., in which it is being observed and described. Thus, it applies to physical things throughout all contexts and situations in which they exist, that they are made up of atoms, electrons, and all the rest, and that atoms and electrons of physical things exist and have precisely the properties, which have been observed and described under scientific circumstances. In this sense, what physics describes applies in general to things in reality and their physical properties; this is what generality within the sciences is about.

But, no matter how true it is that electrons exist in all situations in which physical things exist, *descriptions*, *propositions* and *theories* about electrons do not exist independently of reference to situations in which they are observed. Thus, we may say that in contrast to physical things and their properties, which exist independently of situations in which they are observed and described by persons and language users, descriptions or propositions about physical things do not exist independently of persons and language users being in particular situations in reality, having particular opportunities of observation, and a language which may be used to describe and refer to these things.

Secondly, neither the *truth* nor the *meaning* of any beliefs or propositions - be they scientific or otherwise - can be determined independently of or without referring to such situations and to things and circumstances in those situations that the beliefs or propositions are about. No belief or proposition about things has a meaning in itself, nor may its truth be determined independently of referring to such situations in which the things exist - be it in ordinary everyday or in scientific situations. In this sense the beliefs and propositions of a scientist investigating electrons are context dependent, i.e. in that their meaning and truth cannot be determined independently of referring to the situations, the experimental set-up, measuring instruments etc., in which his investigations and measurements are carried out.

This means, thirdly, that although the truth of scientific propositions may be both general and eternally true, no propositions or beliefs about things exist without persons and language users to whom they are *known* to be true, and who know the situations or contexts in which their meaning and truth may be determined. Unless, of course, we are prepared to say that propositions and beliefs about matters of scientific interest do not involve the cognition of persons, nor situations or contexts in which their meaning and truth may be determined. Rather, I think we shall have to contend that no descriptions and propositions exist, which may be true or false about anything - be it things in scientific or everyday situations - independently of persons and language users, and without presupposing that persons and language users are "the sorts of thing", which may have *knowledge* and *beliefs* about such things; knowledge and beliefs, moreover, which cannot be accounted for independently of referring to contexts, opportunities for observation, interests, and so on; indeed, independently of presupposing that persons and language users may find themselves in many different situations in reality, having different kinds of knowledge and beliefs about the same things, which may be true or false.

Now, let a cognitive theory about mental states be granted, e.g. along the lines suggested in Stich's Syntactic Theory of Mind (cf. p. 208), which may explain and predict how such states affect the behaviour of people. Just as is the case for a physical theory about electrons, their properties and behaviour, in order for such a theory to have any scientific credibility and to make any sense, it must be possible to specify experimental circumstances in which behavioural effects predicted by the theory may be observed, and unambiguously be attributed to the presence of the mental states described by the theory. And just as we cannot talk in well defined ways about electrons and their properties, nor develop theoretical generalizations about their behaviour outside experimental situations without referring to experimental situations in which the meaning of terms for electrons and their behaviour and properties are defined, we cannot talk in well defined ways about the existence and properties of mental states of persons – or, for that matter of the "abstract syntactic objects" onto which such states may be mapped - nor develop theoretical generalizations as to their behavioural effects, without referring to experimental situations, in which the meaning and terms for both such states and behaviour are defined. In this sense the meaning and test of a cognitive theory about mental states is just as dependent on context and circumstances as is the meaning and test of a physical theory concerning electrons. However, in contrast to what is assumed in the above quotation by Stich – there is yet another way in which experimental tests of cognitive theory is "context dependent", which makes the conditions for such test fundamentally different from the tests of a physical theory. For not only do experimental tests of a cognitive theory, and the ascription to subjects of the psychological states it concerns, require that reference be made to actual contexts and circumstances in which to observe and determine these psychological states. They also require reference to, indeed *rely* on the ascription of *beliefs*, *intentions* and *interests* on the part of both the experimenter and the subjects taking part in the experiment concerning those situations and how they influence the subjects' behaviour. Here is why:

Controlled experiments on cognition, which are carried out according to stringent methodological criteria, often require that the cognitive scientist tells his subjects what features of the experimental situations they are supposed to attend to, and what sort of responses are relevant in the situations. For example, in an experiment of the kind outlined by Stich, which aims to test predictions and generalizations of his Syntactic Theory of Mind, subjects are instructed to memorize a list of propositions, and subsequently to check which on another list of propositions they believe comply with the ones they have memorised. Now, to carry out such experimental tests presupposes, first, that the cognitive scientist may trust that the subjects have understood his instructions, and that their behaviour and responses in the experimental situation are determined by their *understanding* of and *desire* to obey his instructions. And it presupposes, secondly, that the experimenter may assume that the propositions on the lists presented to the subjects have some particular implications and meaning for them, and that it applies to his subjects as it applies to himself that part of understanding the propositions involves understanding in what situations and contexts their meaning and/or truth may be determined. These conditions and assumptions granted, then, and only then, is the experimenter in a position to begin to test whether the theoretical generalization of his model of psychological states applies to his subjects, and to test their predicted effect on their behaviour.

If so, we may contend, first, that neither are the conditions for, nor the assumptions involved in experimentally testing theoretical generalizations about the subjects' psychological states and their behavioural effects, different from the general assumptions of FP about the conditions for attributing beliefs and other propositional attitudes to oneself or others, nor from the assumptions of FP of the ways in which they may affect our own behaviour and those of others. And neither are the terms used by the cognitive scientist to account for the conditions for carrying out the experiment (i.e. of the material presented to the subjects or of his instruction to the subjects), nor are the terms he uses to describe the assumed behavioural effects of the psychological states of the subjects during the experiment, etc., different from the "acute context sensitive" terms encountered in "folk psychological content ascription".

So, far from being methodologically suspect, the very conditions for the experimenter's attribution of psychological states to his subjects during the experiment, and his test of the theoretical generalization about their behavioural effects, necessarily involve and rely on reference to context, beliefs, intentions, interests and other psychological phenomena denoted by the terms of FP. If so, there is no reason whatsoever, nor any possibility that a cognitive psychologists who intends to test a theory of the mental states underlying our behaviour (along the lines suggested in Stich's Syntactic Theory of Mind) should or indeed *could* avoid such reference. Nor is there any reason why anyone seriously concerned with issues of human cognition and behaviour should obey Stich's autonomy principle, let alone could possibly restrict themselves to describing the cognition and behaviour of persons in the same abstract, non-intentional and context-free terms and language used to describe the properties and relations between "abstract syntactic objects" on which this cognition and behaviour is supposed to supervene. If psychologists, despite the conditions for formulating any sensible and scientifically testable theories concerning such objects and supervenience, would maintain the possibility of describing the cognitive states of persons and their behavioural effect solely in such terms, then they could rightly be accused of being un-scientific about their subject matter. For, they would have ignored the conditions for and ways in which the existence of such mental states and their behavioural effects may be put to scientific test, i.e.

conditions and ways characterizing and defining experimental subjects as persons who act according to their beliefs, interests, desires and understanding of the situation in which they find themselves.

10.5 Some differences between ascribing beliefs to people and properties to objects

A person is someone who, in virtue of having a body has physical properties, and who has knowledge and beliefs about his or her body as well as about many other things, and who therefore is someone to whom both physical properties and beliefs may be attributed. And just as our propositions about the properties of things and persons may be true or false, so our propositions about our own beliefs and those of others may be true or false. Thus, we may falsely attribute beliefs to other persons or to ourselves that they or we do not have. Conversely, we may correctly attribute beliefs to a person which are false. In other words, we may correctly or incorrectly attribute to other people - or to ourselves - beliefs which themselves may be correct or incorrect, true or false. This clearly makes ascription of beliefs to persons both different from and more complicated than ascription of properties to things. But in spite of this complication our belief ascription track record is not significantly inferior to that of ascribing physical properties to things or persons. On the whole it works pretty well. As mentioned earlier, the existence of societies and social institutions speaks so loudly as to be almost deafening about how good we actually are in tackling these complication. We may of course sometimes be wrong or uncertain in our ascription of belief to other people - and even to ourselves - indeed we may realize that we are wrong and uncertain - just as we may sometimes be wrong and uncertain in our ascription of properties to things - and realize that we are wrong and uncertain.

Now, the fact that we may sometimes be wrong and uncertain when ascribing physical properties to physical things - in ordinary everyday as well as in scientific situations - has not (recently) made anyone suggest that we should stop ascribing properties to things. Nor has it been suggested that, because the propositions and explanations about physical matter as expressed in either everyday or scientific terms may be wrong and even sometimes *underdetermined*, then it may safely be concluded that reality and its physical things do not exist. However, in the case of belief ascription and ascription of propositional attitudes, claims have been made to the effect that the notions of beliefs and of ascribing belief to others - and even to oneself - are so indeterminate as to be unsuitable in scientific investigations of the cognition and behaviour of people (see e.g. Quine, 1960, 1970). Indeed, it may safely be concluded that psychological idioms about beliefs and propositional attitudes are void, and what they denote does not exist. Other arguments have it that because the propositions in which belief ascriptions are expressed suffer from logical "peculiarities" or "anomalies" - i.e. when analysed from the point of view of symbolic logic in the same way as propositions within the natural sciences about things and their properties may be - they do not fulfil the conditions for scientific descriptions. In what follows I shall argue that the "peculiarities" encountered do indeed exist, but also that these "peculiarities" are only poorly dealt with and understood if we fail to realize that the conditions and procedures for determining the implications and truth of propositions about *things* and *their properties* are indistinguishably the same as the conditions and procedures for determining the implications and truth of propositions about our *knowledge* and *beliefs* about these things and properties.

First, I shall try to thrash out what it means and requires to ascribe beliefs to somebody and the problems it may entail, and to outline the assumptions we necessarily have to make about beliefs and about someone to whom beliefs may be ascribed.

In order to ascribe to someone, let us call him Sam, a belief expressible in the proposition, say,

1. Sam believes that the cat is on the mat

we, or who ever else makes the ascription, must know what the belief being ascribed to Sam means, and thus what the proposition

2. 'the cat is on the mat'

means. That is, we or they must know what the proposition in (2) implies and know the conditions under which it is true or false. If we ourselves did not know the implications and correct use of the proposition 'the cat is on the mat', we could not possibly ascribe a belief to anyone to this effect. But neither can we ascribe this belief to anyone, for example to Sam, unless we presuppose that Sam knows what we know, and that Sam has concepts of 'cats' and 'mats', of 'is' and 'on', which have the same implications and uses for him as they have for ourselves, and that he knows to what those concepts apply. And like we ourselves he must know the conditions under which it is true or false to believe that the cat is on the mat. If we did not presuppose this knowledge on the part of Sam and ourselves, we could not ascribe to Sam the belief 'the cat is on the mat' - and the belief statement

3. Sam believes that the cat is on the mat

would be meaningless.

We may contend then, that in order meaningfully to ascribe to Sam the belief "the cat is on the mat", and thus for the belief statement (1) to be meaningful, it must be presupposed that Sam, being a user of language like ourselves, is capable of putting forward first person propositions in the same language that we ourselves speak. And it has to be presupposed that being able so to do implies that in this language propositions may be put forward, which have perfectly determinate meanings, and that together we are able not only to determine what we are talking about, but also to carry out appropriate acts in a procedure for determining the truth of the implications of the propositions as well as their correct applications.

Now, it may be that the intentional depths or widths of Sam's belief in the actual situation are different from ours and, thus, that some of the implications of Sam's beliefs about the cat on the mat are different from ours. It may be that Sam knows more about the cat than we do. For example, he may know that the cat in question belongs to his neighbours, and that the mat being referred to is the one he got as a birthday present from his aunt. We may not know this - and the belief being ascribed to Sam will lack these extra implications. Conversely, we, in our capacity of being cat specialists, may know something about the cat, which Sam does not know, for example, that the cat in question belongs to some rare Siamese family. But this real or potential discrepancy of implications of the belief (i.e. between the person to whom it is ascribed and the person making the ascription) does not mean that the implications of our own beliefs and those of Sam's are irrevocably indeterminate or indeterminable, nor that ascribing beliefs to someone is therefore impossible. On the contrary, we could not talk in a well defined way about the existence of any such discrepancies of belief, nor point out any concrete cases of discrepancy, without presupposing that their implications are indeed determinate and determinable.

What it does mean is that when ascribing beliefs to somebody, we have to bear in mind that such potential discrepancies in knowledge may exist which in the actual situation may make a difference. So, when we ascribe to Sam the belief "the cat is on the mat", we may have to ensure that we are not ascribing more to Sam than he would be willing to commit himself to. In order to ensure this, it may be necessary to discuss with Sam whether in addition to the basic implications of the belief being ascribed to him (i.e. that it is about a small furry animal with four legs, a tail etc., being at a particular place at a particular time etc.), Sam or we have different implications in mind about the actual cat or mat.[4]

[4] In psychological investigations, properly carried out, the investigator takes such problems into account. Indeed, a substantial amount of work in preparing a psychological

Hence, no particular belief could be ascribed to Sam without presupposing, and thus at the same time ascribing to Sam, a vast amount of general knowledge and propositions about reality, ourselves and other people, which we share with Sam. Otherwise, it would not be possible together with Sam to carry out the necessary procedures for determining the precise implication of what Sam is actually believing, let alone determine whether any difference exists between the implications of his and our own beliefs. Please note: the existence of such procedures are part of the general conditions for communication and action between language users, not only in ordinary everyday situations with our everyday possibilities of observation and description, but also in scientific situations with the possibilities of observations and descriptions, which such situations afford.

Differences may of course exist in the sets of implications and meaning of beliefs held and propositions put forward by different persons and by the same person in different situations; in this sense it is right to say that beliefs and propositions are prima facie indeterminate. However, that we may talk about the existence of and, thus, are able in a well defined way to account for different sets of implications and meanings of beliefs and propositions, relies on the fact that each of these sets does have determinate implications and meanings; this is the condition for determining any differences between our own beliefs and propositions and those of others, or for determining differences of our own beliefs and propositions in different situations (*pace* Quine).

Now, when we ascribe beliefs to other people we may be mistaken. Thus, we may be mistaken in ascribing to Sam the belief that the cat is on the mat, which makes the belief statement

1. Sam believes that the cat is on the mat

a false statement.

This belief statement is false if Sam does not believe, or is not willing to propose anything of the kind. Indeed, we may, granted that the necessary conditions for ascribing beliefs to Sam holds, be able to determine and come

---

investigation - no matter of what kind - is put into ensuring that uncertainties regarding the subjects' understanding of what is being said to them by the investigator - and vice versa - do not invalidate the data of the investigation. However, such precautions are not always taken. A case in point is the work by Nisbett and Wilson and their co-workers on "verbal reports on mental processes" (Nisbett and Wilson, 1977; for a critical review of the work of Nisbett and Wilson, see e.g. Ericsson and Simon, 1980, Praetorius and Duncan, 1987). Unfortunately, these methodologically flawed investigations are widely used as "empirical evidence", particularly by philosophers who, like Stich (op.cit.) and Dennet (1991), attempt to argue that people can't talk sense about their beliefs.

to know that we have made a mistake, and that our statement about what Sam believes is false.

But no matter how we may find out what Sam believes or does not believe, the statement expressed in (1) is true if Sam believes that the cat is on the mat, and it is false if he does not so believe.

Conversely, the proposition of which the belief statement may be said to be a component, i.e.

2. 'the cat is on the mat',

may turn out to be false. In other words, for some reason or other Sam may wrongly believe that the cat is on the mat. However, this does not affect the truth of the ascription of the belief to Sam that the cat is on the mat. Thus, the belief statement

1. Sam believes that the cat is on the mat

may be true or false independently of whether the statement

2. 'the cat is on the mat'

is true or false.

As has been correctly argued, this means that from the truth of a belief ascription statement we cannot draw any conclusion about the truth of what is being believed. No matter how true it may be that Sam believes that the cat is on the mat, the proposition 'the cat is on the mat' is not true if the cat is not on the mat. Conversely, from the fact that the statement, 'the cat is on the mat' is true, it cannot be concluded whether the belief statement, "Sam believes that the cat is on the mat", is true or false. Similarly, it cannot be concluded from the truth that someone believes that a particular cat is on a particular mat that either the cat or mat in question exist.

In this respect it is arguably correct that belief statements are logically "peculiar" and differ from statements used in the natural sciences and biology to ascribe properties to things. The language of physical and biological science, as pointed out, is largely extensional and can be formulated approximately in familiar predicate calculus. One feature of symbolic logic is that the truth of an expression only depends on what the expression refers to (its extension), not its meaning (intension). Thus, in so called extensional expressions we can substitute a term with one meaning with a term having another, which both refer to or stand for the same thing, without changing the truth value of the sentence. For example, in the sentence, "the cat is on the mat", the term 'cat' can be substituted with 'Jerry's pet' if they both refer to the same cat. Another feature of extensional expressions is that they are truth-functional, which means that the truth value of a statement which is composed by other statements, e.g., 'the cat is on the mat and it is raining', can be ascertained simply by knowing the truth value of the component sentences.

According to these arguments a belief ascription sentence may be analysed as a compound statement which exhibits a "logical anomaly" in that it fails the extensionality conditions. Thus, the belief sentence, "Sam believes the cat is on the mat", may be true whereas, "Sam believes that Jerry's pet is on the mat", may be false, i.e. if Sam does not know that the cat on the mat is Jerry's pet. Likewise, the belief statement, "Sam believes that the cat is on the mat", may be true - and yet its component statement, "the cat is on the mat", may be false. By the same token, a belief statement may be true; however, no conclusion can be drawn as to the actual existence of what is believed. It is widely claimed today by philosophers and logicians that intensional sentences cannot be equivalently rephrased or replaced by extensional sentences - and that, therefore, the propositions used to describe psychological states, such as beliefs and other propositional attitudes, cannot be scientific.

Now, if we look at the sort of compound sentence into which a belief sentence may be rephrased, e.g.,

3. 'Sam believes that the cat is on the mat, and the cat is on the mat'.

it is quite clear that no conclusion can be drawn about the truth or existence of its components - or vice versa. In this respect a belief ascription statement is similar to statements ascribing other propositional attitudes, such as, "Sam hopes that the cat is on the mat", or, "Sam fears, or wishes, or thinks, or imagines .... that the cat is on the mat". However, as we normally use the terms "believe", "hope", "fear", "wish", "think", and the like, it simply does not follow from someone believing, hoping, or wishing something to be true or false, that it *is* true or false, or that what is hoped for, believed etc. exists. Indeed, it could be argued that the whole point of using terms like 'hope', 'fear' and 'believe' is precisely to leave open, and thus uncertain, the truth and existence of what is being hoped for, feared, believed, etc. If someone says, "I hope that the cat is on the mat", or, "I believe the cat is on the mat", it means just that, i.e. I hope or believe it - but I am not certain: or I believe or hope it is the case, but I do not know it.

This fact, in itself a triviality, has unfortunately been vastly complicated by philosophers who - not least in their attempt to help psychologists and cognitive scientists to sort out notions of cognition - use the notions of 'belief' and 'knowledge' as if they were synonymous concepts. *They are not*, and as the next example illustrates, it should come as no surprise that when analysed in terms of propositional logic, statements which contain the verbs 'to know' and 'to believe' behave quite differently. Thus, if instead we had ascribed to Sam,

4. Sam knows that the cat is on the mat

the logical "peculiarity" and "abnormality" encountered in the case of belief ascription do not occur. Thus, if the knowledge ascription sentence in (4) is true, it *does* follow that its component sentence, 'the cat is on the mat', is true and that the cat exists. If we said otherwise, we would quite simply be using the term 'to know' incorrectly. We cannot say without contradicting ourselves that the statement ascribing the knowledge to Sam that the cat on the mat is true, but the cat is not on the mat - nor that the cat which Sam knows about does not exist. Conversely, we have to say that the conditions on which the *statement*, 'the cat is on the mat', is true are the same as the conditions for Sam's *knowing* that the cat is on the mat. These conditions granted, we may just as safely ascribe the knowledge to Sam, "the cat is on the mat", as we may put forward the proposition 'the cat is on the mat'. Indeed, if the argument is correct that no proposition is true or false in itself or independently of being put forward by persons and language users who know and may determine that it is true or false, there is no difference in principle in saying, "the cat is on the mat", and saying "someone *knows* that the cat is on the mat". In other words, to assert that the proposition, "the cat is on the mat", is true, implies asserting that it is known to someone that it is true that the cat is on the mat, i.e. someone who may determine that it is true.

Now, if we adopt a Quinian line of reasoning the conclusion which could be drawn from the above example is that a cognitive theory, which restricts itself to being about the *knowledge* of people may be scientific - as opposed to one which also wants to include the *beliefs* and other propositional attitude of people. But apart from the fact that such a theory could hardly be a comprehensive theory of cognition, a much more obvious conclusion could be drawn from the logical similarity between psychological propositions ascribing *knowledge* to persons and biological or physical propositions ascribing *properties* to things. This similarity, as pointed out above, is due to the fact that what both propositions state to be true or false is, implicitly or explicitly, *known* so to be. That is, in both cases it is known to someone who has a language and terms in which reference may be made to that which is known, in the form of propositions which may be true or false. Indeed, in both cases the very notion of 'true' and 'false' implies and necessitates a notion of "knowledge".

In contrast, in the case of ascribing hopes, fears and beliefs to others (or oneself), it is left open or uncertain whether what is believed, hoped etc. is in fact known to exist or to be true or false. What is *not* left open or uncertain, however, is that anyone to whom it may be ascribed that "he or she hopes, believes, or fears that the cat is on the mat", knows what a cat and a mat is, and knows how to determine whether or not some particular cat is on some particular mat. And in order to ascribe any of these propositional attitudes to anyone, it has to be presupposed that the one to whom they are ascribed knows how to determine whether the proposition 'the cat is on the mat' is true or false, and thus knows how it may be determined whether or not his or her hopes or fears come true, and whether or not his or her beliefs are true. And to ascribe to a person that he or she *believes* as opposed to *knows* 'that p' presupposes that both we and the person know what it means to believe 'that p', as opposed to knowing 'that p'. This cannot itself be merely a matter of belief, hope, and the rest.

The point is not only that there are good reasons why we should distinguish between the notions of 'belief' and 'knowledge', i.e. that these notions are not synonymous, but also that the ascription to persons of any propositional attitudes - beliefs included - at the same time necessarily presupposes and implies ascription of *knowledge*. Hence, a cognitive theory cannot have a notion of 'belief' or of any other propositional attitudes without, or which is not firmly based on, a notion of knowledge; nor can it make do without a notion of persons having this knowledge, and who know how to distinguish between and refer to different kinds of cognitive states they may be in, and to whom such states may be ascribed.

Now, the point is well taken that people do not act or reason solely on the basis of what they know, but also sometimes on the basis of what they believe (or hope, desire, think, or fear). If so, it may sometimes be more important for psychologists, in their attempts to account for the action and reasoning of people, that they get right what people actually believe, rather than whether what they believe is true or false. What is of importance for psychologists is to make sure that their belief ascriptions are true, and to make sure that they correctly distinguish between cases and conditions in which people believes 'that p' as opposed to knowing 'that p'.

It may well be that the psychologist's business of attributing beliefs or knowledge to people is more complicated and uncertain than the physicists business of attributing properties to physical matter. But this, as I have argued, does not leave the implications and use of the idioms of psychology irredeemably uncertain or indeterminable, nor does it rule out the existence of the psychological states to which they refer. In particular, it does not do so any more than similar uncertainties in describing material things and their properties rule out the existence of such thing and properties. These idioms form part of our everyday language and terms, which we use to describe and communicate to others our cognition and action, and without which we could not together determine - in ordinary every day or scientific situations - what are true and false propositions and knowledge about matters of fact, as opposed to what is merely believed, hoped, feared, imagined and so on.

If, furthermore, the argument is correct that propositions about matters of fact, which may be true or false, are propositions put forward by someone who knows about these facts, and thus someone who may determine their truth or falsehood, then the business of ascribing *knowledge* and *beliefs* about matters of fact, need be no more indeterminate than to put forward *propositions* about matters of fact. If it is similarly true that it is a condition for defining precise, determinate implications and correct uses of descriptions and propositions about matters of fact - in scientific as well as in everyday situations - that the speakers involved share knowledge about these facts, as well as procedures and a language in which together they may determine what they know about them, then the conditions and procedures for determining in well defined and precise way what people *know* about matters of fact are exactly the same as those which apply for establishing precise, determinate implications and correct uses of *descriptions* and *propositions* about such facts. If we said otherwise, if we said, for example, that although language users - in scientific as well as in everyday situations - are able together arbitrarily precisely to define the implications of propositions about matters of fact, and to determine their correct applications and truth, propositions concerning what these same people *know* and *believe* about these same matters of fact in these same situations, are both indeterminate and indeterminable, we would quite simply be talking nonsense. We would be saying in effect that although we can use language in determinate and correct ways to put forward assertions ascribing properties to things, we cannot use language in determinate and correct ways to put forward assertions ascribing to ourselves and others knowledge of the matters about which we may put forward true and determinate assertions. If we do say so, we would in the final analysis be using language to question the very conditions for language users to assert anything determinate about anything, and thus question the very possibility that language may be used to say anything determinate about anything.

As has hopefully been made clear so far, to analyse statements ascribing beliefs and other propositional attitudes to people solely according to the logical criteria which pertain to statements ascribing properties to things, would be to miss important points about psychological state ascription. In particular, it would be to miss the crucial differences which exist between ascribing knowledge, beliefs and propositional attitudes to people, i.e. knowledge and beliefs which may themselves be true or false, as opposed to ascribing properties to physical things, which may or may not exist, but which cannot themselves be true or false. Conversely, the consequence of denying the possibility of determinate propositions ascribing cognitive states to persons, including ascriptions of the knowledge and beliefs of persons about properties of things, on which are based their descriptions and assertions about these properties of things, would inevitably also be to deny the possibility of determinate proposition ascribing properties to things.

Chapter 11

Propositions about real as opposed to fictitious things

11.1 Brentano's thesis about the intentionality of the mental

It would be fair to say that cognitive psychology is way behind the natural sciences when it comes to developing rigorous terms, methods and theories to describe and account for the mental states and phenomena falling within its subject area. This is partly due to the immense complexity of cognitive phenomena, and partly due to the misconceptions, reminiscent of the Mind-Matter bifurcation, that they are phenomena which may be conceived of as some sort of "psychological objects", the properties of which - just like the properties of physical objects – may be studied scientifically *in isolation*, or *sui generis*. According to this view, cognitive phenomena, such as beliefs and propositional attitudes of people (i.e. their thoughts, hopes, intentions, desires, fears, etc.) are some sort of *mental states*, the content and properties of which can be treated and accounted for independently of that which it is about[1] - despite the fact that neither the content nor the properties of "aboutness" or intentionality of such states are well defined without reference to something existing independently of these states.

This misconception has probably been exacerbated by Brentano's thesis of what distinguishes – and separates – mental phenomena and states from physical things and states. Mental states, says Brentano, are *intentional* in

[1] In much the same way as it is assumed that we may talk about the percepts of people, i.e. talk about *what* people see independently of talking about *that* which they see (cf. Chapter 4).

that they are directed towards or refer to something. This intentional property, Brentano insists, makes mental phenomena fundamentally different from purely physical phenomena, and precludes mental phenomena being studied with the tools of the physical sciences. However, he also insists that what makes mental phenomenon intentional is that they "*contain an object intentionally within themselves*"- and thus all mental phenomena are intentional in precisely this sense - whether or not their objects concern real existing, physical things (such as the chair I am sitting on) or non-existing or fictitious things (such as Santa Claus or a unicorn). Hence, it is the fact that mental states involve *presentation* of objects towards which they are directed, which constitutes their intentionality, and distinguishes them from physical phenomena. In Brentano's own words:

> "Every mental phenomenon is characterized by what the Scholastic of the Middle Ages called the intentional (or mental) inheritance of an object, and what we might call, though not wholly unambiguously, reference to a content, directed toward an object (which is not to be understood here as meaning a thing), or immanent objectivity. Every mental phenomenon includes something as object within itself, although they do not all do so in the same way. In presentation something is present, in judgement something is affirmed or denied, in love loved, in hate hated, in desire desired and so on. This intentional in-existence is characteristic exclusively of mental phenomena. No physical phenomenon exhibits anything like it. We can, therefore, define mental phenomena by saying that they are those phenomena which contain an object intentionally within themselves." (Brentano, 1874/1973, p. 88)

Since this intentional property applies whether the content of mental events concerns things existing in material reality or something non-existing or fictitious, the relation between mental states and their "intentional object", is a relation which may lack *actual relata* - but which may still "point". Brentano characterizes the special "psychical relation" involved in mental states thus:

> "In the case of other relations, the Fundament as well as the Terminus must be an actual existing thing. ... If one house is larger than another house, then the second house as well as the first house must exist and have a certain size. ... But this is not at all the case with psychical relations. If a persons thinks about something, the thinker must exist but the objects of his thoughts need not exist at all. Indeed, if the thinker is denying or rejecting something, and if he is right in so doing, then the object of his thinking must not exist. Hence the thinker is the only thing that needs to exist if a psychical relation is to obtain. The Terminus of this so-called relation need not exist in reality. One may well ask, therefore, whether we are dealing with what is really a relation at all. One could say that we are dealing with something which is in a certain respect similar to a relation,

and which, therefore, we might describe as something that is 'relation-like' [etwas 'relativliches']." (1879, quoted in Chrisholm, 1967, p. 149)

Thus, it would seem that a mental phenomenon is intentional in virtue of being directed towards something – and also that for this intentional property or "psychical relation" of a mental phenomenon and its object to exist, only takes a thinker who thinks of it. However, it would seem that since mental phenomena are thus intentionally, i.e. are directed towards an object within it self, there can be no distinction in the mind of objects presenting real as opposed to non-existing things. If so, one may say that Brentano's thesis of what makes mind and mental states like beliefs, hopes, desires etc., irreducibly intentional, at the same time fundamentally *separates* mental states from physical matter - for both mental states and the objects towards which they are intentionally directed are firmly positioned in the "realm" of the mental.

Well, if this reading of Brentano is correct, the consequences are serious. For how, for example, would Brentano's thinker, equipped with this kind of intentional mind, be able to distinguish himself as someone who truly exists, from something non-existent presented in his mind in precisely the same way? And how could the thinker - if all his mental states could be intentionally directed towards are objects presented within themselves - possibly be either right or wrong in "denying or rejecting that the object of his thinking must exist"? (op.cit) Not only would there be no way in which Brentano's unfortunate thinker could distinguish beliefs about things which exist, from beliefs about things which do not - and thus distinguish facts from fiction - but there would be no way either that he could distinguish between *beliefs* about things from the *things* themselves - be those things real or fictional. For his beliefs and the objects of his beliefs would be one and the same thing. In other words, no distinctions could be made, or indeed could exist, between beliefs about things and the things themselves, be they actually existing physical things or non-existent fictitious things. Indeed, there could be no distinction between mental states and physical things, and thus nothing "physical" from which the "mental" could be distinguished or separated.

If we accept Brentano's conception that mind and its states may be directed intentionally, albeit the objects towards which they are directed are invariably objects of the mind, we would be compelled to admit that none of the objects about which we have beliefs concern objects in the world. And we would be compelled to admit that mental states entail an intentional relation, although the objects that this relation concerns may not exist. In the words of Richardson,

"On the one hand, if we maintain that there are non-existing but real objects which are the objects of our thoughts, we are compelled to admit that none of

> our objects [the objects about which we have beliefs, etc.] are in the real world. We are debarred of thinking of the concrete. On the other hand, if we admit that mental acts are not really relational, we are led to the conclusion that mental acts cannot really direct us to objects in the world (or out of it either). Our thinking does not direct us to the world. In either even such acts can hardly be viewed as intentional." (Richardson, 1981). p. 177 - 178)

However, Richardson also contends that we cannot really sacrifice either claim if we are to deal adequately with intentional phenomena. But not to sacrifice one or the other inevitably leads to conflicting assumptions. For, on the one hand, it is assumed that the mind has the capability to "reach out towards" and present objects in the world, i.e. physical objects existing outside and independently of the mind. Indeed, without so assuming, no "thinker" could possibly talk sense about objects which "must" or "must not" exist, let alone distinguish such objects from fictitious objects. On the other hand, it is also assumed that, somehow, in the process whereby the mind "reaches out towards" objects existing in the world, the objects being "reached" are *internalized* or *absorbed* into the mental landscape - and transformed into purely mental objects, i.e. objects of the kind that the mind and its mental states may include intentionally within themselves. Consequently, all that the mind actually is and can be intentionally directed towards and related with is something of its own (mental) kind.

These conflicting assumptions have led to the problem, first, that since the intentional directedness of our thoughts and beliefs may *indiscriminately* concern both actually existing and non-existing or fictitious objects, there can be no way of accounting consistently, nor scientifically, for the intentionality of our thoughts and beliefs. Secondly, the insuperable gulf between the mental and the physical created by Brentano's thesis and definition of mental states, both seems to legitimate and necessitate the assumption that the mind and its states, such as beliefs and other propositional attitudes, their content and intentionality, may, indeed, only *can* be viewed *sui generis*.

One solution aiming to solve the problems arising from Brentano's thesis of intentionality without sacrificing either of the assumption it involves, has been to suggest that the content of mental states and, thus their intentional objects, are *representations*, and furthermore, that *veridical* mental representations, i.e. representations of objects actually existing in the world, are related to the world by a causal connection to those objects.[2] However, if the intentional objects of mental states are (mere) *representations* of things and events in the world, no account of how these representations are caused by

[2] This conception of mental states being representations forms part of models developed within the branch of cognitive science endorsing a computational view of cognitive processes. In the next chapter I discuss Fodor's arguments for this view.

things in the world may explain how mental states may be about and refer to actual objects in the world - as opposed to representations of them. Similarly, if *all* we have in our minds are *mental representations*, and nothing in the mind itself distinguishes *veridical* representations of things actually existing from representations of non-existing things, then no causal explanation could ever be developed and explain how some of our mental representations connect our mind to real existing things, while others do not.

Another solution to the same problems has been to suggest that since mental states are describable in terms of propositional attitudes (Russell, 1940), they may be analysed as linguistic sentences using verbs such as "believes", "hopes", "desires", "thinks", or "fears", i.e. sentences of the form, "Mary believes that her lamb fell over the cliff". In this form the verb expresses the *attitude* of the mental state, while the proposition following "that" is the *object* towards which the attitude is directed. However, the problem of this conception of mental states is the same as before. For, if *propositions,* so the argument goes,[3] are the objects towards which our attitudes are intentionally directed, then (by definition) mental states are about propositions rather than about objects in the world. In short, viewing the content and intentional objects of mental states as being either *representations* or *propositions* does not solve the original problems to which Brentano's thesis about intentionality give rise; the problem has merely been re-phrased in terms of representations and propositions, without, however, providing a distinction between veridical as opposed to non-veridical representations, nor between propositions which may be true or false about real existing things as opposed to being true or false about non-existing things.

In the next section I shall argue that only if we give up the assumption that mental states are objects which may be analysed *sui generis*, i.e. as some sort of "mental kinds", the constituents of which may be determined independently of or without referring to - among others - the "natural kinds" that their content may concern, only then will there be any hope of ever finding an alternative to the prevailing Brentanion view of intentionality and the problems it inheres.

11.2 Brentano's thesis of intentionality reconsidered

As Brentano rightly points out, it is a feature of our mind that it allows us to direct ourselves intentionally toward things, of which we know and may assert that they do *not* exist in physical reality (such as imagined and fictitious things), and even towards our own beliefs, be they of fictitious or real things, and thus that the objects and content of our beliefs may concern

[3] See e.g. Bechtel (1988).

something "mental", i.e. something within the mind itself. And he was also right in asserting (cf. the quotation on p. 236) that a "thinker" equipped with this kind of reflexive mind must be someone who exists - and thus is someone who may refer to and identify him or herself as someone existing in some concrete situation in physical material reality. That is, someone who knows that he or she exists as a person with a body being part of the physical reality and things in reality of which he or she may have beliefs. This, I think, must be a brute fact about someone being able to reflect on and direct one's mind intentionally towards things which he or she *knows* do not exist in physical reality, for example, direct ones mind intentionally towards and reflect on one's beliefs about such things. However, contrary to what Brentano also asserts, to be able so to reflect and know must inevitably be to know that some of our beliefs can be intentionally directed towards things existing in physical material reality *as well*, and thus that the objects and content of our beliefs can indeed concern physical things, and hence that our mind is capable of being directed intentionally towards objects not "within itself" - i.e. not being part of the mind itself.

Equally, I think we can agree with Brentano that it is an irreducible property of beliefs *in general* that they may be intentionally directed toward, refer to, or be about something. However, again in contrast to Brentano, I think we shall also have to agree that it is a property of beliefs about things existing in physical reality *in particular* that they are about things, which exist *independently* of persons and the beliefs that persons may have about them. They are the kinds of things we may observe and act on, just as we may observe our own bodies, arms and legs carrying out those acts - and they exist just as independently of our beliefs about them as our bodies, arms and legs, exist independently of the acts we carry out with them. Hence, in the case of beliefs about physical things (including our bodies), we shall have to agree that for a "psychical relation" to obtain between such beliefs and that which they concern, requires not only that we ourselves exist, but also the things toward which our beliefs are intentionally directed – *pace* Brentano (op.cit.). Furthermore, we cannot even begin to characterize or describe these physical things independently of assuming that they exist as things we may have beliefs and knowledge about, i.e. that they are things that our beliefs and knowledge may concern, and thus be intentionally directed towards. And, equally importantly, although we can certainly distinguish between *beliefs* about physical things and the *things* they concern,[4] we cannot describe or characterize our beliefs about physical things without or independently of referring to them. For example, I cannot characterize or describe the content of my belief or knowledge about the chair I sit on, inde-

[4] - just as much so as we may distinguish between the acts we carry out on things and the things on which we act.

pendently of or without at the same time describing the chair; nor, conversely can I characterize or describe the chair without assuming that the chair exists as something that I may have knowledge and beliefs about - and thus that the chair is a thing that my beliefs and knowledge concern. In other words, we shall have to assume that a *necessary* relation exists between our beliefs and knowledge of physical things and those things.

Indeed, without *both* assuming the existence of physical things independently of our beliefs of them, *and* assuming a necessary relation between these beliefs and the physical things they concern, the difference and distinction between physical things and mental phenomena on which Brentano insists, would not be possible, nor well defined. However, these assumptions immediately invalidate and rule out the possibility of a *separation* between mental phenomena and physical things of the kind suggested by Brentano in his thesis of intentionality.

Now, the point made by Brentano that beliefs and other mental phenomena are intentional in that they may be about or refer to something, is certainly correct, and so is the point that this feature of beliefs and other mental phenomena uniquely distinguishes beliefs and other mental phenomena from physical things. Moreover, it is certainly correct that things in physical reality exist independently of persons and the beliefs or knowledge of person, but also correct that *beliefs* and *knowledge* about these things do not exist without or independently of persons who have them. Persons, that is, who are able to distinguish beliefs and having beliefs about physical things and the things their beliefs concern. Physical things, states or events, on the other hand, cannot have beliefs, nor be about or refer to something, let alone imagine or have memories of their own states, or of those of other physical things. Neither can physical things be right or wrong, true or false about themselves or other things or states of the world, nor have desire or hope vis à vis such things or states. The chair I am sitting on is not about anything. It may be causally related to other things or states in the world, which may cause physical changes to it, just as the chair may cause physical changes to other things in the world. But it cannot be intentionally directed towards nor have desires and beliefs about, or imagine such other things or changes in the sense that the minds of persons, their imagination, beliefs or desires may be intentionally directed towards things and changes. Nor is there any way in which intentionality, beliefs and other sorts of "mentation" may be explained in terms of causal physical relations or processes.

Brentano was therefore right in pointing out that what is present in mind – and how it gets there - cannot be accounted for or explained in terms of physics, and the causal laws applying to physical things. Although the presence of a chair in my field of vision via some physically describable causal processes in my visual system may be necessary for the occurrence of my perception of the chair, this perception and thus my awareness of the presence of the chair, cannot be explained in terms of the causal physical process between the chair and my visual system; nor can my *assertions* about the chair or the fact that these assertions may be true of false. For there is no way in which a causal physical process can give rise to *knowledge* or *awareness* about that which caused it, nor give rise to *assertions* about it. A billiard ball may move when hit by the cue, but it does not result in anything, not even remotely, which could be described as knowledge about what hit it, nor in awareness of having been moved from one place to another as a consequence of having been hit.

However, the fact that beliefs and other mental states are not accountable in terms and laws applying to physical things - and vice versa - does not warrant the kind of separation between mental states and physical things suggested in Brentano's thesis about the intentionality of mental states. On the contrary - as the analysis above and in earlier chapters should have made clear. In the next section I shall try to show how, following the assumptions concerning *both* the differences *and* necessary relation between beliefs and the "objects" they concern, the problems inherent in this thesis of distinguishing beliefs about fact from beliefs about fiction may be overcome.

11.3 Beliefs about real and fictitious things

So far, I have dealt only with beliefs and other propositional attitudes of persons about *physical* things - but not with beliefs about non-existing or totally fictitious things. For good reasons. For, it makes good sense to start a discussion of how persons may distinguish between beliefs about real existing things and fictitious things by making clear the conditions applying to our cognition and beliefs about ourselves as persons, and the physical reality of which we are part. In particular, to make clear the nature of the intentional directedness, and hence the *necessary* relation which exists between beliefs about physical reality and the reality towards which they are intentionally directed. Indeed, it would seem reasonably to assume that only after having made this clear, will it be possible to begin to thrash out any well defined similarities and differences, which might exist between beliefs about real existing things and non-existing or fictitious things, as well as similarities and differences in the ways in which either type of belief may be intentionally directed towards them.

So, let me begin the discussion about some of these differences and similarities by repeating some of the arguments of Chapter 7. According to these arguments, part of our cognition and beliefs about physical things is knowing what we may or may not do to them (or they to us), and what will happen to things (or to ourselves) as a consequence of the acts we carry out on them. In this sense, our cognition and beliefs about physical things entail *projecting*, *anticipating* and *imagining* changes in the properties and features of things as a consequence of the acts we carry out on them. And in this sense, to have knowledge and beliefs about things involves imagining properties and features of things as yet not present. Furthermore, it is part of our knowledge and beliefs of things subsequent to the changes furthered by our action on them that they concern the *same* things, which had different properties and features before we acted on them. Consequently, *memory* of how things were before such changes must also be part of our knowledge and beliefs of things, upon which we may act and induce changes. Indeed, it would seem that a lot of "mentation" in the form of imagining, anticipating and remembering features and properties of things not being present, is involved in our knowledge and beliefs of them. To the above may of course be added that it is also part of our beliefs and conception of physical things that by carrying out investigations on them, we may come to discover that they have properties and features which, prior to such investigations, were totally unknown to us.

Now, let us assume that this capacity of our minds to imagine, anticipate and remember changes in properties and features of physical things, and even to imagine that we may discover properties and features of things not yet known, is an essential and necessary part of our cognition and conceptions of real existing physical things. If so, it would not be unreasonable further to assume that this "mentation", which is involved in our cognition and beliefs about real existing physical things, could be both a prerequisite and constitutive for our minds to form beliefs about non-existing or fictitious things. Indeed, it would seem that for us to take the further step of imagining changes and occurrences of features and properties of things which, literally, are beyond belief in the case of real existing things, is not very far - and thus that the creation of beliefs about non-existing or fictitious things, having non-existing properties and features, are "born" out of the same sort of "mentation" which is part of our cognition and beliefs about real existing physical things. Likewise, it would be reasonable to assume that the fabrications of beliefs about possible or fictitious worlds, inhabited by these non-existing things and their non-existing properties and features, are born out of the same sort of "mentation", which is a necessary part of our cognition and beliefs of real things existing in the physical world.

But what about the relation between our beliefs of non-existing or fictitious things, and the things they concern? Well, as argued above, in the case of belief about physical things it is obvious to assume that a *necessary* relation exists between such beliefs and the independently existing things they concern. But when it comes to non-material or fictitious things, such as Santa Claus or unicorns, would it not then be right to say that we are talking about something, which is entirely mental in Brentano's sense, i.e. to say that such beliefs are mental phenomena, which are directed towards or contain an intentional object within the mind itself? I do not think so. For one thing, in the case of having a belief about Santa Claus[5] and making assertions about this belief, is not to have a belief about something mental, nor to make assertions about something mental *any more* than when we are having beliefs and make assertions about physical things in material reality. What we are having beliefs of and make assertions about is a *fictitious figure* existing in some collectively shared fictional world of the kind narrated in adventures and folklore; adventures, which are passed on from generation to generation within our culture by story tellers, or printed in books for all who can read to share. For another, in the fictional worlds of such adventures, unicorns and Santa Claus exist as things about which something is the case and something else is not the case, and about which true and false assertions may be made, just as is the case for real existing things. That is, they exist as fictitious figures, the existence and identity of which are firmly based in those narratives or fictitious contexts. According to such contexts it would be false to say about e.g. Santa Claus that he is only three inches high, and that he has a body like a lion, but we do have to say that he looks like a man with a long, white beard, and all the rest. If beliefs about Santa Claus could not be true or false, and assertions about Santa Claus did not have necessary implications and correct uses, there would be no beliefs about Santa Claus, nor could we talk about him - not even as a fictitious figure; indeed, there could be no fictitious figures or events to talk about - in contradistinction to physical figures and events in material reality.

In *contradistinction* to, indeed, for there are differences between beliefs about fictitious things existing in fictional or possible worlds and beliefs about physical things existing in material reality - just as there are differences between persons having beliefs about fictitious things and worlds, and the fictitious things and worlds they may have beliefs about. Thus, persons who may have beliefs about such non-existing things or worlds are persons who necessarily - as argued above - at the same time know that they themselves exist in concrete situations in material reality. That is, persons who know that they have physical bodies which may move around in real time and space, and who may carry out real physical acts with real physical things - but who at the same time know that they may not physically "enter into" or find themselves in situations in fictional or possible worlds, nor carry out physical acts with or manipulate the sort of things which inhabit fictitious worlds. Persons, furthermore who know that they cannot carry out empirical investigations in the world of fictitious things, e.g. in order to test whether

[5] Or any other fictitious figure or phenomenon from our common heritage of adventures and folklore.

fictitious things or figures have the properties and features they are believed to have, let alone carry out such empirical investigations in order to discover as yet unknown features and properties of them. Although the features which "make up" Santa Claus and unicorns are those of an old white bearded man and a small horse-like creature with a horn on its head respectively, it is part of being a person capable of fabricating and holding beliefs about such fictitious things that one knows that one cannot ask whether Santa Claus or unicorns have digestive systems like human beings or small horses, and hope to get an answer by empirical investigations of the sort which brought about knowledge of digestive systems in human beings and small horses. However, persons capable of inventing stories involving beliefs about Santa Claus and unicorns can elaborate on the fictitious features and properties of these figures, and decide whether they have digestive systems resembling those of human beings and horses, albeit fictitiously. In this sense beliefs about fictitious things as well as of their features and properties are *par excellence* inventions, being fixed by conventions by people sharing these beliefs.

However, as often noticed,[6] the language and concepts used to describe narrative or fictitious figures and events, is exactly the same language and concepts that we use to describe events in material reality. When describing Santa Claus as a figure looking like a man with a long white beard, or a unicorn as a figure which looks like a small horse with a very long horn growing out of its forehead, the meaning and correct use of these terms are exactly the same as when used to describe real men, beards, horses, horns, length, and the like. And so are the notions of 'true' and 'false' when we describe fictitious figures and their features. We do not have a notion of "truth" applying to descriptions or beliefs about real existing things, and another applying to descriptions or beliefs about non-existing things. Indeed, we know and may correctly determine the implication and correct uses of terms used to refer to things in fictitious circumstances *in virtue* of our knowledge of the language and meaning of terms we use for describing real existing things in material reality; and we may so do, because the notions of 'true' and 'false' are the same when used to describe fictitious and real things.

But there are differences here as well, namely differences concerning procedures for determining the truth and correct applications of descriptions and assertions of real versus fictitious things. Indeed, to be able to hold and fabricate beliefs about fictitious things and worlds, and thus to distinguish descriptions and beliefs of such things and worlds from descriptions and beliefs of things in physical material reality, requires that we know these differences of procedures. One of these differences is that, whereas the im-

[6] See e.g. Tye (1992).

plications and truth of an assertion or a belief about real existing objects may be determined by checking whether it holds true of some materially producible thing, the determination of the implications and truth of a belief or an assertion about a fictitious thing can only be carried out by *verbal* means. - Two people discussing whether the horns of unicorns are one meter or rather two meters long, or whether they are made of ivory or some other "stuff", will, as part of their knowledge of unicorns as fictitious figures, know that the dispute cannot be settled by going to the Zoo - nor by searching anywhere else for a materially producible exemplar. Nor will they be able to determine the truth of any description or assertions about unicorns by confronting the assertion with a unicorn so produced. The only way to settle the dispute will be to consult a storyteller, or an expert folklorist, who knows how unicorns are described in transcripts of original stories or depicted in paintings or ancient tapestries.

Let me sum up. The invention of collective fictitious worlds and the figures which inhabit them, such as Santa Claus and unicorns, relies entirely on conventions among persons who at the same time know that they themselves exist in material reality; persons sharing a language in which they may make conventions about such fictitious worlds and figures. However, this language is no other than the language we use to describe and communicate our beliefs and knowledge of things existing in physical material reality. And, far from the popular view that the fabrications of fictitious figures and worlds are the products of primitive minds being unable to distinguish fact from fiction, the capacity to invent and fabricate narratives of fictitious worlds and things requires a highly sophisticated language and knowledge by its users of how language may be used correctly in radically different circumstances, and about radically different things. Indeed, to be able to distinguish fact from fiction, and thus to determine the implications and truth of beliefs and assertions about real as opposed to fictitious phenomena, requires that one can distinguish between something which may *exist* and about which something is *true and false in material reality*, and something which may *exist* and about which something is *true and false in a fictitious or abstract context*. And it requires knowledge about the difference in procedures for determining the truth of beliefs and assertions in real versus fictitious contexts. Furthermore, it requires the knowledge of persons that, unlike material things which exist independently of their beliefs and the circumstances in which they find themselves in material reality, fictitious things and events do not exist independently of persons and the circumstances in which they are *made* to exist as *fictitious* things and events by persons.

According to the foregoing arguments the relation between beliefs about things and the things they concern, be they real or fictitious, is a *necessary* relation. Thus, it applies just as much to beliefs and descriptions of fictitious things as to real things that they may be correct or incorrect, true or false, i.e. they have determinate and determinable implications and correct applications. Otherwise there could be no beliefs about fictitious things, and no fictitious things to talk about. However, unlike things in material reality, which exist independently of and as something different from our beliefs and descriptions of them, things inhabiting collectively sustained fictitious worlds only exist in virtue of our beliefs and descriptions of them, and of the conventions which determine the content or objects towards which they are intentionally directed or refer to.

11.3.1 Conclusion

Thanks to the versatility of our mind unambiguously to direct itself intentionally towards things existing in physical reality, as well as towards fictitious and "abstract" things and phenomena; and thanks to the versatility of our mind to direct itself intentionally towards our beliefs about such radically different things, and to determine how they differ: and, not least, thanks to the fact that our mind may even direct itself intentionally towards and uncover the logical structure of our beliefs and thinking - whatever objects and phenomena those beliefs and thinking concern - it has been possible to develop systematic logics and mathematics, which may further improve our thinking and reasoning, as well as our understanding and description of reality - and whatever else our minds may be directed intentionally towards. However, all this is only possible because there is no, and can be no *general* problem of reference of beliefs and description in any of these special cases, i.e. the things and phenomena in these cases exist as things and phenomena towards which our beliefs and descriptions may be intentionally directed, and thus are things and phenomena about which something determinable and determinate may be the case, true or false. This, together with our knowledge of the different procedures by which to determine the truth of beliefs and assertions about real existing and fictitious or abstract things, must be the point of departure for any further investigations into any *special* problems of reference, which may exist in these different cases.

Now, nothing has been said (neither here nor anywhere else) which may explain how it comes about that we have a mind and mental states that may be intentionally directed towards objects existing in physical reality, or towards fictitious and abstract things – nor of how it comes about that our mind may be intentionally directed towards objects or states within the mind itself. In short, nothing has been said which may explain that mental phenomena may refer to or be about something – whether inside or outside our mind. For, nobody knows how to explain this feature of our minds. In particular, nobody knows how to explain the mentation and logical structure of our beliefs and other propositional attitudes in terms of something more elementary, *which does not presuppose these features and structures*. On the contrary, any attempt scientifically to explain or account for these features and logical structures of our mental states both relies on and must take these features and logical structures for granted. Indeed, such explanations and accounts are necessarily themselves characterized by and originate from the very same intentional features, mentations and logical structures as those involved in our ordinary everyday knowledge and beliefs about things - fictitious as well as real.[7] If so, *reasons of principles* exist why neither such accounts, nor the intentional features and logical structures they concern, can be explained in terms of or be reduced to something more basic or "scientific" (cf. Chapter 8).

Furthermore, we do not yet have an inkling of what has caused the development of minds in human beings to have these intentional features and logical structures, even less of what has motivated human beings further to develop and make use of these features and logical structures to direct their minds intentionally towards a wide variety of objects in as many different kinds of contexts or "worlds" - encompassing the real as well as the worlds of fiction, myth and religion, and of logic and mathematics. Hence, to commit ourselves to the explanatory paradigms of the natural sciences in advance of a better understanding of these issues would not only be untenable for principled reasons, but also vastly premature. It would be just another highly unfortunate reflection of present days scientistic telescopic vision - a vision which blatantly ignores that there are no differences in principle between the basic conditions which apply for knowledge and description of everyday and scientific matters, and thus ignores the impossibility scientifically to account for or explain these conditions - without presupposing them.

In the next chapter I shall discuss Fodor's Causal Theory of Content (Fodor, 1987). By analyzing this concrete example, it should be clear why no general problem may be formulated and no general account or explanation may be given of how it comes about that our beliefs and their content are intentionally directed towards that which they concern - without presupposing that it is in fact the case. Because of this presupposition, I argue, neither persons, nor their minds or any of the features of their minds may be explained in terms of or be reduced to any of the things that persons may have knowledge and beliefs about.

[7] So if (to go back to the claims by Stich and many others), these features and structures cannot be considered scientifically "kosher", too bad for science.

Chapter 12

Why there still cannot be a causal theory of content

Epistemological consequences of Fodor's solution to the problems of intentionality, meaning and truth

12.1 Introduction

In his now classical paper, *Fodor's Guide to Mental Representation* (1985), he observes that no adequate semantic theory exists for mental representations and predicts that providing such a theory will "be the issue in mental representational theory for the foreseeable future" (Fodor, 1985, p. 96). And so it has. In the introduction to their collection dedicated to this issue, Stich and Warfield (1994, p. 4) contend that "In one guise or another, the project of providing a semantic theory for mental representations (or a "theory of content" as it is often called) has been centre stage in the philosophy of mind for much of the past decade. Many writers now view it as the central project in the philosophy of mind".

Fodor's own solution to the problem, the so called *naturalized, causal* theory of content (first published in Fodor, 1987, and with "refinements" in Fodor, 1994), aims to explain, within a Computational and Representational Theory of Mind, how it comes about that beliefs come to denote and veridically "mean" things in the world - in virtue of causal connections between properties of the things in the world and the mental states that represent them. More specifically, it is Fodor's aim to show that since this causal relation between mental states and things in the world is accountable in purely non-intentional, non-semantic terms, viz. in purely naturalized terms, then the 'intentionality', 'meaning', 'veridicality' and 'truth' of beliefs and mental states are themselves accountable in purely naturalized terms.

However, the "naturalized" causal theory of content proposed by Fodor has been met with a series of objections - ten of which he discusses and rejects in a recent paper (Fodor, 1994).[1] The aim of the discussion in this chapter is two-fold. It is to show, first, that still other reasons exist why Fodor's solution to the problems, which hampers not only his own but any naturalized, causal theory of content, does not work. And to show, secondly, that these *are reasons* of principle why no naturalized theory of intentionality, meaning and truth would work.

But first, a broad historical outline of what we need to know about the Computational and Representational Theory of Mind supported by Fodor if we are to follow his arguments for a causal theory of content.[2] According to the *Computational* view, the human mind may be conceived of as a symbol processing computer in which symbols are stored and manipulated in essentially the same way as happens in a digital computer (i.e. a Turing machine) (Fodor 1980). Just like real computers have input facilities, the mind's computer is equipped with transducers ("sense organs"), providing access to information about the environment which may be stored in the memory. However, as in a real computer the symbols of the mind are manipulated according to rules internal to the mind without any consideration as to what is represented by the symbols, or whether what they represent corresponds to anything outside the mind. Mental processes, says Fodor, are *symbolic* in that they are defined over representations, and *formal* because they apply to representations in virtue of the syntax of the representations. That mental processes are computational, and thus formal, means that it does not matter to such processes whether the input functions are transducers faithfully mirroring the states of the environment, or whether they are merely the output end of a typewriter manipulated by a "Cartesian demon bent on deceiving the machine" (ibid. p. 65). Formal mental processes only have access to the formal properties of the representations of the environment that the senses provide. Hence, "they have no access to the semantic properties of such representations, including the property of being true, of having referents, or indeed, the property of being representations of the environment" (ibid.) What is of importance for psychology is the formal structure of the symbols of mind which, according to Fodor, both determines the content of a person's propositional attitude and distinguish the content of different

[1] Additional objections appear in the Stich and Warfield volume among others by Millikan, Horgan, Stich, and Adams and Aizawa. To these objections must be added the critique by Michael Tye of Fodor's and other "brands of reductive naturalism" (Tye, 1992).

[2] This Computational *cum* Representational Theory of Mind, together with the parts of a causal theory of content proposed by Fodor and discussed in this chapter, is the Background Theory on which Fodor's recent theory of Concepts is developed (Fodor, 1998).

representations. What needs to be known, moreover, are the rules by which the symbols are manipulated and operations carried out.

According to the *Representational* view, propositional attitudes are to be construed as relations that organisms bear to mental representations. Thus construed, the symbols of the mind's computer assume the role of propositions in propositional attitude discourse and serve to represent what one is thinking and having beliefs about; hence, they are referred to as *mental representations*.

Now, since 1975 Fodor has argued that psychology cannot explain rational human behaviour without assuming that human reasoning is couched in mental representations of a language-like kind, and he proposed that the symbolic representation of the mind is a "language of thought". (For reasons which I shall come back to in Chapter 15, he assumes that this "language of thought" or "Mentalese" is innate.) Thus, any explanation of rational behaviour must allow for organisms to consider the consequences of the actions they intend and may think on. This requires, says Fodor,

> "that agents have means for representing their behavior to themselves; indeed, means for representing their behavior as having certain properties and not having others. [...] It is essential to the explanation that the agent intends and believes the behavior he produces to be behavior of a certain kind [...] and not some other kind. Give this up, and one gives up the possibility of explaining the behavior of the agent by reference to his beliefs and preferences." (Fodor 1975, p. 30 - 31)).

Now, according to the Computational theory (cf. above), Fodor also argues that the only thing which can influence our behaviour is what is *syntactically* represented in our minds; what exists outside in the world does not affect our behaviour - unless it affects the internal structures - so, "it's what the agent has in mind that causes his behavior" (Fodor, 1980, p. 67), not what these mental states refer to. So, although he thinks that mental representations allow an agent to consider the consequences of its behaviour, such mental representations need not refer to things in the world that the behaviour concerns, nor for that matter to the actual behaviour carried out and its actual consequences (cf. above). By taking this position, Fodor obviously endorses a *solipsist* theory of mind, in which - so it would seem - the thinking by people about their behaviour and all other things in the world, is totally removed from this action and world. However, for a theory to explain the intentionality of mental states obviously requires that it explains how mental representations connect with objects in the world. A computational theory of the mind, like any other theory positing (merely) intentional objects of mental states, does not do the job (cf. the discussion in the previous chapter).

12.2 Naturalizing intentionality and the content of beliefs

In order to overcome this problem, Fodor, in 1987, proposed a theory which explains how mental representations are about features of the world in virtue of their causal connections between mental states and states of the world. The strategy adopted by Fodor is to show how intentionality, or "aboutness", and the content of beliefs may be naturalized, i.e. reduced to and identified with causal relations between states of the mind and states of the world, either of which are themselves nonintentional and nonsemantic. For, as he says in the introduction to what he calls a *Crude Causal Theory of Content* (CCT), the intentional and semantic are not real properties of things and, by implication, cannot be real properties of mental states either:

> "I suppose that sooner or later the physicists will complete the catalogue they've been compiling of the ultimate and irreducible properties of things. When they do, the likes of spin, charm, and charge will perhaps appear upon their list. But *aboutness* surely won't; intentionality simply doesn't go that deep. It's hard to see, in fact of this consideration, how one can be a Realist about intentionality without also being, to some extent or other, a Reductionist. If the semantic and the intentional are real properties of things, it must be so in virtue of their identity with (or maybe of their supervenience on?) properties that are themselves *neither* intentional *nor* semantic. If aboutness is real, it must be really something else." (Ibid,, p. 97)

The problem of naturalizing aboutness and other intentional categories, and thus in general to explain what it is for a physical system to have intentional states, says Fodor, is to explain in nonintentional terms that propositional attitudes have conditions of semantic evaluation, e.g. that beliefs have *truth conditions*. Now, according to his Representational Theory of Mind one generates such conditions by fixing a context for the tokenings of the "Mentalese" symbols, which jointly constitute a system of mental representations. For example, one fixes "a context for tokenings of the (Mentalese) expression 'this is water' by specifying, among others, that in the context in question the symbol 'water' expresses the property H_2O, or the property XYZ, or whatever" (ibid., p. 98).

According to Fodor's scheme of things, then, the problem of naturalizing intentionality, and thus to "articulate in nonsemantic and nonintentional terms the sufficient conditions for one bit of the world to be about, represent, or be true of another bit", boils down to saying (in nonsemantic and nonintentional terms) what it is "for a primitive symbol of Mentalese to have a certain interpretation in a certain context" (ibid., p. 96). Well, although Fodor admits that he does not have a clue *how* to carry out such a program, he thinks it is plausible that the interpretation of Mentalese symbols "is determined by certain of their causal relations. For example, what makes it the case that (the Mentalese symbol) 'water' expresses the property H_2O is that tokens of that symbol stand in certain causal relations to water samples" (ibid., p. 98).

This in broad outline is how Fodor's causal "story" of content goes in its original version (Meaning and the World Order, 1987), as well as in the latest revisions of it (A Theory of Content, II. The Theory., 1994). In the sections which follow I shall present Fodor's main arguments for the CCT and discuss some of the problems they raise.[3]

So, let us start, as Fodor does,

> "with the most rudimentary sort of example: the case where a predicative expression ('horse', as it might be) is said of, or thought of, an object of predication (a horse, as it might be). Let the Crude Causal Theory of Content be the following: In such case the symbol tokenings denote their causes, and the symbol types express the property whose instantiations *reliably cause* their tokenings. [Italics added] So, in the paradigm case, my utterance of 'horse' says *of* a horse that it *is* one (ibid. p. 99). [...] Suppose, for example, that tokenings of the symbol 'A' are nomologically dependent upon instantiations of the property A; viz. upon A. Then, according to the theory, the tokens of the symbol denote A's (since tokens denote their causes) and they represent them as A's (since symbols express the property whose instantiations cause them to be tokened). But symbol tokenings that represent A's as A's are ipso facto veridical. So it seems that the condition for an 'A' token meaning A is identical to the conditions for such a token to be true." (Fodor, 1987, p. 101).

"Reliable causation", says Fodor, requires either that instances of properties actually cause tokenings of symbols, or that they would *were they to occur*, or both. And he supposes that it is sufficient for such reliable causation that there be a nomological, lawful, relation between certain (high-order) properties of events - in the above case between the property of being an instance of the property horse and the corresponding property of being a tokening of the symbol 'horse'. The intuition that underlies the Crude Causal Theory, says Fodor, "is that *the semantic interpretations of mental symbols are determined by, and only by, such nomological relations*". (ibid., p. 99, italics added).

This immediately prompts the following comments. First, no matter how lawful the relation between purely physical, non-intentional and non-semantic properties of tokens belonging to human symbolic systems and

[3] The presentation is based mainly on Fodor's 1987 text; however, reference will also be made to the 1994 text - partly in order to acknowledge the revisions introduced in this later text to arguments in the original version, and partly to show that, despite these revisions, the consequences of his causal theory of content remain unchanged.

similar physical, non-intentional, and non-semantic events in the world may turn out to be, it does not - by any definition of causation between physical events - explain how it comes about that events in the world may cause other non-intentional, non-semantical events in physical systems to denote or be true about them. To suppose, as Fodor does, that it is sufficient for reliable causation that there be a nomological - lawful - relation between the property of being an instance of the property, say, *horse* and the corresponding property of being a tokening of the symbol 'horse', is merely to state what it takes for relations of denotation and meaning to obtain between tokenings of symbols and instances of properties, i.e. to state that such relations in fact exist. But this statement in itself does not of course amount to explaining how instances of properties of the world cause or "make" non-intentional, nonsemantic "items" or tokens of a physical system denote and *mean* those instances. Nor does it amount to explaining what it is about the purely physical and non-intentional, non-semantical "items" or tokenings belonging to human symbolic systems, which "make" them acquire the function to represent - as opposed to the non-intentional and non-semantic "items" of other physical systems.

Well, nothing has been said so far about the states of the human symbolic system and its non-intentional, non-semantic "items" (i.e. ortographic-phonetic sequences or strings of syntactic items, or the like), which distinguishes this system and its states and items *in principle* from other physical systems and their non-intentional, non-semantic "items". On the contrary, it is the whole point of the naturalization project that the CCT is designed to carry through, that there *is* no difference. So, if what according to the CCT applies for the human symbolic system and its items *in virtue* of this system being a purely physical system, it would seem to apply to *any* physical system and its non-intentional states or "items" that by being causally related to physical states in the world, sufficient conditions exist for these states and "items" to "mean" and be true about what causes them. In other words, it would seem to apply for physical systems in general that they are *intentional* systems, the states and items of which not only "denote what cause them", but also assume truth values in virtue of being causally linked to events in the world (cf. Millikan (1994, p. 244) for a similar argument).

Now, it so happens that it is assumed that a distinction *can* be made between human symbolic systems and other physical systems - a distinction which precisely concerns the fact that, in contrast to those other systems, human symbolic systems are *representational* systems in that their items or tokens may (come to) denote and be true about things in the world to which they are causally linked. However, if these items or tokens are not different in principle from, and thus cannot be distinguished from, the non-intentional, non-semantic items or states of other physical systems *prior* to having been "causally interpreted", it would seem that the items or tokens of the human symbolic system can only be distinguished from items and states of other physical systems *after* they have acquired their denotational and other semantic properties. That is, it would seem that they can only be distinguished from such other items and states by their causally acquired semantic properties. If so, distinguishing syntactic items or tokenings of the human symbolic system from items of other physical systems cannot be done without invoking *semantic* notions. And if what distinguishes different syntactical "items" or forms of a symbolic system from one another in *representationally relevant ways* is precisely their semantic, denotational content, then a distinction between different *syntactical* forms cannot be made without invoking semantic notions. Now, if this means that the very business of establishing that reliable causation obtains between particular symbolic items belonging to the human symbolic system and particular (corresponding) states in the world requires that these items have already been identified as items having the semantic property of denoting and being about the states in the world which cause them, then there is no way that a causal theory of content may avoid circularity.

The circularity in question becomes clear in Fodor's story (in the Causal Theory of Content, II) about what God (who can see what goes on inside peoples' heads) could tell about the content of a mental state just by looking at its actual causal relations. Here are the main points of the story (Fodor, 1994, p. 212 - 14):

> "For simplicity, I assume that what God sees when He looks in your head is a lot of light bulbs, each with a letter on it. A mental-state type is specified by saying which bulbs are on in your head when you are in the state. A token of a mental-state type is a region of space time in which the corresponding array of bulbs is lit. [...] Let's suppose that here is how it looks to Him in a particular case; say in *your* particular case. There is a light bulb marked *c* that regularly goes on when there are cats around; and there is light bulb marked *s* that regularly goes on when there are shoes around. We can assume that the right story is that *c*'s being on means *cat* (i.e., it constitutes your entertaining as a token of the concept CAT), and *s*'s being on means *shoe*". [...]

But God can't really assume this - for

> "it turns our that, though some of the *c* tokenings in your head are caused by cats, it's also true that some of your *c* tokenings are caused by shoes. Moreover, like the cat → *c* causal pattern, the shoe → *c* causal pattern supports counterfactuals; there are circumstances in which shoes cause *c*s *reliably*". [So] "What's in your head doesn't determine content and actual causal relations don't determine content".

However, the story continues,

> "if God has a look at *both* the actual causal relations of your mental state *and* the surrounding space of counterfactual causal relations, *then* He can tell the content of your state".

In the case in question there are two possibilities,

1. "you could have gotten the shoe-caused *c* tokens even if the cat → *c* connections hadn't been in place. However, then shoe-caused *c* tokens can't mean *cat* - for no symbol carries information about cats unless its tokenings are somehow nomically dependent on cats". So, God can only take shoe-caused *c*-tokens to mean *cat* if He is prepared to give up the basic principle that ... "the content of a symbol is *somehow* dependent on the lawful causal relations that its tokens enter into".

Because Fodor reckons that God will not give up this principle, the only relevant possibility left is that

2. "shoes wouldn't cause *c*s if cats didn't cause *c*s". So, "if even "shoe-caused *c*s are causally dependent, on cats - in the sense that if cats didn't cause *c*s then shoes wouldn't either - then it's OK for God to read a *c*-token as meaning *cat* even when it's caused by a shoe".

Two crucial assumptions are made in Fodor's story. First, that God shares with him the "foundational intuition that a symbol means cat in virtue of some sort of reliable causal connections that its token bears to cats" (Fodor, ibid., p. 215) and, second, that God, *in virtue* of this intuition, is capable of sorting out the relevant kind of counterfactual causal relations - i.e. is capable of "picking out semantically relevant causal relations from all the other kinds of causal relations that the tokens of a symbol can enter into" (ibid. p. 181-82, Fodor's italics). However, for God to be able to sort out the semantically relevant causal relations and counterfactuals, and thus to get from (mere) *information* of tokenings to their *meaning*, necessarily requires the assumption, thirdly, that He is able to distinguish not only cs carrying information about cats from ss carrying information of shoes, but also *cats* from *shoes*. Without this assumption, and hence that God is able to distinguish cases in which cs are caused by cats, and thus *veridically* denote *cats*, from cases in which they do not, He could not possibly determine that *c*s actually mean cat (whatever their cause); nor, for that matter, would He ever be facing the problem of whether *c*s mean *cat* as opposed to *cat **or** shoe*. So be it. However, with these assumptions of how God may square the problem of what are the semantically relevant lawful relations between tokenings of symbols and things in the world, from which the meaning of tokenings may be determined, the story being told by Fodor can only be told in a circular way. He just did.

If this is not immediately obvious, it may help to consider that in contrast to the omniscient God in Fodor's story, the human owner of the brain He observes, for example Fodor, only has access to his *mental states*, and thus may only distinguish *c*-tokens carrying information about cats from *s*-tokens carrying information of shoes, but not cats from shoes. Consequently, for him it would not make the least difference whether *c*s were caused by shoes or cats, and thus whether they denote or mean cats or shoes, or both; indeed non-veridical *c*s (i.e. *c*s caused by shoes or any other non-cats) would go quite unnoticed - and so would any other non-veridical tokening. Hence, he would never be in a position to figure out whether his *c*s mean cats or shoes, or both - or something quite different; nor would he be in a position to settle the question, should it ever arise. (This point is further elaborated in the section: *Errors in the Crude Causal Theory*).

We may contend, therefore, that the fact that lawful and semantically relevant causal relations obtain between tokenings and items of the human symbolic system and corresponding properties of states in the world, both means and requires, i.e. by definition of precisely such relations, that they are tokenings of symbols which *denote* and *express* these properties. If so, it is the circular intuition underlying the CCT that the semantic interpretations of mental symbols is determined by, and only by, the relation between the semantic interpretations of mental symbols and instantiations of properties in the world which cause these interpretations.

I am inclined to think that the circularity of causal explanations of "aboutness" and content - be it of tokens of symbols, mental representations, propositions or beliefs - is inevitable. The reason may be stated in general terms thus. Although things in the world and their properties exist independently of beliefs, propositions, and mental representations of them (or of tokenings of symbols which may come to "mentally represent" them), such beliefs, propositions and mental representations do not exist *as such* without or independently referring to the things they are about. Thus, we cannot talk about *what* we believe or assert ('horse out there!') without at the same time talking about *that* which it is about or means (the *horse* out there). But nor can we meaningfully say that our beliefs or propositions are about - let alone are *true* about - horses or any other things, which exist "out there", without presupposing that those horses and things exist as something about which we may have beliefs and put forward propositions, which may be correct or true. Without these presupposition. neither our notion of horses and things existing "out there", nor of propositions, beliefs and mental representations about them, or of "true" and "correct", would have well defined meanings. So, if we cannot talk about the content of beliefs independently of or without referring to and talking about the things they are about, and cannot talk about these things independently of or without referring to the beliefs we may have about them, then we cannot talk about *either* without or independently of talking about the other.

I think we shall have to agree, furthermore, that meaning, "aboutness", truth and veridicality are properties of beliefs, propositions and mental representations, but not of physical things or states. Physical things and states exist, and it may be true - or false - that they exist; but physical things and states cannot be true of false, nor be "truth functional".[4] So, although beliefs, propositions and mental representations about things do not exist without these things, the fact that it is true that things exist in the world necessarily requires that beliefs, propositions or mental states exist that may be true about them. Conversely, no meaning, aboutness or truth exist in any well defined way independently of being the meaning, aboutness or truth of beliefs, propositions or mental representations. That is, the asymmetrical independence of things on beliefs notwithstanding, interdependencies or *necessary* relations do exist between, on the one hand, "beliefs", "propositions", and "mental representations" of things, their "meaning", "aboutness", "truth" and, on the other, "things in the world and their properties". It is because of these interdependencies that the circularity in CCT encountered above is inevitable, indeed is inevitable in any causal account of how "aboutness" and meaning of beliefs, mental representations or propositions come about. (I shall come back to this point in a later section).

There are other problems as well. It would seem that if a semantic interpretation of a mental representation is conditional on its being causally connected to states in the world, then the CCT - like its logical positivist forerunners - is confined to accounting for how mental representations may come to denote or be true about things *only* in cases where it makes sense to talk about lawful, causal relations between mental representations and states in the world. This, of course, leaves out a fair amount of the things that our mental representations or beliefs actually denote and may be true about (among them things which are not and never have been part of states and properties of the world), and ways in which they come thus to denote and be true.[5] In particular, it seems to leave out and be unable to explain how we

[4] Indeed physical things are precisely the kind of things which may be distinguished from beliefs and propositions in that they may be described - and can only be correctly described - in non-semantic, non-intentional terms.

[5] Michael Tye (1992) has pointed out that the component properties of imagined, non-existing things such as golden mountains (i.e. being golden, being a mountain) are "environmentally connected to thinkers" and thus that each of those properties is an external reality which may be environmentally located. In this sense beliefs about imagined things may be said to be "parasitical" to beliefs about existing things. However,

come to use language *reflexively*, i.e. to talk correctly about the meaning and truth of both beliefs and propositions, and thus how we come to use language to talk correctly about beliefs and use of language.

However, let us for the moment with Fodor disregard what seems to become progressively unacceptable consequences of the CCT, and turn to some of the problems he *does* consider. Thus, the CCT says that a symbol expresses a property if it is nomologically necessary not only that *all* but *only* instances of the property causes tokenings of the symbol (ibid. p. 100). However, there are problems with both clauses; not all horses cause 'horse' tokenings (a lot of horses live and die unnoticed), and not only horses cause 'horse' tokenings (sometimes cows cause 'horse' tokenings, e.g. when they are mistaken for horses). In Fodor's view these problems are the worst threats to the CCT; however, he also thinks that they have natural and appealing solutions. Thus, the problem presented by the "only-clause", may be solved if it can be shown that not only veridical representations, but also misrepresentations and nonveridical representations "get into the causal picture". What needs to be shown in this case is why a causal theory does not need to require that "*only* horses do consonant with 'horse' meaning HORSE". The problem of the "all-clause" may be solved by showing that, although a CCT does not require that only horses and all horses cause 'horses', it is nevertheless "in virtue of the causal connections between horses and 'horses' that 'horse' means what it does" (ibid., p. 101). These two problems, as well as the problems to which their proposed solutions give rise, will be presented and discussed one at a time in the following paragraphs.[6]

---

this does not, of course, explain how beliefs about non-existing things are caused, even less how a distinction between beliefs about existing and non-existing things comes about. Thus it does not, as Tye seems to suggest, get us any nearer a solution to the problem of this distinction, which arises from the assumption of mental states being representations. Merely to treat beliefs about non-existing things as "parasitical" to beliefs about existing things won't get us out of the problem with this distinction.

[6] One of the main revision of Fodor's 1994 version of the CCT, is that naturalizing intentionality no longer requires necessary and sufficient conditions, but only the production of naturalistically specifiable sufficient conditions for a physical state to have intentional content. In view of the "relaxation" of the all and only clause just mentioned, however, it could well be argued that this change of condition for the CCT is already inherent in his original version. However, in his Theory of Content, II Fodor introduces conditions for a theory of meaning which are different from the original conditions; these will be presented and their consequences discussed where appropriate.

12.3 Errors in the Crude Causal Theory

The only-clause seems to imply that no misrepresentations or false representations can occur, nor be accounted for by the CCT. However, misrepresentations do occur; it does happen that B's (e.g. cows) sometimes cause token of 'A' ('horses'). However, not to worry. For, since 'A' tokens express the property of being A, then 'A' tokens that are caused by B's, represent B's as A's - and ipso facto are *misrepresentations* of their cause; for the same reason 'A's thus caused are ipso facto nonveridical.

This, says Fodor, is how the story about misrepresentations may be told according to the CCT. However, the story just being told makes for an imperfect causal dependence of 'A's upon A's - since not only A's but also B's are *sufficient* to cause 'A's; thus 'A' tokens may be caused by *either* A's *or* B's. However, if symbols reliably express the properties whose instantiations reliably cause them, then it seems that 'A' expresses the *disjunctive* property of being (A v B) - in which case B-caused 'A' tokenings are veridical (because B's *are* A v B). Thus, it seems that the CCT does not deliver a theory of misinterpretation after all, and thus cannot account for the problem of nonveridical or false representations. In more familiar terms, Fodor states the problem thus,

> "I see a cow which, stupidly, I misidentify. I take it, say, to be a horse. So taking it causes me to effect the tokening of a symbol; viz., I say 'horse'. Here we have all the ingredients of the disjunction problem, [...]. So, on the one hand, we want it to be that my utterance of 'horse' means *horse* in virtue of the causal relation between (some) 'horse' tokenings and horses; and, on the other hand, we *don't* want it to be that my utterance of 'horse' means *cow* in virtue of the causal relation between (some) 'horse' tokenings and cows. But if the causal relations are the same, and if causation makes representations, how can the semantic connections not be the same too?" (ibid., p. 107).[7]

At this point Fodor introduces the observation by Plato that falsehoods are ontologically dependent on truths in a way that truths are not ontologically dependent on falsehood. Thus, the mechanisms that deliver falsehoods are somehow "parasitic" on the ones that deliver truth. This means, according to Fodor, that "you can only have false beliefs about what you can

[7] The disjunction problem is a "robust" problem which arises in one or another guise in any causal theory of content, so Fodor contends; some even argue that the problem that causal theories have with misrepresentations is inherent and ineliminable. I agree (for reasons which should become clear in the following), just as I agree with Fodor that none of the solutions to the problem that have been suggested in the literature (among them Dretske's information transmission solution, and Stampe's and Fodor's own teleological solution), which Fodor discusses in the paper at some length, are satisfactory. (Stampe, 1977; Dretske, 1981; Fodor, 1990.)

have true beliefs about (whereas you can have true beliefs about everything that you can have beliefs about)" (ibid. p.107). To Fodor this "intuition" points the way out of the disjunction problem. For, if there is an asymmetric dependence on falsehood of beliefs upon true ones, then "any theory of error will have to provide for the asymmetric dependence of false tokenings upon true ones" (ibid., p. 110). So, here is Fodor's route out of the disjunction problem,

> "From a semantic point of view, mistakes have to be *accidents*: if cows aren't in the extension of 'horse', then cows being called horses can't be *required* for 'horse' to mean what it does. By contrast, however, if 'horse' didn't mean what it does, being mistaken for a horse wouldn't ever get a cow called 'horse'. Put the two together and we have it that the possibility of saying 'that's a horse' falsely presupposes the existence of a *semantic set-up* for saying it truly, but not vice versa. Put in terms of CCT, and we have it that the fact that cows cause one to say 'horse' depends on the fact that horses do; but the fact that horses cause one to say 'horse' does *not* depend on the fact that cows do.
>
> So, the causal connection between cows and 'horse' tokenings is, as I shall say, *asymmetrically dependent* upon the causal connection between horses and 'horse' tokenings. So now we have a necessary condition for a B-caused 'A' token to be wild: B-caused 'A' tokenings are wild only if they are asymmetrically dependent upon non-B-caused 'A' tokenings. [...] If B-caused 'A' tokenings are wild - if they falsely represent B's as A's - then there *would be* a causal route from A to 'A' even if there *were no* causal route from B's to 'A's; but there would be no causal route from B's to 'A's if there were not causal route from A's to 'A's." (Op.cit. p. 108).

This is how the disjunction problem is solved, according to Fodor. For,

> *"you don't get the asymmetric dependence of B-caused 'A' tokenings on A-caused 'A' tokenings in the case where 'A' means A v B.* [...] In the case of disjunctive predicates, what you get is symmetrical dependence. Asymmetric dependence thus does what it's supposed to do if it's necessary for wildness; viz., it distinguishes wildness from disjunction. I am inclined, however, to think that asymmetric dependence is both necessary *and* sufficient for wildness." (Op.cit. p. 109).

Now, in his Theory of Content, II, the route out of the disjunction problem is a bit different. According to this theory, solving the disjunction problem requires not a theory of *error*, but a theory of *meaning*; if the theory is any good, says Fodor, the conditions for disjunction meaning should fall out as a special case.

Such a theory of meaning is one in which a syntactic item "X" means X, if

(1) "Xs cause 'X's" is a law

(2) Some "X"s are actually caused by Xs

(3) For all Y not = X, if Ys qua Ys actually cause "X"s,

then Ys causing "X"s is asymmetrically dependent on Xs causing "X"s (Fodor, 1994, p. 224).

Used on the horses and cows above, the syntactic item "horse" means *horse* if (1) *horses* lawfully cause 'horse' tokens, and (2) some "horse" are caused by *horses*, and (3) if *cows* cause "horse", then *cows* causing "horse" is asymmetrically dependent on *horses* causing "horse". In other words, among the conditions for the syntactic item "horse" to mean *horse*, is that *cow*-caused "horse" tokens are asymmetrically dependent upon *horse*-caused "horse".[8] This would seem to imply that in addition to the sufficient conditions listed above, "X" means X if

(4) There are some non-X-caused "X"s

that is, "horse" means horse if there are some *cow*-caused "horse".[9] While these conditions of Fodor's theory of meaning may be sufficient to cover both the case of *disjunction meaning* and that of *misrepresentations*, they cannot in themselves solve the problem of *distinguishing* between disjunction meaning and misrepresentations, i.e. errors. Hence, "the theory is residually verificationist in assuming that if cows-on-dark-nights actually do cause "horses", either "horse" means something disjunctive or it is nomologically possible to distinguish horses from cows-on-a-dark-night, (i.e. the residual verificationism is required so that tokens of "horse" that are caused by cows on dark nights can fall under condition 3.)" (Fodor, 1994, p. 210).

[8] Cf. also Fodor p. 182: "Cow" means cow because, as I shall henceforth put it, non-cow-caused "cow" tokens are asymmetrically dependent upon cow-caused "cow" tokens").

[9] In the 1987 version of the causal theory of content, the asymmetrical dependence spelled out in condition 3 was not part of the sufficient conditions for "horse" to mean horse; nor did it figure as a condition that "Horse" means horse if there are some cow-caused "Horses" (4). On the contrary, in his earlier version Fodor was adamant, as we saw in the quotation above (p. 261) that "if cows aren't in the extension of 'horse', then cows being called horses can't be required for 'horse' to mean what it does", and "the fact that horses cause one to say 'horse' does not depend on the fact that cows do".

Now, whether one endorses the earlier or more recent version of Fodor's attempt to solve the disjunction problem, and thus to account for errors and misinterpretations within a causal theory, they incur the same difficulties. For it applies to both versions that, given circumstances in which *sufficient* conditions may obtain for cows to cause 'cow' tokens, i.e. in virtue of cows being there to cause them, then sufficient conditions would *also* obtain for the same cows to cause 'horse' tokenings - if only a causal route exists between 'horse' and *horse*. So why do misrepresentations and errors not happen all the time - or, say, fifty percent of the time? And if the existence of a causal route between 'horse' tokenings and horses is a sufficient condition for cows mistakenly to cause representations of horses, then what prevents the same conditions from being just as sufficient for anything else in the world (e.g. tables and chairs, mountains and trees, elephants and mice) to be misrepresented as *horses* - and vice versa? So, why does it not happen all the time - or just by chance - that chairs and tables, mountains and trees, or elephants and mice are misrepresented as horses - or vice versa? Indeed, how do we ever come to spot the *difference* between things, and thus to distinguish one thing from another, if sufficient conditions obtain *willy-nilly* for a thing to be represented as any other thing?

It is here, I suppose, that Fodor's "residual verificationism" becomes required. However, the verificationism needed is of a kind which makes the story very different from what we are told. It is a story which right from the beginning should have been told this way: A 'horse' token caused by cows is nonveridical or false *first and foremost* because it makes *a false representations of cows*, and not only because a 'horse' token, given the right circumstances, happens to represent something else (and to do so veridically). Indeed, it is only if there is a causal route between 'cow' and cow, and thus only 'cow' veridically represents cows, that it would ever occur to anybody that cow-caused 'horse' tokens are nonveridical, wild or false. Otherwise, misrepresenting cows as horses would go quite unnoticed. Thus, *determining* that B-caused 'A' tokenings are false or wild, does not rely on an asymmetrical dependence of non-B-caused 'A' tokenings *alone*, but *also* on B-caused 'B' tokens - for precisely the reason given earlier by Fodor: "you can only have false beliefs about what you can have true beliefs about" (op.cit).

More importantly, however easily overlooked, 'horse' tokens can only be false about cows (as can other tokens not being part of the extension of 'cow') if cows are the sort of things about which something may be *both* true *and* false. And this takes us right back to Plato's observation about the ontological dependence of falsehoods on truths which - Fodor's interpretation of its asymmetrical nature notwithstanding - must go both ways. Thus, to say "this is a false belief or proposition about this thing" not only requires beliefs or proposition which are true about the thing, but also that *about* this thing something may be *false*; indeed infinitely many beliefs and propositions may be false about it. So, if being a thing, i.e. this particular thing, implies that something is the case or true about it - and that something else is not the case or false about the same thing - then it must somehow figure in the extension of a notion or belief about it *both* what is true *and* what is false about it; and by implication, it must figure in the extension of the mental representation or the tokenings of the symbols by which it is expressed. In this respect part of having true beliefs or notions about a thing is just as much to know that something else is false about the same thing.

The dependence of falsehoods on truth, and vice versa, becomes obvious when we consider that saying, "this is true about p", means that it would be false to deny it. Aristotle even made a logical principle out of it; he called it *the principle of contradiction*, and it states that p and non-p cannot both be true. This is what true means and what it means that a belief or proposition is true about something. And this is what it takes, i.e. a principle of contradiction, to have a concept of truth. Indeed, to say that something is true about a thing only makes sense provided that something else would be false about it, including the contradiction of what is true about it. Without falsehood, "true" would not mean a thing. Hence, we cannot have true beliefs about a thing independently or without believing that something else is false about the same thing.

Now, misrepresentations, false beliefs and propositions may be so in more than one way. A representation, a belief or a proposition about a thing and its properties may be quite simply false - full stop - i.e. without necessarily being true about something else. It is simply not correct, as Fodor claims in support of the CCT that "... any theory of error will have to provide for the asymmetric dependence of false tokenings on true ones" (op.sit). Just imagine the case in which, stupidly, I misrepresent a horse - I take it to be a horse with five legs (although it only has four just like any other normal horse). If the CCT and Fodor's claim were correct, then for a five-legged-horse representation in the actual case to be nonveridical or false, and thus for a 'five-legged-horse' tokening to be wild, would require that a causal route already exists between instances of horses with five legs and their veridical 'five-legged-horse' tokening. (The reader will probably be able to find more appropriate examples).

But is there really any possibility for the CCT (or the Theory of Content, II) to account for beliefs about a thing which are just plainly false, let alone to distinguish this type of falsehood of beliefs from misidentifications, i.e. beliefs or representations which would be correct of other things - given the right kind of circumstances? Well, if I am right, to determine either false beliefs or misidentifications of things *as such* requires that something be not only true, but also false about the thing being misidentified or misrepresented. However, it would seem that the CCT (or the Theory of Content, II) has no credible means of accounting causally for how beliefs of things may quite simply be false about them. If the CCT - or any other causal theory - were to account for the fact that having true beliefs about a thing logically implies knowing also that something else is false about it, it would have to be stipulated that, somehow, lawful causal connections could exist between false beliefs about things and some missing or non-existing properties or feature of the things - i.e. properties or features that the thing does not have. This, I suppose, suffices to make clear why a causal theory of content cannot account for the notion of truth - nor, therefore, for faults or misrepresentations.

Now, it so happens that we can distinguish between these different cases of falsehoods and wildness of beliefs and propositions about things. If we could not, we could not possibly have notions about "the same thing" vs. "different things", nor any distinction between "same" and "different". The reason why such distinctions and their importance does not occur to Fodor, is probably due to the fact that although beliefs and propositional attitudes are thought to occur in a "language like" form, the syntactical forms which constitute the content of mental states are talked about in terms more suitable to *representations* than to *propositions*. However, if having beliefs and putting forward propositions about a thing imply knowing both that something is true and that something else is false about it, then there is a fundamental difference between having beliefs or putting forward propositions about things - and having representations of them.

That a difference exists between, on the one hand, a representation or a symbol and, on the other, a belief or propositional attitude, is already reflected in how we talk about representations and beliefs or propositions. Representations are representations *of* something, while beliefs and propositions are *about* something. When we say of e.g. a map that it represents a country, or that the smoke is a symbol of the fire - or about Fodor's Graycat that its behaviour in front of its bowl represents its wanting some food to be put in its bowl - such symbols or representation *cannot* be false in the sense that a belief or proposition about something can be. They do not and cannot in virtue of being representations of something convey what that something is *not* - only what it is. The smoke, in virtue of being a symbol of the fire, can "stand" for fire in the sense that fire may be reliably inferred from the smoke; but the smoke cannot be wrong about the fire. The same holds for Graycat's behaviour. It may be interpreted as Graycat's wanting food in his bowl - and granted that interpretation, the pattern of behaviour displayed by Graycat cannot be false about what the behaviour signifies. (Even if Graycat were able to fake his wishes, the behaviour he displays would still signify: "food in the bowl, please".)

Likewise, a representation, say a map of a country, does qua the rules of cartography "stand" for something about the features being represented, such as the relative size of towns, the distances between towns, the length and directions of rivers etc. Granted that we have used the production rules of cartography correctly, the map is a correct cartographical representations of the country. If we have not, the map cannot in any sensible way be said to be a representations of the country in question, albeit an *incorrect* one. The map cannot be mistaken or false about the country it represents in the sense that beliefs or propositions about the country may be; i.e. it cannot represent features of the country, which it does not have, nor represent those which it has incorrectly, and *still* be a representation of it. There is no way that e.g. a map of Denmark can be a mistaken representation of Sweden; nor does it ever happen that sometimes a map of Denmark appears as a result of our drawing maps of Sweden.

Well, propositions and symbols of a linguistic kind may be used as representations. But to entertain a propositional attitude towards a thing, i.e. to assert something about it which is true - and, by implication, that something else may be false about it - is not just to have a representations of it. The propositional attitude being expressible in an assertion, say, "this cup is red", implies at the same time a whole range of other assertion about the cup, which would be false, including, for example, assertions to the effect that the cup is green, blue, violet etc. However, representations and symbols do not have false implications or meanings about that which they represent or symbolize; indeed they do not have any implications at all *unless* someone sees them *as* representations, i.e. sees them as representing something else. That is, unless the rules of interpreting the representation are *known* to that someone, who therefore knows the whole story about that "something else" which is represented as well. This story is necessarily told in a language which is *richer* than the language and signs of the representation; and, necessarily, it is the (very same) language in which the story is told about the relation between the symbols or signs of representations and what they represent. Now, we may of course say about a representation that it is a veridical representation of what it represents. However, that it may be so relies on the truth of the determination and description of the very things on which the construction of the representations of them are based.[10]

[10] But what, then, about the so called "internal mental" representations we may have of things, i.e. in the form of perceptions or of memory images of things we have perceived? Well, I think we shall have to agree, firstly, that we cannot talk about our perception of things without at the same time talking about what we perceive, i.e. the things, and thus cannot talk about a perceptual representation of things without taking about it as that which we perceive. Secondly, I do not deny that we may have memory images of things which we have perceived; but to say that they are representations of things in our mind makes no more sense than saying that the (original) perceptions of the things on which they are

I contend, therefore, that you cannot develop beliefs or propositions about things in the world, nor a language in which both such beliefs are expressed and things described, from *representations* (be they "language like" or not). In particular, I contend that beliefs about things being particular things of which something may be both true and false - as opposed to being about some other things, of which something else may be true and false - have nothing in common with the kind of "contextualized", "interpreted" representations of things which ensue from their causal link with the things being represented. Moreover, specifying the nature of the "interpretations" and "contextualizations" of representations both requires and relies on a language in which such specifications may be expressed in terms which are logically related to assertions about the things being represented. This language can only be *natural language* being as intentional, semantical, and contextual as any language could be. This will be all the more clear when, in the sections which follow, we go to Fodor's solution to the problems arising from the CCT's "all-clause".

First, a step-by-step account will be given of the arguments on which this solution is built.

12.4 The CCT's psychophysical explanation of content

The idea that a Mentalese symbol, say, 'horse' express the property of being a horse only if all instantiations of the property cause the symbol to be tokened in the "belief box", says Fodor, is preposterous on the face of it. What is needed for the CCT to hold, he contends, is (merely) a plausible sufficient condition for 'A' to express A such that, in at least *some* cases, 'A's express A because that condition is satisfied (ibid., p. 111). Consequently, the viability of the CCT only depends on its being

> "able to specify (in naturalistic vocabulary, hence in nonsemantic and nonintentional vocabulary) circumstances such that (a) in those circumstances, 'horse's covary with horses; i.e., instantiations of horse would cause 'horse' to be tokened in my belief box (i.e., would cause me to believe *Here's a horse*!) were the circumstances to obtain; and (b) 'horse' expresses the property *horse* (in my idiolect of Mentalese) in virtue of the truth of (a)."

But what are those circumstances? The problems of the "all-clause" thus specified are still very deep; indeed, says Fodor, "it's here that the plausibility of a serious causal theory of content will be tested". (ibid., p. 111). The "sketch" he offers for their solution is rather long, however, the points being

---

based, are but representations of those things which may be referred to and talked about independently of or without at the same time referring to and talking about the things themselves.

made are few and easily put into a much shorter version. Fodor starts his sketch by noticing that,

> "There are circumstances in which beliefs about observables do seem to force themselves upon one in something like the way that the Crude Causal theory requires. For example: Paint a wall red, turn the lights up, point your face towards the wall, and open your eyes. The thought 'red there' will occur to you. Just try and see if it doesn't." (Op.cit. p. 112)

Such circumstances are *psychophysically specifiable*, says Fodor, indeed psychophysics is precisely in the business of telling us - in nonsemantical, nonintentional vocabulary - how much of the wall has to be painted red, how red it has to be painted, and how close to the wall one has to be, and how bright the lights have to be, etc., in order for you to think 'red' if your eyes are pointed toward the wall and your visual system is intact. And if psychophysics can specify the circumstances in which red instantiations control 'red' tokenings, Fodor sees no reason why a suitably extended psychophysics should not be able to do just as much, say, for HORSE, PROTON, or CLOCKTOWER. Of course there are instantiations of red, horse, clocktower and proton that do not affect the contents of one's belief box. But this is, according to Fodor, only because one doesn't occupy a *psychophysically optimal viewpoint* with respect to those instantiations; what these are can be specified by psychophysics. Thus, according to Fodor, psychophysics provides exactly what a naturalized semantics needs, i.e. circumstances in which instances of a property are guaranteed to cause tokens of the symbol that expresses it. Indeed,

> "The *psychophysical cases* are close to ideal from the point of view of a causal theory; for they're the ones for which it appears most clearly possible to enumerate the conditions in which (reliable, causal) *correlation makes content*" (ibid., p. 114, italics added)."

But alas, it doesn't work. For although psychophysics can guarantee circumstances in which one will see a horse and, presumably, have the appropriate horse experience, it cannot guarantee the *intentional content* of the mental state that one is in in those circumstances, i.e. it cannot guarantee that one sees the horse as a horse; for, seeing a horse as a horse requires applying the *concept* HORSE to what one sees. However, it is perfectly possible even for an intact observer not to have a concept of HORSE, and therefore it is perfectly possible for someone to be in psychophysically optimal relation to a horse, and yet the thought "here is a horse!" does not occur to him or her.

In this respect HORSE and PROTON differ from RED. What makes RED a special psychophysical concept, says Fodor, is that the difference between merely seeing something red and succeeding in seeing it as red vanishes when the observer's point of view is psychophysically optimal. One cannot, so Fodor claims, see something red under psychophysical optimal viewing conditions and not see it as red. Hence, applying concepts such as RED perceptually are independent of, i.e. not mediated by, the perceiver's background of cognitive commitment (ibid. p. 117).

But unfortunately, it only applies to a very small set of mental representations that the distinction between *seeing* and *seeing as* vanishes in psychophysically optimal circumstances; such psychophysical concepts cannot provide a reduction base for the rest of Mentalese. But nor can the psychophysical model be extended in any obvious way to concepts which are inferentially mediated, such as HORSE or PROTONS.

End of the CCT story? Not at all. For, according to Fodor, there might be a less obvious way in which the psychophysical model *can* be extended to such concepts as well. So, here is the continuation of Fodor's story: although horse isn't a psychophysical property, instantiations of *horse* very often are causally responsible for instantiations of what *are* psychophysical properties. It is because "Dobbin" is a horse, says Fodor, that it has a "horsy look" - and it is plausible that that look reduces to having some or other bundle of psychophysical properties - although *being a horse* does not.[11]

So, what is needed is a kind of connection between horses (or protons) in the world and 'horses' (or 'protons') in the belief box, which is such that because this connection obtains, 'horse' means horse (and 'proton' means *proton*). All that the CCT requires is that this connection be a causally reliable covariation. Now, although such covariance cannot be mediated by brute psychophysical law - because neither HORSE nor PROTONS are psychophysical concepts - there is no reason, says Fodor, "why instantiations of psychophysical properties should not be *links in a causal chain* that reliably connects horses and protons with 'horses' and 'protons' respectively" (ibid., p. 119). Thus, for observational concepts such as HORSE it applies that,

> "'Horse' means *horse* if 'horse' tokenings are reliably caused by tokenings of psychophysical concepts that are in turn caused by instantiations of psychophysical properties for which instantiations of *horse* are in fact causally responsible. The causal chain runs from horses in the world to horsy looks in the world to psychophysical concepts in the belief box to 'horse' in the belief box. 'Horse' means *horse* because that chain is reliable. [...] [Hence, in the case of observational concepts], the nomologically sufficient and semantically relevant conditions for their tokenings are specifiable 'purely externally'; viz. purely psycho-

[11] It's a bit different with protons which have no observable looks like horses have - but it is possible to construct environments in which instantiations of proton do have characteristic psychophysical consequences as well, i.e. experimental environments which involve deployment of "instruments of observation".

physically. And [...] all the other semantically relevant symbol/world linkages run via the tokening of observation concepts." (ibid., p. 122).

As far as Fodor is concerned the CCT story has now completed its job of presenting the sufficient conditions for instantiations of *horses* and *protons* to cause 'horse' and 'proton' to be tokened in one's belief box, and hence cause one to believe "Here is a horse!" or "There is a proton!" respectively. What has been shown, more specifically, is that to explain in causal, naturalized terms that 'horse' means and is true about horses is just a matter of "enumerating" in nonsemantical and nonintentional vocabulary the conditions "in which (reliable, causal) *correlation makes content*" (op.cit).

In the section which follows I shall argue why the CCT has not explained or shown anything of the kind, indeed why *no* naturalized causal theory can explain how the content and intentionality of beliefs come about. Such explanations are invariably attempts to explain how the meaning, "aboutness" and truth of our beliefs of things may be reduced to (some of) the features of the things that we may have beliefs about. So, more generally, I shall argue why none of our beliefs about anything can be reduced to any of the things or their properties, which we may have beliefs about.

12.5 Why a causal theory of the intentionality and content of beliefs does not work

I shall begin my arguments against the possibility of a causal theory of content and intentionality by pointing out some of the inconsistencies and glosses in Fodor's account of his Causal Theory of Content.

In his specifications of the conditions which must be fulfilled for the CCT to work, Fodor has helped himself, as he puts it, to the notion of 'one event being the cause of another', and to the notion of 'an intact organism'. And as the attentive reader has noticed, he has also helped himself to the notion of 'optimal circumstances' (even optimal points of view) in which one event may cause another to occur in an intact organism. With these helpings the story being told by the CCT and its causal chain of events begins to smack more of a story about "teleology" (or just plain circularity) than a story about causation. For, an *optimal* (psychophysical) condition is one in which one event, say, a horse, cause another event, i.e. 'horse', to be tokened in the belief box of an *intact* organism, whilst conversely, a perceptually *intact* organism is one in which, under *optimal* psychophysical condition, horses causes 'horse' to be tokened in the belief box.

Equally noteworthy, Fodor has helped himself to the notion of 'psychophysical concepts', and even to the notion of 'psychophysical properties', both of which are introduced without further ado - or at least with very little attempt to illuminate the reader as to their content, or to what they refer. All we have been told is that RED is a *psychophysical concept*, because the difference between merely seeing something red and seeing it *as* red vanishes when the observer's point of view is psychophysically optimal. And about *psychophysical properties* we have been told that, in the case of horse, its having a "horsy look" may (plausibly) be reduced to having some or other psychophysical properties - for which instantiations of horse are causally responsible.

However, this does not make us any wiser as to what, more precisely, a psychophysical property is. On the one hand it seems to be a property that a thing, e.g. a horse, may have; just as it has 4 legs, a body, a head and a tail, and so forth, it also have the property "horsy look", which, unlike its legs, tails, head and other physical properties, is not only physical, but also partly psychological. But if this "horsy look" is really a property of a horse on a par with other properties constituting a horse, it cannot be psychological, not even partly. For, who has ever heard about properties of physical things in the world being partly psychological?

On the other hand, we are also being told that this psychophysical 'horsy look' property is something for which instantiations of horse is causally responsible. So, a "horsy look" cannot really be a property of a horse either; for, who has ever heard about properties of things being caused by the things which have them? For example of the legs, head, body, and so forth, of a horse being caused by the horse, or of instantiations of water causing H_2O?

However, if what is being referred to as *psychophysical properties* (e.g. "horsy looks") are not the properties of things (*in casu* of horses), but something else, which is caused by instantiations of things (i.e. of horses); something, moreover, which is partly physical, partly psychological, then I would simply not know what to look for, nor where to go looking for it. Going to psychophysics itself will definitely not help us solve the puzzle of what may possibly be meant by psychophysical *properties*, nor what the notion psychophysical *concepts* refers to. For such concepts or properties figure nowhere in psychophysics or its vocabulary. What we do have in accounts of psychophysical investigations are descriptions of experimental settings in which things are presented under varying conditions, and in which measurements are carried out of physically specifiable features of the things and conditions. This is the *physical* part of the psycho-physical investigation. And we have observers who inform the psychophysicist what they experience under these varying conditions; this is the *psychological* part of a psycho-physical investigation. The result of such an investigation may be that a (lawful) covariance is found between, on the one hand, the physically specifiable features of the circumstances and things presented to the observers and, on the other, their experience of them.

Now, it is just conceivable, but only just, that a psychophysicist, when presenting a horse to an observer under various experimental conditions (e.g. varying conditions of illumination, or conditions in which various part of the horse are visible whilst others are not) may be able to detect certain features or properties of a horse which are sufficient to convey to an observer a "horsy look". Thus, the observer, being in those circumstances and exposed to those properties, may say: "what I see has a "horsy look", or more likely, "looks like a horse" (in contrast, say, to a cow, an elephant, or a chair). If so, it may just conceivably make sense to say that these properties are the properties of a horse which, when presented in the right kind of circumstances, are sufficient for a horse to have this "horsy look" for someone observing it. Our psychophysicist, having specified (in appropriate physical, naturalized terms, and all the rest) these properties of the horse and conditions in question, may then formulate a psychophysical law about the covariance between specifiable physical features of a horse and an observer's experience of "horsy look".

Very well. But granted that the properties thus psychophysically specified are the properties of a horse, which to an observer make it have a 'horsy look', how then could this 'horsy look' figure as a "link in a causal chain which reliably connects horses with 'horse'? (op.cit.). For, however else this psychophysically specifiable horsy look gets into the observer's belief box (if indeed it does), it should not, according to the first part of the CCT story, cause 'horse' tokens in his belief box. Thus, if we stick to this first part according to which A's causes 'A', and thus causes 'A' to denote its cause, then the properties unveiled, specified and described in purely naturalized terms by our psychophysicist and presented to an observer, should cause 'horsy look' tokenings *only* in the observer's belief box, but not 'horse' tokenings. If it did, we would have a case of misrepresentation on our hand. But nor would it help much to say that such a misrepresentation of *horsy look* depends on an already existing causal route from *horse* to 'horse' tokenings (as the first part of the story has it). For, according to the psychophysical part of the CCT story with which we are now concerned, the causal route goes in exactly the opposite direction, i.e. from 'horsy look' to 'horse'. However, this does not really make sense either; for, how could anyone have an experience of 'horsy look', who did not first have a belief or notion of a horse? And so on.

Now, there is of course a third alternative, namely that the term 'psychophysical properties' is introduced to indicate that, according to the CCT-*cum*-Representational Theory of Mind, a psychophysicist - just like the rest of us - only has access to *mental representations* of horses; in which case his specifications of the physical features of a horse that account for its "horsy look", are merely specifications of features of a *mental representation* of a horse (and ipso facto of something psychological), and not of the real, tangible horse "out there". But that would clearly not make sense either, for in that case the psychophysically specifiable features themselves would also have to be merely mental representations of physically specified features of mental representations of a horse ... and so forth in an infinite regression of mental representations. And this would leave us with a quite different story, i.e. one about how some psychological or mental events cause other psychological or mental events. Thus, it would leave us with a story which, for reasons of principle, would forever defy naturalization or physical reduction. For, notions such as "physical things", "out there", or "real horses" would have no meaning at all; if so, nor would notions about mental representations of such things.

So, in order to keep the right kind of causal story going, i.e. one which does not prevent its being naturalized even before it has got started, it would have to be assumed that we do indeed have access to the things "out there" in physical, material reality - and not only to mental representations of them. However, so to assume would obviously render both the RTM and a causal theory of content not only totally superfluous, but also meaningless. For who, in his right mind, would suggest that our cognition of things in the world is a matter of merely having *mental representations* of things caused by some features of things, if telling the story which explains all that presupposes that we do indeed have knowledge about and may correctly describe both those features of the things and the things to which they belong?

Well, we knew it all along. When we see a horse out there and go to stroke it, we not only stroke a mental representation of a horse, nor are we having a mental representation of stroking a mental representations of a horse. And it simply does not make the least bit of sense to say that some particular sample of partly "naturalized", partly psychological features of a horse via some inferential mechanisms cause beliefs about a tangible horse to occur in our minds, let alone cause a tangible horse to emerge out there in the world. For again, how could anyone presented with samples of psychophysically specifiable "horsy look" properties make the inference, "horse out there!", who did not first have a notion of the *whole* horse, and thus knew and could determine to what these features apply? In particular, how could a psychophysisist possibly claim that the physically specifiable features of "horsy look" are properties of *horses*, and thus that his description of these features applies to horses, without presupposing ordinary everyday determinations and concepts of horses?

The above attempt to untangle the notions of psychophysical concepts and properties to which Fodor has helped himself in order to get his causal story of content going, already points to some of *the reasons of principle* why his causal theory - as well any other naturalized theory of content and "aboutness" of beliefs and propositions - does not work.

Let me start the further analysis of these reason by first mentioning some facts about beliefs and propositions and what they are about, which are so elementary as to be almost self-evident. In material reality physical things exist, such as horses, protons, wavelengths and the like, which - by definition of physical things - are *nonintentional*, *nonsemantical*, and all the rest. In contrast, beliefs and proposition about such things *are* intentional and semantical - by definition of beliefs and propositions; that is, they have a meaning, refer to, and may therefore be true or false about such things. This applies of course whether the beliefs and propositions about the things and their nonintentional properties are expressed in ordinary everyday or in so called "naturalized" scientific terms. To say otherwise would be nonsense.

Well, Fodor does not, of course, maintain that descriptions, beliefs and propositions are non-intentional and non-semantical just in case they are about non-intentional and non-semantical things. However, he *does* maintain that describing in "non-intentional, non-semantical, viz. naturalized terms", the circumstances in which instantiations of things in the world covary with beliefs about them, amounts to explaining, in purely naturalized terms, how one event causes another to be intentional and true about what causes it - thereby explaining how intentionality itself reduces to something purely "naturalized". Indeed, this is the whole point of the naturalization and reduction project of the CCT (cf. Fodor's remarks quoted on p. 252).

But, as already pointed out at the beginning of this chapter, the implications of the notion of 'one event causing another to be about it', to which Fodor has helped himself, figure nowhere in the definition of causation in the natural sciences. Thus, no account of the causal relations which exist between physical things, and no definition of causation applying to events accounted for within those sciences, can be brought to bear on the claim that "enumerating" the condition in which psychophysically specifiable features of things in the world covary or correlate with beliefs about these things, amounts to specifying *"reliable, causal correlations" that "make content"* (cf. ibid. p. 114).

Well, the aim of the CCT, so we were told at the outset, was to present a naturalized, causal theory which precisely explains how "one bit of the world may be about another bit". However, at the end of his account, Fodor seems to think that it is not part of the solution to the naturalization problem for semantics to explain how correlations "make" content - it is enough that they do, or to claim that they do. Thus, in a footnote Fodor says,

> I take it [...] that it's OK for an astronomer to say 'a meteor was the cause of the Great Siberian Crater', knowing that he means more by this than 'no meteor, no

crater'. Well, if he can help himself to cause without further analysis, so can I. I propose to.

> It is *simply unreasonable* that a solution to the naturalization problem for semantics should also provide an account of causal explanation. Semantics is respectable if it can make do with the same metaphysical apparatus that the rest of the empirical sciences require; it doesn't *also* have to incorporate a theory of that apparatus. (Op.cit. p. 165)

But is it really the case that the kind of naturalized semantics that Fodor proposes, can make do with the same metaphysical apparatus that the rest of the empirical sciences require? Is it not precisely because of the *anomalies*, the incomplete causal dependencies which "make" veridical as well as non-veridical representations, that his attempt at naturalizing semantics has shown itself *not* to be able to make do with that metaphysical apparatus of the natural sciences and *their* notion of causation? And is it not precisely because terms such as "veridicality" and "nonveridicality", "truth", "meaning" and "aboutness" is not part of the vocabulary used in expressions about causal relation between physical things within the natural sciences that Fodor's or any other semantics cannot make do with the metaphysical apparatus that they require?

Needless to say, we accept an astronomer's story about how the Great Siberian Crater was caused by a meteor, and thus that a causal relation may obtain between the crater and a meteor, because a *more* general explanation exists within his science about a reliable, causal correlation between a moving object (of a certain mass and velocity), and the impact it makes (of a certain size) on other objects (of a certain mass, etc)., when it hits them. In the case of the story being told by the CCT, however, we simply *lack* such a further story. To repeat, what we lack is precisely a story about how nonintentional, nonsemantic, physical things and events in the world may cause other nonintentional, nonsemantic, physical events in the world to become truth functional and to "denote", "express" and "be about". In short, we lack a notion of 'causation' and of being 'the cause of' by which to account for how a non-intentional event may "make" other events be about it.[12]

But there are other compelling, i.e. *logical*, reasons why a causal story about the content of beliefs cannot be told, let alone be substantiated empirically, by any alliance between semantics and psychophysics. These reasons have to do with the fact pointed out earlier that psychophysical specifications

[12] As it now stands the CCT might be said to explain too much. Thus, as pointed out earlier, if the notion 'being the cause' of in the other empirical sciences did refer to the kind of causality that Fodor wants to help himself to, then the CCT might as well explain how the causal relation or correlations between meteors and craters on the earth causes these craters to represent and be true about meteors.

of the physical features or properties of things in the world, for example of horses, presupposes and relies on ordinary everyday descriptions and determinations of horses, and thus on ordinary everyday knowledge and concepts of horses. Without such determinations and descriptions, a psychophysicist could not claim that his observations and measurements of the physical features in question apply to horses. This means, firstly, that the description of the psychophysically specifiable physical features of horses *rests on an abstraction* in the sense that it rests on an another description, i.e. an ordinary everyday description of horses. Therefore, in the causal chain running from horse in the world to 'horse' in the belief box,

> horse → physically specifiable horsy look → 'horsy look' in the belief box → 'horse' in the belief box[13]

the determination of horse which figures as the initial condition is necessarily identical with the description of 'horse', which figures at the end of the causal chain. It better be; for, if the psychophysicist did not presuppose, secondly, that his initial description of horse were identical with the description of 'horse' in his and his subject's belief boxes, the conditions for carrying out the psychophysical investigations would not exist. Indeed, if we did not presuppose *in general* that our descriptions of things in the world - be they ordinary everyday descriptions of the things or descriptions of their physically specifiable features - are identical with the descriptions of our *belief* about these things and features, then the conditions for carrying out investigations of anything in the world would not exist. For these reason the causal chain above cannot be a chain; on the contrary, it has be a circle - on all levels of description.

This could also be said as it was said earlier in this chapter, as well as in previous chapters: logically, we cannot talk about the *content* of a belief ('horse out there!', or 'physically specifiable features of the horse out there') without referring to *that* about which it is believed (i.e. the horse out there, or its physically specifiable features). Although we can certainly distinguish between having beliefs about things and the things themselves, and distinguish between beliefs of things and the things themselves, no beliefs about things exist which may have a meaning and be true or false independently of referring to the things that the belief is about. Otherwise, the notions of 'beliefs', their 'meaning' and 'truth', would be ill-defined. Conversely, without presupposing that our beliefs about things in the world concern and are about precisely the things which exist in the world, and presupposing, therefore, that things in the world exist as things about which we may have beliefs which are true, our notion of both 'things' and 'exist' would be ill-defined. That is, for any of these notions to be well-defined, we shall have to

[13] Op.cit, p. 122.

presuppose that a *necessary* relation exists between beliefs and what they are about. However, because of this presupposed necessary relation, any attempt to give causal accounts of the content and "aboutness" or our beliefs would be circular.

We may contend, therefore, that the determination of the physical, non-intentional, nonsemantic features and properties of things which, according to Fodor, cause beliefs about them, relies on descriptions within the same semantic, conceptual apparatus that we use to describe both our beliefs about these properties and the properties themselves. This applies irrespective of the level of description.

And this leads us right to the reasons why a naturalized reduction of content, "aboutness", and truth does not work: The whole naturalization and reduction project relies on the possibility of ignoring, first, that the determination of the physically specifiable features of things which (are claimed to) cause beliefs about these things, presupposes and is based on descriptions and beliefs about these things which, therefore, cannot at the same time be reduced to something purely physical. Secondly, it relies on the fact (if indeed it is a fact) that ... "the nomological sufficient and semantically relevant conditions for the tokenings of observations concepts are specifiable 'purely externally'; viz. purely psychophysically". However, this cannot possibly mean that identifying, observing and specifying these conditions does not involve *concepts*, nor terms and descriptions which are semantical and intentional. So to say would be to say, in effect, that the business of determining and specifying the conditions giving rise to the content of observational concepts, does not involve *cognition*; or it would be to say that beliefs and description of observable non-intentional, non-semantic physical features, are themselves not intentional, nor semantic - both of which is nonsensical. So, if we do not want to talk nonsense, we had better assume that which reductionist, causal theories of content will have to ignore in order to get their causal story going: the observation and determination of scientifically specifiable nonintentional features and properties of things *does* involve the cognition of persons being in intentional states, as well as descriptions of these things and features, which are intentional and semantic. Hence, no accounts can be given for how 'beliefs', 'knowledge' and 'propositions' about things in the world are derivable from or reducible to something purely non-intentional and non-referential, without using terms and propositions and having beliefs about this non-intentional and non-referential something, which *are* intentional. That is to say, beliefs about this non-intentional and non-referential something, which are described in the very same intentional, semantic vocabulary in which the *propositions* and *statements* about it is put forward.

It may be easier to appreciate the impossibility of causal theories of content if we consider that even if accounts could be given of how purely naturalized, non-intentional and non-semantic features of some particular things in the world cause beliefs about those things, such accounts could not be about how *all* beliefs and propositions of states in the world can be reduced to something nonintentional. Hence, the CCT and other causal theories can only solve the naturalization problem if it is ignored that the business, say, of psychophysically specifying features and properties of things requires that this business itself be specifiable in (some other) psychophysical terms. That is, if it is ignored that whenever we manage to explain how it comes about that, e.g. horses in the world, via some specifiable psychophysical circumstances in the world and physical states in our brains, give rise to the cognition of horses, we are inevitably left with yet another problem; namely, the problem of explaining how still other psychophysically specifiable properties of the world and states in our brains may give rise to the cognition of the psychophysical circumstances and states in our brains, giving rise to cognition of horses - and so on in an infinite regression of specifying psychophysical conditions for cognizing psychophysical conditions for cognizing yet others ...

Where does this leave us? Well since, generally, there is no conceivable way of talking correctly about and meaningfully accounting for that which is supposed to "cause" the notions of truth, content and aboutness, and to which these notions are supposed to be reducible, *without* using these notions, and independently of implying the existence of these notions, I think we are left with having to accept that these notions are not thus reducible. On the contrary, they are notions which have to be taken for granted, must be "primitives" in any theory and accounts of cognition and language - and that which cognition and language is about.

Hence, the general result of the above analysis seems to be that

1. We cannot reduce any of our beliefs about things, nor that of having beliefs or knowledge which may refer to and be true – or false - about things, to any of the things or features of the things of which we may have knowledge and beliefs.

2. Nor may we derive the notions of truth, content and "aboutness" from any of the features of the things, about which we may have knowledge, and of which we may put forward propositions which may be true or false, i.e. from features which themselves do not imply the existence of these notions.

This, I take it, must be a matter of principle, a fundamental condition for our cognition, language and use of language. We shall have to agree with Fodor's supposition that the notions of intentionality and semanticality will not figure in the completed catalogue compiled by physicists of the ultimate and irreducible properties of things; indeed, *reasons of principle* exist against it (cf. also Chapter 8). However, these are reasons which at the same time preclude a reconciliation between the natural sciences and cognitive science of the kind suggested by Fodor, as well as in general by any other kind of reductionism.

12.6 Conclusion

In the summary of arguments for his Causal Theory of Content, Fodor contends that the theory provides a sufficient condition for one part of the world to be semantically related to another part; it does so in nonintentional, nonsemantical, nonteleological, and in general, non-questions-begging vocabulary. However, he admits that the theory relies on the claim that, whatever may be the right unpacking of the notions of 'one event being the cause of another' and 'intact organism', it won't smuggle in intentional/semantic notions. On the other hand, Fodor continues, should it turn out that

> "INTACT ORGANISM and THE CAUSE OF AN EVENT are indeed covertly intentional/semantic, then it looks as though belief/desire psychology isn't, after all, the only science whose practice presupposes that intentional/semantic categories are metaphysically kosher. That the organism is (relevantly) intact is part of the background idealizations of practically all biological theorizing; and (so I believe) we need to distinguish between the cause of an event and its causally necessary conditions *whenever* we do the kind of science that's concerned with the explanation of particular happenings (cf. the formation of the great Siberian Crater, the extinction of the dinosaur, and the like).
>
> So if INTACT ORGANISM and THE CAUSE are indeed intentional/semantic, then there is nothing special about belief/desire psychology after all. The availability of intentional apparatus would then be quite widely presupposed in the special sciences, so its deployment by psychologists for the explanation of behaviour would require no special justification. In which case, I've just wasted a lot of time that I could have put in sailing." (ibid. p. 127).

Well, to this I argue, firstly, that causal reductionist theories about the content and intentionality of mental states of the kind proposed by Fodor and others, do indeed require that notions be "smuggled in" of BEING THE CAUSE OF which imply that causation itself be something intentional/semantic. Secondly, the practice of any science, natural or special, necessarily requires "descriptive apparatuses" or "systems" which are intentional/semantic, that is to say apparatuses or systems about which it is presupposed and taken for granted that they allow true propositions to be put forward about that which is observed. Presupposed and taken for granted, furthermore, is that the things being observed exist as things about which we may have knowledge and put forward propositions which are true. Things, furthermore, about which more knowledge may be acquired by observing them - under various new or different circumstances, about which we may also acquire knowledge - and which may be described in "descriptive apparatuses" or "systems" that allow propositions to be put forward about them, which are true. Without these presuppositions, no further investigations could sensibly be made of anything.

That in this sense the descriptive apparatuses or systems of any science are "intentional/semantic" does not, of course, make *that* which is described, known or believed within those sciences intentional/semantic - covertly or otherwise. Thus, physical things and their causal relations studied by the natural sciences are nonintentional through and through; they do not have "contents", nor may they be true or false in the sense that propositions and beliefs about them may be - by definition of physical things and their causal relations. Such things and relations which the physical world is full of exist, and they exist independently of whether or not we describe or have knowledge about them, and independently of the propositions being put forward about them. In contrast, descriptions, beliefs and propositions about things, which exist may denote, have a content and be true or false about these things, only exist *as such* in virtue of their *necessary* referential relation to those things. These are the crucial dependencies and the asymmetry which have to be presupposed between things in the world and propositions and beliefs about them. Indeed, everything we say, not only about material reality, but also about language and knowledge of material reality, as well as about the intentionality and truth of propositions and knowledge, hinges on our understanding of the importance of these presupposed dependencies and this asymmetry.

However, these dependencies and this asymmetry are ignored in any attempts to develop causal, naturalized theories of content and intentionality. The inevitable consequences of such theories are their futile attempts to navigate between the The Scylla and Charybdis of these alternative assumptions: *Either* the intentionality and truth of beliefs and propositions may be reduced to something being as physical as the physical things and their causal relations, which they are about - in which case beliefs and propositions - just like those things and relations - are not really the sort of thing which may denote or be true or false. *Or* physical things and their causal relations may denote and be true and false in the *same* sense as beliefs and propositions about them may be. Granted either assumption, neither our concepts of 'physical things' and their 'causal relations', nor our concepts of 'knowledge', 'proposition' or 'truth' could have any well-defined meanings.

Therefore, in conclusion, this is what needs to be said. Just like any other sciences, the practice of psychology requires a descriptive apparatus or system of which it is presupposed that descriptions and propositions may be put forward that refer to and are about, and thus may be true or false about the things being observed. Hence, it is part of the background presuppositions for psychology - as it is for any of the special and natural sciences - that beliefs and knowledge, as well as descriptions and propositions about the objects being observed, whatever their nature, are intentional and semantic. However, what crucially distinguishes the objects of a psychology, studying beliefs, knowledge, action and their causes, from the objects of the natural sciences, is *precisely* that the objects of psychological studies are *themselves* intentional and semantic - as opposed to the objects of the natural sciences. Indeed, it is part of the general background presupposition of psychology, as well as of any other science, that they are.

Therefore, attempts by a Cognitive Psychology (or by a Philosophy of Mind) causally to explain how beliefs and other mental states come to refer to and be true about things in the world, would at best be circular; they would be attempts to justify or explain that which logically forms part of the background presuppositions for investigating and substantiating the explanation or justification - be those investigations of an empirical or of a conceptual kind. At worst, attempts to explain the intentionality, aboutness and truth of mental states and beliefs by reducing intentionality, truth, and the rest, to something purely nonintentional and nonsemantic, would be just as purely selfcontradictory. For, again, they would be attempts to reduce away that which logically forms part of the background presuppositions for investigating and substantiating such reductionist explanations - be those investigations of an empirical or of a conceptual kind.

Maybe we should all go sailing more often, to clear our minds and thoughts.

Chapter 13

The relation between language, cognition and reality I

Summary and consequences

13.1 Epistemological and Ontological assumptions and their inter-relatedness

Ryle, in *The Concept of Mind* (1949), observes that many people can talk sense with concepts but cannot talk sense about them. They may know in practice how to operate with concepts, at least inside familiar fields, but without being able to state the logical regulations governing their use. They are, says Ryle, like people who may know their way about their own parish, but without being able to read a map of it, much less a map of the region in which their parish lies. However, to be able to talk sense about concepts requires knowledge about the "logical geography" of the propositions in which they are "wielded", and thus knowledge of what propositions are consistent and inconsistent with what other propositions, and what propositions follow from them and from what propositions they follow. Otherwise, says Ryle, we risk making logical mistakes when talking about concepts (ibid. pp. 9 - 10).

It is Ryle's aim to "rectify the logical geography" of propositions about the mind by exploring the sets of ways in which it is legitimate to operate with concepts about mental states and processes. In particular, it is his aim to show that the Cartesian polar opposition of *Mind* and *Matter* (or Body) involves a breach of logical rules, i.e. in that it allocates the concepts about the mind to the same logical category as the concepts about matter (or body).

This breach of logical rule, says Ryle, has led to a "double life theory", representing a person as a "ghost mysteriously ensconced" inside a physical machine.

Now, when two concepts or terms belong to the same category it is proper to construct *conjunctive* or *disjunctive* propositions embodying them. Thus, a purchaser, says Ryle, may say that he bought a left-hand glove and a right-hand glove. But he cannot say that he bought a left-hand glove and right-hand glove *and* a pair of gloves. Nor may he say that *either* he bought a left-hand glove and right-hand glove, *or* he bought a pair of gloves. For that would be to commit the logical mistake of constructing conjunctions and disjunctions between terms of different logical types. However, the dogma of the Ghost in the Machine does just that, says Ryle, i.e. by assuming that it makes sense to say that there exist *both* bodies *and* minds. Now, Ryle does not deny that bodies exist, nor that mental processes, such as "doing long divisions or making jokes", exist. What he denies is that the statement, "there occur mental processes" *means* the same sort of things as, "there occur physical processes". The two statements do not indicate two different *species* of existence, but two *senses* of 'exist' - somewhat as 'rising' has different senses in, "the tide is rising", "hopes are rising", and, "the average age of death is rising". And for the same reason that it would not make sense to say, in the same tone of voice, "the tide is rising", and, "hopes are rising", it does not make sense to conjoin the statements, "there exist mental processes", and, "there exist physical processes". But nor does it make sense to disjoint these statements as happens in Materialism and Idealism. The reduction of material world to mental states and processes, as well as, conversely, the reduction of mental states and processes to physical states and processes, presupposes the legitimacy of the disjunction, "Either there exist minds or there exist bodies (but not both)". However, so to say would be like saying, "Either he bought a left-hand and a right-hand glove, or she bought a pair of gloves (but not both)" (ibid. p. 24).

In my analyses of the concepts we use to account for *language*, *knowledge* and *reality*, I have been walking in the foot-steps of Ryle in the sense that I have attempted to reveal the logical implications of propositions in which these concepts are "wielded", and to show the ways in which they may be legitimately used. Thus, I have attempted to show with what other propositions and concepts they are *logically* related, and what follow from the logical relations which exist between these concepts and propositions. In this pursuit both my overall aim (i.e. to "dissipate the polar opposition between Mind and Matter"), and the strategy used to accomplish this aim (i.e. of using *reductio ad absurdum* arguments to "disallow operations of concepts implicitly recommended by the Cartesian myth") is the same as that of Ryle's. However, both the result of my analysis and the consequences which follow are radically different from those suggested by Ryle. The principle reason for this discrepancy is that, in my view, there is not just *one*, but *two* sets of problems incurred by the Cartesian Mind-Matter dualism which needs to be addressed and resolved. The one is the set of *Mind-Body* problems which arises from the different ontological nature of Mind and Matter, among them problems of how Mind and Matter (or Body) may be reconciled.[1] The other is the set of problems of an epistemological nature, i.e. the *Mind-Reality* problems, which arises from the Cartesian assumption of the possibility of a polar division of Mind and Matter into two separate "entities" or "realms", each of which may be talked about and characterized independently of referring to the other.[2] It has been the aim of the chapters in this Part II and in Part I to show, first, that the very possibility of formulating and determining an ontological distinction between Mind and Matter precludes *epistemological* Mind-Reality dualism, i.e. precludes the assumption of Mind and Matter as two independently determinable "entities" or "realms". Secondly, it has been the aim to show that the epistemological conditions for talking in well defined ways about and distinguishing Mind from Matter, at the same time implies *and* necessitates the assumption of *ontological* Mind-Body dualism, which precludes that the properties and phenomena of Mind that uniquely distinguishes Mind from Matter, be reduced to, derived from, or explained in terms of Matter - and vice versa. A short summary of the first part of the argument, i.e. the impossibility of an epistemological dualism between Mind and Matter, goes this way.

First, obvious as it may be, no such thing as a *Mind* exists, nor has anybody ever encountered, found or observed a Mind - be it introspectively or otherwise. What exists and what we may find or observe, are mental events or phenomena, such as thinking, feeling, imagining, or mental acts such as perceiving and cognizing something, and use of language to put forward propositions about what we perceive, imagine or think. Similarly, no such thing as *Matter* exists, nor has anybody ever encountered, found or observed Matter. What exists and what we may find or observe are physical things or phenomena existing in material reality, such as tables and chairs, mountains and trees, molecules and elementary particles. The concepts 'Mind' and 'Matter' are just shorthand for such mental events or acts, and physical things or phenomena respectively.

[1] To these belong problem such as "how does the will of the Mind cause spatial movements of arms and legs and other physical effects", and conversely, "how do physical causes such as physical changes in the optic nerve cause the Mind to perceive a flash of light", or "how do physical changes in C-fibres cause sensations of pain".

[2] To these belong problems such as, "if we only have direct access to what goes on in our Minds, and thus to something mental, how can we be sure that our perception and knowledge of reality is *true*, and that what exists independently of and external to our Minds is *material*".

Now, it would seem just as obvious that we cannot talk in well defined or sensible ways about things in physical, material reality, i.e. what they *are* or are *not*, without presupposing that we may indeed perceive and have knowledge of them, and a language in which we may put forward true propositions about them - and thus without presupposing that physical reality and those things exist as things which our cognition, perception and description may be about. And obvious, furthermore, that we cannot talk in well defined ways about and characterize our cognition, perception and descriptions of things in reality, without at the same time referring to and describing reality and these things. This suggest that an interdependence or *necessary* relation exists between Mind or Matter, i.e. we cannot consistently talk about either without or independently of talking about and referring to the other.

The fact of this interdependency between experience, description and perception of things in the world, and things existing in the world, may become clear when we consider that an account or description of one's perception, say, of the glass on the table in front of oneself, would not be different from an account or description of the glass on the table. That is, given I am asked to give a description of my *perception* of the glass, it would not be different from my description of the *glass*; indeed, I would quite simply not be able to distinguish between a description of my *perception* of the glass and a *description* of the glass. This is just another way of saying that we cannot describe *what* we perceive (our perception of things), without at the same time talking about and describing *that* which we perceive (the things) – and thus of saying, once more, that a *necessary* relation exists between perceptions of things and the things perceived.

Now, the above interdependency does not mean that we cannot *distinguish* between perceiving, knowing or describing things, and the things perceived, known or described, nor between having perceptions and knowledge of and putting forward description of things, and the things "themselves". We can certainly make such distinctions. For one thing, perceiving, knowing and describing things have properties which are fundamentally different from these things – and vice versa. I can eat and get nourished from eating a bread roll, but I cannot eat nor get nourished by a description or perception or my knowledge of a bread roll. Conversely, whilst my perception, knowledge and description of a bread roll may concern or be *about* the bread roll, indeed even be a *true* or *false* perception or description of the bread roll, the bread roll cannot concern, be about or refer to anything, let alone be true or false about anything. The bread roll may exist, and it may be true or false that it exists, but the bread roll itself cannot be true or false about anything.

Another, equally important, although often overlooked distinction, has to do with the fact that although perception, knowledge and descriptions of things in the world depend on persons who perceive, put forward descriptions and have this knowledge about things, the *existence* of things in the world does *not* depend on these things being perceived, known or described. Indeed, the existence of these things does not depend any more on the cognitive, perceptual or linguistic acts of persons than their existence depends on the physical acts we may carry out on them with our bodies, our hands, fingers, and so on. And such things exist just as independently of the notions and terms of our perception, knowledge and description applying to them, as our bodies, hands and fingers, exist independently of the things on which we may act with our hands and fingers. Or, would we really be prepared to say that we may move around in the world and be at various places relative to things, and also that we may carry out physical acts with them, manipulate and investigate them, indeed even eat and get nourished from eating them - things, therefore, of which we necessarily have to say and assume that they exist independently of ourselves, our bodies and the acts we carry out with them, etc. – yet these *same* things do not exist independently of our experience, knowledge and description of them?

To sum up, although we have to assume that things in reality exist just as independently of the cognition, description and action of persons, as bodies of persons exist independently of the things with which they may carry out acts, we cannot talk sensibly about these acts nor about reality and these things, i.e. talk about what they *are* or are *not*, without referring to persons, who perceive and have knowledge and notions of reality and these things and acts, nor therefore *without assuming that reality and these things exist as things persons may experience and have knowledge and notions of*. That is, it has to be assumed that a *necessary* relation exists between the experience and knowledge of persons, and the reality they experience and have knowledge about. However, we also have to assume that this relation is *asymmetrical,* in that the things in reality we experience, have knowledge of, and with which we may carry out acts, exist *independently* of our experience and knowledge of them and action with them.

Both assumptions are necessary in order to attain epistemological consistency, and thus consistent concepts of reality, and of cognition and descriptions of reality, of reference and objectivity. However, the consequence of assuming a necessary and asymmetrical relation between reality and our cognition and description of it, must inevitably be that the Cartesian dualism or polar opposition between Mind and Matter, i.e. between Mind and Reality, as two "entities" or "realms", each of which may be characterized and determined *independently* of referring to the other, is impossible.

In the next section I shall argue that it follows *logically* from the epistemological assumption of a necessary relation between Mind and Reality, and thus between *language* and *reality*, that none of the notions we use to characterize cognition, language and reality may be reduced to, deduced from or explained in terms of any of the others, nor be reduced to, deduced from or explained in terms of *that* to which these notions refer. Thus, I shall argue why the impossibility of epistemological Mind-Reality dualism implies and necessitate the assumption of ontological Mind-Matter dualism (or Mind-Body dualism) - which precludes that the properties and phenomena which uniquely distinguish Mind from Matter be reduced to, derived from, or explained in terms of Matter - and vice versa.[3]

13.2 Arguments for the necessity of ontological Mind-Matter dualism

It is a fact of natural language that we may use it to talk about language itself, for example use language to talk about the conditions for using language and linguistic expressions correctly. Thus, we may use language to discuss what particular linguistic expressions mean and refer to, i.e. what their correct implications are and to what they may be correctly applied. Indeed, this reflexive or so called "meta-linguistic" potential of natural language is so fundamental that, as Roy Harris puts it in one of my favourite quotations, "unless we grasped that, we could no more be participating members in any linguistic community than we could play chess without understanding our role as players" (Harris, p. 163, 1996).

However, this reflexive potential of language has apparently led to the mistake, perhaps obvious, that language may be studied and accounted for independently of its use by speakers to talk about themselves and other things existing in the physical material reality in which they find themselves. According to this belief, language may be viewed as and accounted for *sui generis*, i.e. as an "immanent" system from which linguistic expressions may be generated, the meaning and structure of which is determined entirely by the syntactical "forms" or "grammar" inherent in the system. However, thus separated from its use by speakers and the communicative situations in which it is used, the problem then arises of how - or even whether - language and linguistic expressions "hook onto", "correspond to" or are in "accordance with" reality and the things that speakers use language and linguistic expressions to communicate about. Not surprisingly, this view puts us right

[3] The implications and consequences of the relation which exist between language and reality presented in the next section, are just as much implications and consequences applying to the relation between our cognition of reality and reality. This is not surprising since we cannot really talk consistently about propositions put forward in language about reality, which may refer to and be true (or false) about reality, without presupposing that we have knowledge about reality, i.e. knowledge of what is the case (or not the case) about it, nor independently of referring to this knowledge.

back to the Cartesian Mind-Matter bifurcation, now in the guise of a dualism and polar opposition between Language and Reality.

What *is* surprising about this dualism, however, is that it is assumed that we can use language to talk correctly about language, and even discuss the conditions for and possibility of using language to talk correctly about things in reality, and at the same time assume that we may not necessarily be able to use language to talk correctly about anything else, *in casu* things in reality. Just how surprising, if not contradictory, this assumption is, becomes clear when we consider that it implies that linguistic expressions, when used to talk about language and the conditions and possibility of using language correctly about things in reality, have correct, i.e. non-arbitrary, implications and uses, and that we have notions of and know correct uses of 'correct' and 'true' when talking about and discussing these conditions and this possibility; *yet* linguistic expressions do not necessarily have correct implications and uses, nor do we necessarily have notions or know of correct uses of 'correct' and 'true' when we use language to talk about that which these conditions and this possibility concern, *in casu* reality and things in reality. What is so surprising is that philosophers discussing these questions have failed to realize that we cannot begin to talk about or determine conditions and criteria for talking correctly about reality, nor talk about propositions and assertions about reality, without at the same time *talking about and describing* reality. That is, they have failed to realize that we cannot discuss *whether*, nor justify or explain *how*, language and propositions or assertions about reality "hook unto", are in "accordance with" or may *refer to* and be *true* about reality - without presupposing that they do.

This is a presupposition, so I have argued, which must be the point of departure, the condition for any further discussion and investigations concerning both reality and language, and for determining criteria concerning the truth or correctness of proposition put forward in language about reality - or whatever we talk about. I have formulated this presupposition about the necessary relation between language and reality in a principle, i.e. the *Principle of the general correctness of linguistic descriptions*. This principle implies that we cannot discuss or determine the truth of any particular propositions about reality, nor determine any criteria for the correct implications and applications of such propositions, without presupposing that this discussion and determination takes place in a language, which we may use to talk correctly about *both* reality and propositions - and thus a language to which the Correctness Principle applies. But most importantly, it is a principle which implies that we cannot sensibly question whether language and propositions put forward in language may be used to refer to and be true about reality - or whatever we use language to talk about. For, if we did question this, we would be using language to question the possibility of using language to talk correctly about anything – and at the same time thinking that we could get away with using language so to do.

Now, to substantiate the presuppositions of a necessary relation between language and reality, formulated in the Correctness Principle,[4] only *reductio ad absurdum* arguments are possible. However, that suffices, i.e. it suffices to show that the necessity of the Correctness Principle cannot be proven, justified or explained, neither can it be doubted or denied, without conceding the principle. In this respect the Correctness Principle resembles the principles of formal logic; that is, try to prove the principle, and you will be proving what you have to presuppose in order to carry out the proof; or, try to disprove it, and you will be disproving what you have to presuppose in order to carry out the disproof.

It follows from the arguments for a necessary relation between language and reality, or between cognition and reality, that a *logical relation* exists between notions we use to characterize language, cognition and reality, such as 'statements', 'knowledge', 'facts of reality', and 'true' and 'false'. This is to be understood in the sense that none of these notions have well defined meanings independently of or without referring to well defined meanings of the others. If so, this obviously precludes any of these notions, and what they refer to, being reduced to, deduced from or explained in terms of any of the other and what they refer to. That is, it follows logically from what has to be a fundamental epistemological conditions for talking consistently about both Mind and Matter that the concepts we use to characterize Mind – and *that* which they refer to - cannot be reduced to, deduced to or explained in terms of the concepts we use to characterize Matter, and *that* which they refer to – and vice versa.

It now remains to summarize the arguments and reasons why the impossibility of epistemological Mind-Reality dualism implies and necessitates *ontological* Mind-Matter or *Mind-Body dualism*, having the same consequence, namely that of precluding that the *properties* and *phenomena* which uniquely distinguish Mind from Matter be reduced to, deduced from or explained in terms of Matter - and vice versa.

As said earlier, "Mind" and "Matter" are shorthand used partly to refer to things and phenomena existing in physical, material reality, and partly to the cognition of human beings of this reality of which they themselves are part, as well as to their use of language to communicate what they know about reality, themselves and these things. Thus specified it is obvious that it does not make sense to talk about the *Minds of persons* without talking about and

[4] And, by implication, the necessary relation between our cognition of reality and reality, which may be formulated in a similar principle about the general correctness of our cognition of reality.

referring to persons having bodies, which they may locate relative to other physical things in space and time in the world of which they themselves are part. Nor does it make sense to talk in this way about the *Bodies of persons*, without referring to persons having knowledge about themselves, their bodies and other things existing in physical, material reality.

Put this way it becomes clear that when discussing whether or how Mind or any of its properties may be reduced to or explained in terms of Matter, and vice versa, the issue being discussed is *either* whether *persons* may be reduced to or explained in terms of what they know about their *bodies* and the physical, physiological and biological functioning and interaction of their bodies with physical, material reality. *Or*, conversely, whether the *physical, material reality* of which the bodies of persons are part, may be reduced to or be explained in terms of the m*inds* of person, and thus in terms of or as products of their cognition, description and other mental activities.

However, there are compelling reasons why such discussions do not make sense. For one things, the fact that we cannot talk about ourselves without referring to *both* our minds *and* bodies means, inevitably, that we cannot be reduced *either* to what we know and may say about our cognition, use of language or any other of our mental capabilities and properties; *or* be reduced to our bodies and what we know and may say about our bodies or about other things existing in the physical material reality with which our bodies interact. Thus, it must be a triviality that we cannot be reduced to what we know and may say about our cognition and use of language, since we could not begin to determine to whom this cognition and use of language belonged, *in casu* ourselves, without referring to ourselves as persons with bodies being located somewhere in physical material reality; still less could we determine or characterize the *content* of our cognition and linguistic propositions without referring to that which this cognition and these propositions are about, i.e. the reality of which we and our bodies are part. Similarly, it must be a triviality, conversely, that we cannot be reduced to what we know and may say about our bodies and the physical material reality of which our bodies are part and with which they interact, since such accounts could not exist, nor be referred to, without at the same time referring to persons having minds, who may have knowledge of and talk about their bodies and the physical material reality of which their bodies are part.

What does *not* seem to have been a triviality, however, is that what we know and may say about our cognition and use of language cannot be *identical* with what we may say about our body and its physical, physiological and biological functions. Conversely, it does not seem to have been a triviality that what we know and may say about our bodies and the physical, material reality of which our bodies are part, cannot be identical with what we may say about our cognition and use of language to describe these things. Now, to start with the first point, I think we shall have to agree that *accounts* or *explanations* of the physical, physiological and biological functioning of our body and its interaction with physical material reality, *can only be expressed in language*, and only be so expressed by persons who have knowledge about and a language in which they may put forward true or false propositions about such physical processes, functions and interactions. However, the fact that this knowledge and these accounts may be *about*, indeed even be *true* or *false* about the physical, biological and physiological processes and functions of our body and its interaction with physical material reality, *cannot* be explained in terms of the functions, processes and interaction they concern. But nor can the fact of this intentionality and truth of such knowledge and accounts, be accounted for in the physical, physiological or biological terms used to characterize these functions, processes and interaction. To claim otherwise, to claim, for example, that the logical properties of intentionality and truth of knowledge and propositions may be reduced to or explained in terms of physical, biological or physiological properties or processes, would be to claim that they may be reduced to or explained in terms of something which is more elementary than and which does not imply the existence of intentionality and truth. However, it is not difficult to show that *reasons of principles* exist why such reduction and explanation is impossible, among them the logical impossibility of accounting for these more elementary physical, biological and physiological properties and processes without describing them, and thus without implying the existence of intentionality and truth

For the sake of epistemological consistency, then, we shall have to assume that the *minds* of persons, i.e. their cognition, use of language and other mental activities, have *logical properties* which crucially, ontologically distinguish them from their bodies; i.e. properties which cannot be explained in terms of the physical, biological and physiological properties of bodies. However, so to assume is at the same time to assume *ontological* Mind-Matter dualism, i.e. *Mind-Body dualism*, which precludes that that which uniquely distinguishes Mind from Matter be reduced to, deduced from or explained in terms of Matter.

And for the sake of epistemological consistency we shall have to assume, conversely, that the *bodies* of persons and the physical, biological and physiological processes and functions of bodies and their and interaction with physical, material reality, have properties which crucially, ontologically distinguish bodies and such processes and interaction from having knowledge about and describing such physical processes, functions and interaction. If so, we shall similarly have to assume that the bodies of persons and the physiological, physical, and biological functions and processes of these bodies and their interaction with physical reality cannot be reduced to or be explained in terms of cognition and description of them. Hence, we shall have to assume *ontological* Mind-Matter dualism, i.e. *Mind-Body dualism,* which precludes that our bodies and other physical Matter, be reduced to or explained in terms of or as products of our Minds.

All this may be put it general terms: As persons and language users we may acquire knowledge of and develop languages to describe ourselves, our Minds and Bodies, as well as other things existing in the reality of which we are part; however, *the cognition, language and use of language of persons cannot be explained in terms of what persons may have knowledge of and language users talk correctly about.* Furthermore, it seems that the price we have to pay for both epistemological and ontological consistency is that we shall never be able to question or prove that we have knowledge of and a language in which we may talk correctly about reality, nor question or prove the fact that the reality of which we have knowledge and of which we may talk correctly, is a reality which exists perfectly independently of our knowledge and description of it, and as something being different from knowledge and descriptions of it. Indeed, without accepting and taking for granted *both* these epistemological *and* ontological assumptions, there can be no epistemological, nor any ontological consistency.

Central to my arguments for these assumptions is that epistemological and ontological issues and concepts are *inter-related.* Thus, any consistent *ontological* determination and distinction between Mind and Matter involves *epistemological commitments,* i.e. presupposes the assumption of a necessary relation between Mind and Reality (or between our cognition and description of reality and the reality being cognized and described). That is, such determination and distinction between Mind and Matter presuppose that we have knowledge about and a language in which we may talk correctly about both Mind and Matter. Conversely, any *epistemologically* consistent account of either Mind or Matter, involves ontological commitments as to the different nature and properties of Mind and Matter. Furthermore, *both* ontological distinctions between Mind and Matter, *and* epistemologically consistent accounts of either Mind or Matter presuppose that a logical relation exists between concepts we use to characterize our cognition and descriptions of reality, and the reality which this cognition and description concern. Hence, any attempt to reduce Mind to Matter - or vice versa - in order on ontological grounds to solve problems of an epistemological nature about the relation between Mind and Reality, would be attempts which violate the epistemological conditions and presuppositions on which such attempts necessarily rest. Conversely, any attempts on epistemological grounds to question the different ontological nature of the properties of Mind and Matter would similarly violate the conditions and presuppositions of an ontological nature on which such attempts necessarily rest.

Chapter 14

The relation between language, cognition and reality II

Summary and consequences

14.1 The incompleteness of our knowledge and description of reality

In Chapter 5 I argued that a description of a thing is always put forward in a particular situation, i.e. with regard to particular purposes, interests, conditions of observation and a limited set of relevant description, i.e. a logical space of descriptions, which defines its meaning and relative to which its truth may be determined. Thus, an everyday description of a particular things or substance, say, the water in the jug on the kitchen table, is put forward with regard to everyday considerations and relevant implications and correct applications of descriptions of water in everyday situations (e.g. water is the kind of transparent, fluid stuff which may be used for making tea), whereas a scientific, say, a chemical description of the same substance in terms of H_2O, is put forward with regard to the possibilities of observation, descriptive system and relevant procedures for determining its truth in scientific, chemical situations.

This means, first, that any given description of a thing or substance both implies and relies on other descriptions of the thing or substance, i.e. those belonging to the logical space of descriptions within which it has been put forward, and in which it has determinate and determinable implications and applications - and hence within which its meaning and truth may be determined. Indeed, so I argued, we cannot discuss the meaning or truth of any description or statements about reality and things without referring to situations and descriptive systems which determine the set of relevant implications and correct applications of the description - and thus, in effect, without presupposing that the set of descriptions to which it is logically related have correct uses. According to this argument, to determine the truth of a particular description can only mean to determine whether it belongs among the possible correct description of the thing in situations in which it is relevant to investigate its truth. Within such situations there is nothing "relative" or "arbitrary" about the meaning or truth of the description. On the contrary, it is because a description has logical implications that it cannot be used arbitrarily. Nor can we give reasons for its truth - e.g. give reasons for why it is correct or incorrect - without referring to other descriptions about which it is presupposed that they have logical implications and correct uses.

Secondly, however, the fact that any description of reality and things in reality is put forward with regard to particular interests, descriptive systems etc. also means that it is *limited*, i.e. in that it only concerns particular properties or features of reality and these things - while ignoring others. This is immediately obvious in the case of scientific descriptions, which are confined to describing well defined physical, biological or physiological properties and features of things in reality while leaving out other properties and features of the same things, i.e. properties and features which are salient in everyday determinations and description of them. In Chapter 8, I used the expression that a scientific description in this sense *relies on an abstraction* in that it ignores what according to the science in question are defined as non-relevant properties and features of the thing. However, the same kind of abstraction, and thus limitation, applies as well to the broad varieties of different everyday description of things put forward in different everyday situations. Thus, we may say that that which makes the meaning and truth of a description well defined and determinable, is at the same time that which makes it a limited, a *partial* description.

Now, that a description of a thing or phenomenon in reality is a limited or partial description of the thing or phenomenon, does not means that it may, so to speak, exist in isolation. On the contrary. As further argued in Chapter 8, a description of a thing or phenomenon also relies on an abstraction in the sense that it relies on and is related to other descriptions of the same thing put forward in other situations. Again, it is not difficult to show this is the case of scientific descriptions. Thus, to the extent it may be claimed that a scientific description of particular phenomena or features (say molecular structures or elementary particles) are the physical properties or features of things existing in the *same* reality, which in ordinary everyday situations have been described in terms of tables and chairs, rocks and trees, a scientific description of the molecular structure or elementary particles of such things *presupposes* and *depends* on everyday descriptions and determinations of reality and these things. Without this assumed dependency on or *logical* relation between scientific and ordinary everyday descriptions of reality and things in reality, no consistent claims can be made about *what* scientific descriptions concern and apply to.

Because of the above limitation of any description and its dependency on other descriptions put forward in other situations, no description of a thing and its properties can be *exhaustive*. However, despite the limitation, incompleteness or partial nature of any description of reality and things, we shall have to assume that a description which fulfils the requirement of having determinate and determinable implications and correct uses, is a *true* description of reality and these things. Moreover, we shall have to assume that any description of reality and its properties which fulfil these requirements are *equally* true descriptions of reality and its properties, no matter in what particular situation and inside what particular descriptive system it has been put forward. Indeed, in view of the dependency of any description put forward in any situation on other descriptions put forward in other situations of the same thing or phenomenon, we cannot question or "relativize" the truth and objectivity of any of those descriptions without questioning or "relativizing" the truth and objectivity of all the others. In particular, we cannot question the truth and objectivity of everyday descriptions of things in our everyday world on which, logically, depend scientific descriptions of the physical properties of those things, without questioning the truth of scientific descriptions. Indeed, we cannot do so without rendering the notion of truth and objectivity ill-defined both in everyday and in scientific situations. Hence, *truth*, what is true about reality and things in reality, cannot be dependent on particular conditions of observation and description; nor can the determination of what exists *objectively* in reality.

That any description of a thing is a limited or partial description and, therefore, cannot be an exhaustive description of that which it describes, means that no description can be *identical* with what it describes. This certainly goes against the dominant Naturalist view within the sciences and philosophy which holds that reality has *essential* properties, namely those described by the natural sciences, and also that these essential properties are the only true and objectively existing properties of reality. Consequently, only descriptions of such properties are *true* and *objective* descriptions of what exists in reality, and hence are identical with what they describe. However, it seems to be logically necessary to assume that no description of anything put forward in any concrete situation can be exhaustive, nor identical with that which it is about - for the reasons just given, which may be further clarified in this way:

First, I think we can agree that we shall always be able to ask new or additional questions about reality and the things and situations in which we find ourselves, and to develop new possibilities of observing and investigating these things and situations, which will give us new information and knowledge about reality. Now, granted that the point of departure for such further inquiries, for example, scientific inquiries of reality and its properties, are everyday determinations and descriptions of reality and its objects as they exist in everyday situations, we shall have to assume that *more* can be said or known about reality than what is already known in these situations. Indeed, it only makes sense to say that scientific investigations may provide new and *additional* knowledge about reality and its properties on condition that we *can* and *shall have more to say* about reality and its properties, than we do *both* in everyday *and* in scientific situations. Thus, generally, any further inquiries and development of our knowledge about reality and its properties - be it in everyday or in scientific situations - rely on descriptions and determination of reality and properties of reality made prior to and in other situations than the ones in which the further or additional descriptions and knowledge are acquired. In other words, such further or additional descriptions and knowledge depend on and, thus, are logically related to these prior determinations of the same reality and to the descriptions and knowledge of properties of reality in those other situations. Hence, neither the additional knowledge and description of reality, nor the descriptions involved in the determination of the parts of reality to which the additional knowledge and description apply, can be identical with, nor exhaustive of what is being described or known.

The main results of the discussions here and in previous chapters of these issues are the following. First, the fact that a description relies on abstraction in that it is put forward in a particular situation and with regard to particular interests, conditions of observation, descriptive systems etc., and thus is a limited or partial description, does not affect the fact that it may be true. Nor does its truth depend on in what particular situation, under what particular conditions of observation and within what descriptive system it has been put forward. Secondly, no description of a thing exists without having logical relations to other description put forward in other situations about the same thing. Thirdly, it seems necessary to assume that both any given description and the descriptions which it implies and to which it is logically related - in the present situation and in others - are all *equally* true. Without these assumptions, we could not talk in well defined ways about *development* and *change* of knowledge and description of reality and its properties. Indeed, it seems that we can only refer to and make meaningful claims about significant development and changes in our knowledge and conceptions of reality and things in reality on the condition that this knowledge and these conceptions are the knowledge and conceptions of persons, who both *before* and *after* the changes in question are able unambiguously to refer to and make true statements about reality and these things. If we remove this presupposition, both changes in and development of conceptions will disappear.

Fourthly, it seems that general, logical reasons exist for assuming that our notion of *truth*, and thus what is true or false to say about reality and things in reality, do not depend on whether we have *total* or *exhaustive* knowledge about reality and these things. That is, the truth and objectivity of our - partial - knowledge and descriptions, and thus our notions of truth and objectivity, do not depend on *how much* knowledge we have, or what kind of knowledge we have. Nor does the fact that any part of our knowledge and description of reality relies on other parts of our knowledge and descriptions mean that for any partial knowledge or description to be true, it has to be related to some sort of holistic system of total or exhaustive knowledge and description of reality. But neither does the fact that the knowledge and descriptions to which any partial knowledge and description is related, is itself a partial knowledge, i.e. can never be exhaustive or total, mean that, then, the knowledge and description we have about reality at any given time may only be considered *relatively* true. If it did mean any of the above things, we could not, as argued, talk consistently about development in knowledge and description. What it does mean, however, is that what we already know will always form the basis of and be logically related to further development of knowledge.

The consequences of the arguments so far presented seem to be that although it is fundamental to the knowledge and description of persons and language users that their partial knowledge and description of reality only exist in virtue of having relations to other parts of knowledge and language, it is equally fundamental that this knowledge and language has relations to other, as yet unknowns. That is, it is fundamental to our partial knowledge that it does not exist independently of logical relations to other parts of knowledge and language - but also that other and as yet unknown logical relations may exist. If so, our notion of what it is to be persons and language users may be further specified in this way:

To be persons and language users is not only to have knowledge about ourselves, other persons and things existing in the situations in which we find ourselves in reality, but it is also to know that there is something which we do not yet know about these things and situations; something about which further questions may be asked, and which we may investigate and come to know more about. Given that this be the case, however, such knowledge will always be logically related to what we already know. To be a person and a language user, then, is to know that *much more* and very different things can be said about the same reality and things, in the actual as well as in other situations.

Let me conclude this discussion of the consequences of the incompleteness of our knowledge and description of reality (or whatever else we have knowledge about and describe) by emphasising, once more, that because *any* questions being asked about our knowledge of any parts or properties of things in reality depend on, and thus are logically related to other parts of knowledge and descriptions of the same reality, it is possible to account for historical changes and development of our knowledge. Furthermore, because any well defined questions about as yet unknown parts or properties of these situations - as well as their answers - rest on objective knowledge and true descriptions of the situations in which the questions are asked, it is possible to refer to such historical changes - without running into any *general* problem of truth and objectivity. In particular, it is possible so to do without running into the problems of *historical relativism* or other kinds of epistemological relativism - of which I shall discuss an example in the next section.

14.2 Putnam's *Internal Realism*

In *Representation and Reality*, Putnam (1988) presents a series of arguments against his brain child, *computational functionalism*, and the philosophical assumptions on which it is based. These are the assumptions of what he calls "metaphysical" or "scientific realism", and characterises as

> "a bundle of intimately associated philosophical ideas about truth: the ideas that truth is a matter of Correspondence and that it exhibits Interdependence (of what humans do or could find out), Bivalence, and Uniqueness (there cannot be more than one complete and true description of Reality)." (ibid. p. 107).

What is so seductive about metaphysical realism, says Putnam, is the idea that *the way to solve philosophical problems is to construct a better scientific picture of the world*, and he continues,

> "In such an outlook, Independence, Uniqueness, Bivalence, and Correspondence are regulative ideas that the final scientific image is expected to live up to, as well as metaphysical assumptions that guarantee that such a final scientific resolution of all philosophical problems *must* be possible." (ibid. p. 107).

Among the assumptions of this outlook is that a scientific theory of the "nature" of the intentional realm *must* be possible if there is anything "to" intentional phenomena at all. However, as Putnam argues, what applies to any physicalist accounts of the world applies just as much to accounts being made within the most recent paradigm of scientific realism, i.e. the computational paradigm: Such accounts of the world are incomplete in that they cannot account for intentionality. Nor can intentional notions, such as 'concepts', 'meaning', 'truth', and 'reference', be reduced to the terms, such as of 'representation' and 'cause', employed within computational functionalism.

However, the fact that no scientific theory has so far proven capable of accounting for "intentional phenomena" does not rule out the possibility that there is something "to" intentional phenomena - nor that the problems about the intentional are far deeper than thought of within scientific realism. Short of intending to present a whole new theory about the intentional, Putnam concludes his examination of some of these problems by proposing two very general *desiderata* for a philosophical picture of the intentional: it should account both for *objectivity* and for *conceptual relativity*. In what follows I shall present Putnam's account of his alternative "perspective" to scientific realism - which he calls *Internal Realism* - and discuss the claim of this perspective to the effect that the *objective* character of intentional notions, such as 'truth', 'reference' and 'meaning', cannot be treated apart from the *conceptual relativity* of those notions.

Putnam begins his arguments for this claim thus:

> "What do I mean by "objectivity and conceptual relativity"? To begin with objectivity: to say that intentional phenomena are "objective" is not to say that they are independent of what human beings know or could find out (it is not to say that they are Objective with a capital "O", so to speak). If we take truth as our representative intentional notion, then to say that truth is objective (with a small "o") is just to say that it is a property of truth that whether a sentence is true is logically independent of whether a majority of the members of the culture *believe* it to be true. And this is not a solution to the grand metaphysical question of Realism or Idealism, but simply a feature of our notion of truth." (Ibid. p. 109).

Now, if for a moment we ignore the curious bits about objectivity spelled with a capital or small "o", I think it is possible to make sense of and reconcile the two seemingly contradictory statements in the above quotation in this way.[1]

Whereas the facts of reality and its objects exist independently of the knowledge and description of human beings, *knowledge* and *statements* about these facts do not exist independently of human beings and language users, who may know of, refer to, and put forward descriptions of those facts. Nor are knowledge and descriptions of facts of reality and its objects

[1] That is, on the one hand the statement: intentional phenomena such as truth etc. are objective - yet dependent on what people know and can find out, and on the other the statement: a sentence is true is logically independent on what any number of humans believe to be true.

independent of the situations in which language users act with and observe these facts, their possibilities of observation, conceptual systems, purposes, intentions and so on. However, this claim is fully compatible with the claim that the knowledge we may have and description we put forward about such independently existing things, may be *true* - and thus that they exist as things to which we may refer, and about which we may put forward true statements. That is, the truth of our knowledge and descriptions, i.e. the fact that we may use language to put forward true descriptions of reality, is *not* dependent on the situations in which we observe things, nor dependent on the possibilities of observation, intentions or purposes, etc., we may have with particular things in particular situations. Granted this interpretation, I think we shall have to agree on the dependencies pointed out by Putnam, and on the independent nature of truth and other intentional phenomena.

Let me give a concrete example to illustrates what we can agree on. The objectivity of the fact that the cat is on the mat is independent of whether there are any human beings around to observe and make statements about it. But the statement, "the cat is on the mat", is not independent of human beings, i.e. it does not exist without human beings and independently of what they know and what - in virtue of being language users - they may put forward statements about. However, that the statement may be true, is not dependent on the agreement of any number of human beings that it is true; the statement, "the cat is on the mat", is true if the cat happens to be on the mat.[2]

Thus having established the sense in which the "feature of our notion of truth" is "objective", it would seem obvious to contend that *truth* and, mutatis mutandis, other intentional phenomena such as reference and meaning, must be *primitives* or key concepts of any epistemology and theory of language. To Putnam, however, it is not at all obvious why it is necessary, as he puts it, "to take the existence of intentional properties as primitive facts". For, in his view

[2] Again, granted this interpretation, I can fully agree with this suggestion of Putnam's: "The suggestion I am making, in short, is that *a statement is true of a situation just in case it would be correct to use the words of which the statement consist in that way in describing the situation*. Provided the concepts in question are not themselves ones which we ought to reject for one reason or another, we can explain what "correct to use the words of which the statement consists in that way" means by saying that it means nothing more nor less than that a sufficiently well placed speaker who used the words in that way would be fully warranted in counting the statement as true of that situation. What is "a sufficiently well placed speaker? That depends on the statement one is dealing with. [...] facts of the form "If you have to tell whether S is true, then it is better to be in circumstances C_1 than in circumstances C_2" [however] are not "transcendent" facts; they are facts that it is within the capacity of speakers to determine, if they have the good fortune to be in the right sorts of circumstances. What are "the right sort of circumstances"? That depends on the statements one is dealing with. ..." (ibid., p 115).

> "there are *other* properties of truth, reference, and meaning to be accounted for than just the objective character of the notions. In particular there is what I call "conceptual relativity." (Ibid., p. 110).

What Putnam means by the "conceptual relativity" of these notions, he illustrates by the following example:

> "Suppose I take someone into a room with a chair, a table on which there are a lamp and a notebook and a ballpoint pen, and nothing else, and I ask, "How many objects are there in this room?" My companion answers, let us suppose, "Five." "What are they?" I ask. "A chair, a table, a lamp, a notebook, and a ball-point pen". "How about you and me? Aren't we in the rook?" My companion might chuckle. I didn't think you meant I was to count people as objects. Alright, then, seven." "How about the pages of the notebook?" (ibid., p. 111).

Discussions of further examples seems to Putnam to indicate that there is no "right" answer to the question, "how many objects are there in this room" - for the notion of *an object* is indeterminate. We may choose a logical notion of an object, according to which any "substance" or "event" we can refer to with a pronoun, is an object. In this sense, all parts of a person or pages of a notebook, as well as all the particles in the rooms, are objects. But what about *groups* of elementary particles - are they objects?

> "For a moment, let us take it that by a group we are to understand a whole with certain parts, not an abstract set (thus a group is what certain logicians call a "mereological sum"). For example, my hand (we may suppose) is a group of atoms; those atoms are groups of elementary particles. What about the group consisting of my nose and the lamp? Is that an object at all? Or is there no such object?" (ibid., p. 111).

To Putnam it seems clear that the question is one that calls for a *convention* - in that the answer depends on what one means by "an object". But this, in his view, has startling consequences. For, the fact that we are free to choose any of a number of implications of the term 'object' (or other existential quantifiers) means that even the notion of *truth* is affected, indeed there can be no clear-cut relation between *truth* and *facts*. Truth, Putnam contends, must be a notion which is to be found on a *convention-fact continuum*. However,

> "The fact that a truth is toward the "conventional" end of the convention-fact continuum does not mean that it is *absolutely* conventional - a truth by stipulation, free of every element of fact. [...] What is factual and what is conventional is a matter of degree." (ibid., p. 113).

Well, we have to agree with Putnam, that the term 'object' in different situations may have different implications, i.e. it may be used to refer to different things. However, it is not at all clear why this fact should affect the notions of truth, reference or meaning, let alone make "conceptual relativity" and "convention" a property of *those* notions. If so, we might well ask, for a start, what sense there would be in maintaining that it is a feature of the notion of truth that what is true is logically *independent* of how many persons agree or believe it is true, *and* at the same time maintain that what is true, *depends* on what people may agree on making conventions about. And we may well ask furthermore, if truth is a matter of degree, somewhere on the convention-fact continuum, does this mean then that the objective existence and independence of things in reality of which true statements may be made, is also somehow conventional, a matter of degree, i.e. something only "relatively" exiting between fact and convention?[3]

In this case, I think, no amount of clarification or re-interpretation of Putnam's claims will put them right; they are nonsensical and self-contradictory - and quite easily refuted. So, before I present Putnam's further arguments for these claims, let me give the following, quite straightforward reason why the conceptual relativity of the notion 'an object' (i.e. the fact that the term is *un-determined* until it has been specified in what situation it is used and to what it refers), does *not* warrant the claim that then *conceptual relativity* must be a property of truth, reference and meaning as well.

In most situations in which people refer to things in the world with the term 'objects', they will know what are the objects being referred to. If not, they will be able to discuss the matter and reach a satisfactory agreement, just as in Putnam's example, i.e. an agreement of *what*, in the particular situation, they are using the term 'objects' to refer to; for example, that it is the tables and chairs and other furniture in the room, but not to things which form part of each individual piece of furniture, e.g. their legs, seats, table tops, or the elementary particles of which they are made up (which, of course, may be the topic of the conversation in some other situation). Even in the event that they are not able to reach an agreement, people will most likely be able to specify *how* they disagree, as, again, in Putnam's example. Without this presupposition, neither people in ordinary situations, nor Putnam would have any examples or clear-cut cases of *differences of meaning* of the term 'objects' to discuss; nor would it have been possible for them or Putnam to raise any sensible questions of what it implies that the term 'objects' may have different meanings.

The fact that both Putnam and people in general are indeed able to discuss differences and disagreements as to what in different situations they use the term 'objects' to refer to, relies on the presupposition, furthermore,

[3] For example is the existence of the cat on the mat, or any of its parts, i.e. its head or tail, its legs or ears, only relative, a matter of degree somewhere on the convention-fact continuum?

that the things or part of things in reality to which the term refers, *actually exist*, and thus that reality and these things may be "carved" up in different, determinable parts, i.e. as parts and things of which something is the case and something not, and about which we may put forward true and false statements. Indeed, it is necessary to assume that all the different statements we may put forward in different situations about these things and parts are *equally* true statements and determinations of what exists, and is being referred to by the term 'objects' in different situations. Again, without this presupposition, there would be no examples of differences or disagreements to discuss over uses of the term 'objects', let alone any way we could *understand* these examples.

Now, *how* in a concrete situations we talk about things in reality, for example whether we talk about things in the room in which we find ourselves in terms of tables and chairs, walls and floors, pictures on the walls, or talk about their parts, say, the elementary particles or groups of elementary particles of which they are made up, etc., *does* rely on conventions, agreements, purposes, interests, and the like. However, so I have argued, this fact does *not* make the notions of reference, meaning or truth "relative" in those situations, nor do those notions differ or vary "in degree" from situation to situation. For example, the notion of truth that a physicist employs when describing some physical facts about elementary particles observed in his laboratory, is not different from the notion of truth he employs when describing everyday facts about his laboratory - e.g. when he describes the chair he is sitting on. So, although the implications of the term 'objects' may be different when applied to elementary particles and to those everyday things, the physicist does not switch from one way of using the term 'true' when he "switches" from describing elementary particles to describing chairs. That is, the *truth* of any particular statement (i.e. this is a chair, or this (same) thing is a particular structure of particles) does not depend on the situations in which it is put forward (i.e. everyday situations, scientific, or any other situations); nor is the truth of a description, let alone the notion of truth, affected ("relativized" or otherwise) by the fact that the thing to which the description refers in some *different* situation may be described by means of some other true description.

Now, if it were the case, as Putnam suggests, that *truth* is a matter of degree and, just like other intentional phenomena, has properties which puts it "somewhere" on a continuum between "fact" and "convention", then it would of course also have to apply to the truth of Putnam's statement: "truth is a matter of degree". That is, this statement would also have to be true to some degree *only*, or only "relatively" true - in which case it would be utterly impossible to determine what this statement states, or whether it states anything about anything at all. - There is of course the possibility that

Putnam, just like other proponents of epistemological relativism, assumes that his claim about the "relativity" of truth does *not* apply in general - that, for example, "conceptual relativity" only applies to the notion of truth when we use language to talk about reality, but not to the notion of truth employed when we use language to talk about language, and the conditions for using language to talk about reality. However, that this possibility would make even less sense becomes clear when we consider that, logically, we cannot use language to talk about the *conditions* for using language to say something which is true about reality and its facts, without presupposing that we can use language to talk correctly about *reality and its facts*. Thus, to insist that one notion of truth is employed when we use language to talk about language, and another when we use language to talk about reality, would amount to insisting that when determining the conditions for using language to talk about reality, we are employing two different notions of truth at one and the same time. So to insist would not only be untenable, but absurd (cf. arguments in the previous chapter).

The false dichotomy and continuum between fact and convention, and the "conceptual relativity" of intentional notions that goes with it, are, as we shall see next, both the *consequences* of Putnam's refutation of scientific metaphysical realism, and the *means* by which these positions are refuted. Thus, from the observation that "the same situation can be described in many different ways, depending on how we use the words" (ibid. p. 114), it follows as an inevitable consequence, according to Putnam, that *things do not exist independently of how they are described.* Here are, in Putnam's own words, the arguments for this assumption of Internal Realism and against Scientific Realism.

> "A metaphor which is often employed to explain [that there are many ways of *using* the notion of an object] is the metaphor of the cookie cutter. The things independent of all conceptual choices are the dough; our conceptual contribution is the shape of the cookie cutter. Unfortunately, this metaphor is of no real assistance in understanding the phenomenon of conceptual relativity. Take it seriously, and you are at once forced to answer the question "What are the various parts of the dough?". If you answer that (in the present case) the "atoms" of the dough are the *n* elementary particles and the other parts are the mereological sums containing more that one "atom", then you have simply adopted one particular transcendental metaphysical picture; the picture according to which mereological sums "really exist". My view - which I called "internal realism" in *Reason, Truth and History* (I would have done better to call it simply *pragmatic* realism) - denies that this is *more* the "right" way to view the situation than is insisting that only the *n* elementary particles (or only the elementary particles and the atoms and molecules, or only the "organic wholes") really exist. [...]

> The cookie-cutter metaphor *denies* (rather than explains) the phenomenon of conceptual relativity. The internal realist suggestion is quite different. The suggestion, applied to this very elementary example, is that what is (by common-sense standards) the same situation can be described in many different ways, depending on how we use the words. The situation does not itself legislate how words like 'object', 'entity', and 'exist' must be used. What is wrong with the notion of objects existing "independently" of conceptual schemes is that there are no standards for the use of even the logical notions apart from conceptual choices. What the cookie-cutter metaphor tries to preserve is the naive idea that at least one Category - the ancient category of Object or Substance - has an absolute interpretation. The alternative to this idea is not the view that it's all *just* language. We can and should insist that some fact are there to be discovered and not legislated by us. But this is something to be said when one has adopted a way of speaking, a language, a "conceptual scheme". To talk of "facts" without specifying the language to be used is to talk of nothing; the word "fact" no more has its use fixed by the world itself than does the word "exist" or the word "object"." (ibid., p. 113-114)

By insisting that the notion of 'objects existing "independently" of conceptual schemes' is a mistake, Putnam misses the crucial asymmetry between *existence of objects* and *description of objects*, i.e. the fact that while objects exist independently of language and descriptions put forward in language about them, description of objects do not exist independently of objects to which they may refer and about which they may be true. If this asymmetry is *not* acknowledged, "internal realism" will inevitably lead to *Relativism*, or as Putnam puts it, to the view that "it's all *just* language". Thus, if denying that objects exist independently of being described implies that *what* exists, and *how* it exists, only exists in virtue of being described, or how it is being described, then we would be saying that the existence of things in reality depend on the existence of language and the situations in which language users may be.

However, as I have tried to argue in this and the previous chapters, no matter how true it is that we cannot say anything about anything without using language (cf. the quotation from Putnam above) - it must be equally true that the existence of language and use of language logically presupposes that there *is* something to be talked about, something which we may use language to *refer* to, and something, therefore, which exists *independently* of language, *and* as something about which true statements may be put forward in language. In this sense and for this reason we *have* to say that the terms 'facts', 'exist' and 'objects' have their use "fixed" by the world itself. Thus, rather than necessitating "conceptual relativity" of the notion of truth, the observation by Putnam that "the same situation and objects may be described in many different ways", of which no one is "*more* about these objects and this situation than the other", logically necessitates the assumption that such situations and objects exist as something about which many *equally* true descriptions may be put forward. Indeed, the observation that there is not only *one*, but several different descriptions of the world, only makes sense as an argument against Scientific and Metaphysical Realism, provided that Putnam is also prepared to commit himself to the assumption that these different descriptions of the world may *all* be *true* descriptions of the world. Without it, "scientific" and "metaphysical realism" would be immune to Putnam's argument.

In conclusion, true descriptions *are* dependent on language and language users and the situations they may be in; but the *existence* of that about which true description may be made in those situations is *not* dependent on language. If it were, it would all be *just* language - in which case, as argued in Chapter 5, there could be no false descriptions, nor any true ones, and thus no notion of truth. *Relativism* - not even the "non-absolute" kind of relativism Putnam proposes - is not the unavoidable consequence of "conceptual relativity", nor the *only* alternative to the assumptions of Scientific Realism or Naturalism. Putnam may not want to promote *metaphysical relativism*, to borrow his own expression (cf. e.g. his criticism of relativism in the constructionist position (Putnam, 1992)); however he may be accused of failing to see that one cannot be "a little bit relativist". As the examples by Putnam shows, there may be many, indeed incalculably many *concrete* problems of objectivity, reference and truth to discuss. However, such discussions are only possible because of the presupposition that there is no *general* problem of objectivity, truth and reference; in effect, no general problem about the relation between language and reality. And this brings us right back to the *Principle of the general correctness of language*.

14.3 Consequences of the incompleteness of our knowledge and description for computational functionalism

Among "the circles of problems" about intentional phenomena which, according to Putnam (op.cit) present serious problems for any philosophical theory of reference, meaning and truth, as well as in particular for computational functionalism, are the problems of "conceptual relativity" deriving from the fact that a person may use different concepts about the same things in different situations, and that different people (e.g. belonging to different cultures, having different backgrounds etc.) may have different concepts about the same things. These differences in concepts and use of concepts by the same people as well as by people belonging to different speech communities, Putnam argues, cannot be accounted for satisfactorily within a functional computational account of the mind. Thus, the fact that people - in spite of these differences - are able to understand each other and determine the differences, may only be explained within computational functionalism by postulating the existence in each of us of a pre-programmed "universal language", which captures all the possible meanings of all possible concepts, and according to which all concepts may be interpreted - and into which they may be reduced. Furthermore, in order to account for the historical and cultural changes of concepts within different speech communities, this postulated universal language would have to cover the meaning and references of all present, as well as past and future concepts. However, the notion of such an all encompassing pre-programmed or *innate* language, the existence of which - according to Fodor - is necessary for anyone to learn a natural language,[4] is an impossibility, says Putnam. Even if such a language existed, no one would be able to know about it, let alone account for it.

Now, it is doubtful whether we shall ever be able to come up with anything like an ultimate explanation of why or how people in different cultures come to develop and use different concepts and words about the same things; or, for example, explain conclusively why concepts within cultures change over time - although we shall probably be able to account for the conditions under which these differences and changes occur. However, for anyone trying to understand, make sense of, or even theorize about such cultural differences and historical changes, it would seem worth while to consider the following. First, the conditions for people belonging to different speech communities to be able to understand each other, and to determine differences in their use of language and concepts, must be that vast similarities *also* exist in their knowledge, and thus in their notions of things in reality to which their different concepts apply. For, were they not able together to identify, and hence to agree on correct determination of the things that their different concepts concern - i.e. agree that they concern *these* particular things about which *this* is the case or true, and this is not - they could not possible make claims about any determinable differences in their use of concepts and words about these things.

However, secondly, there is no need of assuming that a pre-programmed, universal language and hence universal knowledge about things in reality are required for this identification and determination of things, nor for understanding differences and change in use of concepts about the same things by people in different speech communities. On the contrary, it is far more reasonable to assume that this understanding is due to the *general* recognition by persons that it is fundamental to being persons that their knowledge about the situations in which they find themselves in reality is *incomplete*, or partial. That is, it is reasonable to assume that it is due to the knowledge of

[4] I shall come back to this in the next chapter.

persons that there is something about the situations in which they find themselves, which they do *not* know, but about which much *more* could be known than is already known - for example by investigating the situations according to other and different purposes, interests, opportunities of observation, etc. - *and that they take this into account whenever they try to understand and determine differences and change in use of language and concepts to describe these situations*. Indeed, it would seem that to be persons, who may communicate and collaborate, requires being in a constant process of acquiring *more* and *different* knowledge about the same situations and events - on the basis of the notions and *partial* knowledge we already have and share, and to which our *new, expanded, changed* and *different* knowledge and understanding necessarily is logically related (cf. the arguments in the previous section).

Without assuming this general recognition by persons of the incompleteness and partial nature of their knowledge and descriptions of things existing in the situations in reality in which they find themselves, there is no way in which we could come to realize - and accept - that reality and these things and situations may indeed be described differently, i.e. when viewed from different points of view and using different descriptive terms. And without assuming, furthermore, that *both* our shared knowledge and description of reality and things, *and* the different concepts we apply to these things that they have *necessary* implications and correct uses, there could no differences nor any changes in use of concepts and words to discuss. Thus, once more, far from implying *linguistic or epistemological relativism*, as suggested by Putnam, the determination of cultural and historical *conceptual relativity* precludes such relativism.

Chapter 15

The relation between language, cognition and reality III

Summary and consequences

15.1 The impossibility of explaining how we become persons and language users

In previous chapters I have argued why we cannot explain or justify that we are persons having knowledge, and language users being able to put forward true - and false - statements about reality, in terms of more basic biological or physiological states which do not presuppose the existence of knowledge and language. In this chapter I shall argue why, for similar reasons, it is impossible to explain how we *become* persons and language users, i.e. explain how language, knowledge and the logical properties of referentiality and truth of language and knowledge, originate out of such more basic states. That is, I shall argue why we cannot explain how, from being at some stage mere biological organisms and physiological systems (say, in early infancy), we at some later stage develop into persons and language users, having acquired the logical notion of reference and truth. And I shall also argue why the *transition* from functioning as biological organisms and physiological systems to functioning as persons and language users, cannot be explained in terms of or be derived from the biological and physiological structures or states, which exist prior to the knowledge, language and use of language of persons.

Well, it would seem that what I am going to argue in this chapter is that there is nothing whatever concerning the development of persons and language users, nor of their knowledge and use of language that we may investigate and account for. However, this is not so. Within psychology, for example, it makes good sense empirically to investigate how children come to acquire knowledge about the world, and a language in which to communicate and express this knowledge, and also to investigate how this knowledge and use of language develops over time. Through such investigations we may come to know about the *necessary condition* for being able to function as persons and language users - to which belong, according to these investigations, biological maturation and development of various parts and function of our brains, as well as the possibility of social relations and interaction with other persons and with the physical surroundings. Good examples of investigations of the necessary conditions for the child's (later) functioning as a person and language user are the studies of Jerome Bruner, Roger Brown, Margaret Donaldson, and Daniel Stern on the pre-verbal communication of children (See e.g. Bruner, 1977; Brown, 1973; Stern, 1985; Donaldson, 1992). These, and many other recent and older investigations have vastly enhanced our understanding of how different forms of communication, as well as the various norms and rules by which they are governed, are gradually being established between mother and infant at various stages in the pre-verbal period, which are vital for the child's later acquisition of language and linguistic rules.

However, it is worth noticing that both these investigations and their interpretation rely on the following fundamental assumptions. First, in accounts of the child's pre-verbal behaviour as being that of *communication* and *interaction* with another person, it is assumed that there (already) are features in the child's way of experiencing and conceptualizing the world in the pre-verbal period, which are not *in principle* different from features of those of the language user's. Thus, in the early give-and-take game between the infant and his mother it is assumed that the child is able to identify the objects and persons involved in the game, and assumed that the child can distinguish between the objects and the persons taking part in the interaction. In the description of this interaction it is assumed, therefore, that the child - be it only in *a very rudimentary sense* - has *knowledge* about the objects and persons involved, and that, therefore, for the child something may be the case or true, and something else may not be the case or false, about objects and persons. In the case of the give-and-take game, that the objects being given or taken are in particular places, and may be given to or taken from another person etc. In short, it is assumed that the child is able to perform *acts*, and that the knowledge of the child may be described as being of a *propositional* nature - whether or not the child itself is able to verbalize what it knows.

These are features of the cognition of the child which, I believe, have to be assumed in all investigations in which it is justifiable to interpret the child's 'answers' as (inter)action and *communication* with and *about* something, and when we say that mother and child develop norms, rules, and conventions for their interaction with and about things; i.e. when we talk about the child being able to refer to things and objects in this way:

> "The objective of early reference [...] is to indicate to another by some reliable means which among an alternative set of things or states or actions is relevant to the child's and mother's shared line of endeavour." (Bruner, 1977).

What is said above, could also have been put this way: because in such investigations we have to use language to describe the child's behaviour and ways of experiencing the world, we ignore the differences which may exist between our linguistic and the child's non-linguistic knowledge and action. And for good reasons. For it makes good sense to characterize at least some of the behaviour of the child in the pre-verbal period as *action*, and to characterize its cognition of the world and objects as knowledge of a propositional kind. The point being made, then, is that we cannot describe the behaviour of the child as *action*, and his or her experience of the world as *knowledge* - and at the same time doubt whether the non-linguistic knowledge of the child shares some of the same logical features as that of the language user's. Conversely, when describing the child's behaviour and experience of reality in this way as *action* with and *knowledge* about objects in reality, we characterize it in the same way, or *as if* it was similar to our own. Because of these assumptions underlying the descriptions and interpretations of the communication of the pre-verbal child, it would be misleading to say that the results of the investigations, in any fundamental sense, answer the question of how it comes about that the child *becomes* a person having propositional knowledge and, hence, having (at least rudimentary) logical notions of "reference", "true" and "false" of persons and language users.[1]

Equally important to the point just made is that forms of behaviour and experience or consciousness of the world, which may exist *prior to* what may legitimately be described in terms of knowledge and action, we are for logical reasons debarred from saying anything about, let alone characterizing. For this reason we are similarly debarred from explaining the process whereby the child *becomes* a person, having propositional knowledge and the logical apparatus which go with such knowledge, as a process whereby this knowledge and apparatus is developed *from* such more elementary forms of consciousness and experience. Any linguistic descriptions of such

[1] Psychologists carrying out research in the development of cognition and communication of children are perfectly aware of the point being made here, see e.g. Donaldson, 1992, Bruner, 1990.

more elementary forms of consciousness would have to imply *either* that we could in fact talk logically about them - which, by definition, we cannot; *or* it would have to be possible to derive the logical notions of reference, intentionality, true and false of knowledge and language from something more basic which does not imply the existence of such logical notions - which it is not.[2]

The further general consequence of this argument is that problems of an epistemological nature, such as, "what is knowledge", or, "how does it comes about that we have knowledge about reality", cannot be solved, as assumed by, among others Piaget (1971), by *converting* these problems into empirical questions about how knowledge about reality is acquired by children, or how this knowledge develops *ontologially*, e.g. out of some biological dispositions. This is to be understood in this sense: To the extent that attempts such as Piaget's to explain the existence of human knowledge rely on the possibility that this knowledge and its logical features of referentiality and truth may be derived from or explained in terms of something more elementary (.e.g. in terms of or from biological processes and states), such attempts are doomed to fail. They are so doomed for the same reasons that, generally, it is impossible to explain human knowledge and its logical features in terms of, or to derive this knowledge and its logical features from states and phenomena which do not imply the existence of such knowledge and features (cf. the arguments of Chapter 8). The same applies, *mutatis mutandis*, to more recent *evolutionary or phylogenetic theories* of the development of mind (Sayre, 1986), as well as older versions (Leontiev, 1977), arguing that human cognition, action, perception, and its logical features, may be derived from more elementary forms of experience and behaviour which do not imply the existence of (human)cognition, action perception and its logical features.

In the next section I shall try to thrash out some of the difficulties of describing the transition from organism to person, and at the same time argue why other attempts to explain this transition, i.e. by assuming the existence in the biological make up of human beings of special "cognitive

[2] Although decisive differences exist in the child's situation *before* and *after* we may characterize it as a person in the way described above, it is hopefully clear that I do not maintain that *becoming* a person is something which happens from one moment to the next. Nor do I mean that one can only be a person provided one has full-blown knowledge of reality, neither that this knowledge necessarily has all the sophisticated logical implications which characterize the knowledge of a more mature person. However, it is not the amount of the child's - partial - knowledge which determines that or when the child is a person, but the logical apparatus and relations involved in its knowledge, including relations to that which the child does not yet have knowledge about. Needless to say, vital parts of this knowledge of reality and these relations, which determine when the child is a person, are knowledge about and relations to other persons.

structuring mechanisms" or innate "ideas" of reality, are so difficult - if not impossible - to defend.

15.2 Some difficulties in accounting for the transition from organism to person

In earlier chapters of this book (cf. Chapter 7), a person has been characterized (minimally) as someone who has knowledge about himself and the situations in material reality in which he finds himself; someone, furthermore, who may carry out acts in those situations, and thus may change his possibilities of action and cognition in those situations; someone who may cooperate with other persons finding themselves in the same situations, and who may reflect on and communicate about the knowledge he and they have of these situations and, thus, who have knowledge about other persons as well. It is generally agreed that the process whereby an infant develops into someone with capabilities thus to function as a person, is a process and a development which *occurs over time*. According to some schools of psychology this process and development is assumed somehow to rely on capabilities of the infant existing prior to the time at which it has become a person. For example, *discriminating stimuli* and other capabilities required for an organism to *react* consistently to stimuli and signals in the environment are the sort of capabilities from which eventually develops a person's *knowledge* about material objects existing in time and space, e.g. knowledge of what *acts* one may or may not perform with or upon these objects.

Now, it is a feature of psychological investigations of this development that the *end-result* or *outcome* of the development is known in advance; indeed, it is the very aim of these investigations to account for how some well specified, future capabilities of experiencing reality develop out of states and capabilities of an organism not thus specifiable. This knowledge of the outcome has the immediate advantage that the investigations may be directed towards relevant bio-physiological dispositions and circumstances thought to serve as necessary conditions for an organism later to acquire the capabilities of a person. However, the danger is that in characterising these dispositions and circumstances, which, for example, could be the abilities of organisms to discriminate and react consistently to stimuli and signals in the environment, we tend to ascribe features to such dispositions and circumstances which "anticipate" the significantly different cognitive and behavioural capabilities, which occur at later stages of the development.

Well, there are scores of reasons why it is so difficult to determine the transition which takes place from the time (t_1) when the infant may be characterized as a (mere) bio-physiological organism which reacts to stimuli or signals from the environment, and the time (t_2) when it may be characterized as a person - however rudimentary. One of the reasons is that it is not an all-or-non business; the borders between organism and person are not clear-cut, but blurred. Different forms of behaviour may appear in the transitional period, which may be correctly interpreted as forms of action which require knowledge of objects in the world by the child - while at the same time the behaviour of the child in other situations may be correctly described as mere reactions to stimuli or signals. The point worth emphasizing, however, is that this transitional period cannot be characterized as a period in which the child *becomes* a person, without referring to the *result* which appears at time t_2, but which was not manifest in the intermediate period. The significance of this point will hopefully become clear in what follows.

First, how can we describe what takes place from the time t_1 to time t_2? May we say that in this period the child *learns* to acquire knowledge about himself and the world around him, or *learns* how to observe and act in the world with objects and other persons, or *learns* to apply the logical apparatus involved in knowing and acting in the world? The way a person has been defined in this book, according to which a person does not exist without having *some* knowledge and, therefore, *concepts* about the world, as well as the logical apparatus involved in knowledge of the world (i.e. intentionality, reference and notions of truth), we cannot say that the child learns any of the above things, without at the *same* time learning the others. We cannot say, for example, that the child can perform acts with things, and thus may know of and make choices between different possibilities of action with the things, *before* they have knowledge about the things. Neither can we say that the child has concepts about objects *before* it starts to act and observe the world, and thus has acquired some experience of the world. To say this would involve a Platonic or Rationalistic assumption of the existence in the child of innate ideas or concepts of the world and objects, which is untenable. For, it would imply that concepts and thus knowledge of objects in the world could exist apart from, independently of and *before* reference may be made to that which the concepts concern. However, as has been argued in earlier chapters, such Platonic or rationalistic assumptions as to the possibility of a separation of knowledge about objects from the objects they are about, would render *both* our notions of knowledge and concepts about objects, *and* of objects ill-defined.

But nor can we say that the logical structures, notions and necessary relations involved in propositional knowledge are present or innate - if only rudimentarily - *before* the child starts to experience and make observations of objects in the world. For that would have to imply that such logical structures, notions and relations could exist in the child's mind some time *before* the child had started to learn to experience and make observations of the world. However, since these logical structures and notions of knowledge, i.e.

intentionality and truth, precisely concern the necessary relation or interdependency which exists *between* knowledge and what knowledge is about (such as the world and its objects), these logical structures cannot possibly be claimed to exist "in" the child before the child gets "out there" and acts and makes observations in the world. So to claim would be to claim that not only *knowledge* of the world, but also *the world itself* could exist for the child *before* the child enters the world. That is, we would quite clearly be dealing with claims and assumptions, which are at least as unacceptable as those of Platonism and Rationalism.

Well, that the child learns to experience and acquire knowledge about reality, would seem to require some activity or other by the child with objects and things in reality. However, learning to experience and acquire knowledge about objects, for example that they are objects existing 'out there' at some particular places in space, and about which something may be the case or *true* and something else not is not the case or *false*, as well as the rest of the logical apparatus involved in such knowledge, is not something which "comes about" in the child as a result of mere manipulation and discrimination of things and objects.[3] Again, what makes it so incredibly difficult - if not impossible - to describe what takes place in the child's first learning period is that it applies *par excellence* to what is learned in this period that it cannot really be described without referring to the result - which was not there during that period. The only thing we *can* say (and probably have to say) is that for an individual, who does become a person, possibilities exist that enable the individual at a given time and under given circumstances, to begin *simultaneously* to learn to experience and acquire knowledge about objects, and to act in reality and use the logical apparatus involved in such knowledge and action. And what we *can* do is to investigate *what* these circumstances are, and *when* conditions for such action and cognition exist.

The point, so difficult to express, is that it is logically impossible to explain the transition from organism to person, but that *in practice* this transition may be accounted for as a process *which takes place over time*. Precisely because the transition from organism to person is a process which takes place over time, it has been so obvious to think that this transition could be explained in terms of a *continuous* process - from biological organism and the bio-physiological dispositions of organisms to discriminate and react, to person and the capabilities of persons to act and have knowledge of themselves and the situations in reality in which they find

[3] Thus, apes and other animals do a lot of discrimination and manipulation with things in the world without, so it seems, ever acquiring the kind of knowledge about themselves and things in the world that children do, let alone acquiring the logical apparatus which goes with this knowledge.

themselves. However, if we do attempt to describe this transition as a continuous process, we are in danger of missing the logical conditions for being and acting as persons, and thus in danger, literally from the beginning, of missing the salient features of human cognition and action which ought to be of concern for psychology.

15.3 Arguments against the assumption of an innate language or linguistic structures

It is not difficult to see that the arguments against the assumption of innate "ideas" or cognitive structures to explain how we become individuals with the capabilities of persons, apply just as well to older as to more recent proposals of the existence of innate languages or language structures and principle to explain how it comes about that we become language users. Among the most well known recent proposals to this effect are Chomsky's theory of an innate *Universal Grammar* (see e.g. Chomsky, 1992) and Fodor's theory of an *Innate Language of Thought* (Fodor, 1976).

According to Chomsky's theory, innate "substantial" and "formal" grammatical structures and principles exist, which explain our understanding of and ability to interpret the meaning of linguistic expressions. The correct application of these principles of interpretation, so Chomsky claims, "are not the result of training, or even of experience" (ibid., p. 639). In Chomsky's own word, here is an example which illustrate how one such principle works.

> "Imagine a child learning English who comes to understand the sentence *John ate an apple*. The child then knows that the word *eat* takes two semantic roles, that of the subject (the agent of the action) and that of the object (the recipient of the action); it is a typical transitive verb. Suppose that the child now hears the reduced sentence *John ate*, in which the object is missing. Since the verb is transitive, requiring an objet, the child will understand the sentence to mean, roughly "John ate something or other". So far everything is fairly straightforward if we assume the simple principle that when a semantically required element is missing, the mind interprets it to be a kind of "empty pronoun" meaning: something or other. Perhaps an empiricist linguist might be willing to suppose that this principle is available as an innate element of the language faculty." (ibid., p. 639)

This and other examples of principles of grammar, the application of which are ... "known by any speaker of English without training, without correction of error, without relevant experience", Chomsky attributes to the computational system of the mind/brain, which, he claims, "is designed to force certain interpretations for linguistic expressions" (ibid., p. 640).

Now, even a superficial analysis of this and other examples of Chomsky's innate linguistic principles and structures reveals that the principles and structures in question are grammatical structures of expressions, we use when communicating in *linguistic terms* how and in what ways we relate to things and other people in the world, or how things in the world may be related to us and other things. Thus, in the example above, the sentence, "John ate an apple", is used to convey the fact that someone, i.e. *John*, carried out a particular activity, i.e. *eating*, and that the object of this particular activity happens to be *an apple*. Now, it would seem obvious that the understanding of the fact thus linguistically conveyed requires experience and knowledge of *eating*, e.g. that eating implies that *something*, e.g. an apple, is being eaten by *someone*, e.g. by a person called John. Without this knowledge and experience of what the elements of the expression refer to, nobody could possibly understand its meaning. Indeed, the expression, "John ate an apple", or just, "John ate", only has a meaning which may be understood, in virtue of referring to particular things and persons, and particular actions carried out by persons on things, which are known both to the producer and recipient of the expression.

Well, if this seems obvious, it must be just as obvious that understanding linguistic expressions about things, persons and acts carried out in the world, requires both experience and knowledge of things and persons and of ways in which persons and things may be related to one another. To claim the intrinsic existence of a language faculty, equipped with grammatical principles and structures capable of enforcing particular ways of parsing sentences and interpreting the meaning of linguistic expressions, is to commit the fallacy of Language-Reality Idealism. That is, it is to assume the existence of a linguistic system in which linguistic expressions may be generated, have a meaning and be understood, in spite of the fact that those sentences, by definition of the system being intrinsic or innate, have no relation to what they are about, i.e. matters of fact in reality. And it is to ignore that a substantial amount of the assumed intrinsic structures and principles of the so called language faculty precisely concern implications of concepts, which only exist and are well-defined due to the relation that the concepts bear to that to which they refer, and thus due to the necessary relation between language and reality. Hence, to claim the intrinsic existence of linguistic principles concerning the implications of and relations between concepts referring to states or events in the world, would be to ignore that these implications and relations necessarily rely on their relation with that to which they refer, and thus on knowledge and experience, of structures and relations between things the world.

However, Chomsky is quite willing to ignore this fact, and even to contend that

> "in the context of the theory of knowledge, our knowledge that expression such-and-such means so-and-so is not justified or grounded in experience in any useful sense of these terms." (ibid. p. 640).

Well, if such a theory of knowledge were to allow that language and linguistic expressions could concern or be about reality, it would have to be assumed that the language faculty itself possessed the relevant experience and knowledge about reality, and hence assumed that not only grammatical principles and structures, but also knowledge of reality, is innate and intrinsic to the language faculty of the mind/brain.

* * * * * *

In *The Language of Thought*, Fodor claims that "the only psychological models of cognitive processes that seem even remotely plausible, represent such processes as computational (Fodor, 1976, p. 27). However, as pointed out by Putnam, the problem with computational models of cognitive processes, including those involved in the use of language, is that they require the existence in each of us of a pre-programmed "universal language", and thus of "universal knowledge", which captures the meaning and reference of all possible present as well as past and future concepts (cf. chapter 14). However, in the book just mentioned, Fodor attempts to justify this assumption by arguing that, "you cannot *learn* a language whose terms express semantic properties not expressed by the terms of some language you are already able to use" (ibid., p. 61). In particular, it is a condition for learning a *natural* language that one knows and can use a language which exists prior to the acquisition of a natural language.

Fodor's argument for the necessity of assuming the existence of a language for learning a natural language are the following.

> "Learning a language (including, of course, a first language) involves learning what the predicates of the language mean. Learning what the predicates of a language mean involves learning a determination of the extension of these predicates. Learning a determination of the extension of the predicates involves learning that they fall under certain rules (i.e. truth rules). But one cannot learn that *P* falls under *R* unless one has a language in which *P* and *R* can be represented. So one cannot learn a language unless one has a language. In particular, one cannot learn a first language unless one already has a system capable of representing the predicates in that language *and their extensions*. And, on pain of circularity, that system cannot be the language that is being learned." (ibid. p. 63 - 64)

The further implication of the above argument is that, "... one cannot learn a language whose expressive power is greater than that of a language that one already knows" (ibid. p. 86).[4]

This assumed language (sometimes referred to by Fodor as "mentalese"), which we are supposed to use as a "vehicle" for thinking prior to and as a condition for learning a natural language, Fodor calls the *Language of Thought* (henceforth LOT). If now it were assumed that LOT is itself a language which is learned, then we would be on the road to an infinite regress, for, "learning the *metalanguage* in which representations of the extensions of object language predicates are formulated, must involve prior knowledge of a meta-metalanguage, in which truth definitions are couched. And so on ad infinitum" (ibid. p. 65). So, Fodor does not assume nor claim that the presupposed LOT is a language which is *learned*, only that it is *known*. - And since it is not learned, it must be *innate*.

However, the danger of an infinite regress still looms, if

> "understanding a predicate involves representing the extension of that predicate in some language you already understand. But now consider understanding the predicate of the metalanguage. Doesn't that presuppose a representation of *its* truth conditions in some meta-metalanguage previously understood." (ibid. p. 65).

However, as Fodor explains, *learning* what a predicate means involves representing the extension of that predicate, but *understanding* the predicate does not. A sufficient condition for the latter, says Fodor, "might be that one's use of the predicate is always in fact conformable to the truth rule" (ibid., p. 5).

Put in computational idioms, the internal, innate language and the relation between this and a learned, natural language, so Fodor contends, may be characterized by the following analogy. Real computers use two different languages: an input/output language and a machine language. Compilers mediate between the two languages in effect by specifying biconditionals whose left-hand side is a formula in the machine code. Such biconditionals can be conceived of as representations of truth conditions for formulae in the input/output language. Although the machine must have a compiler if it is to use the input/output language, it does not also need a compiler for the machine language. What avoids an infinite regress of compilers is the fact that the machine is built to use the machine language. That the machine is built to use the machine language means that its formulae correspond *directly* to computationally relevant physical states and operations of the machine. What takes the place of a truth definition for the machine language

[4] It would seem to follow, *mutatis mutandis*, that one cannot acquire any knowledge about reality unless one already has even more extensive knowledge of reality.

is simply the *engineering* principles which guarantee this correspondence (ibid. p. 65 - 66). Following the above analogy, in the case of the internal code this correspondence is determined, so Fodor speculates, by the innate structure of the nervous system (ibid. p. 78).

Now, if the relation between the postulated internal language of thought and a learned natural language is thus analogous to the relation between the machine language and the input-output language of a computer, it would seem that not only *consistent*, but also *correct* (i.e. truthful) use of natural language is guaranteed in virtue of this relation. Unless, of course, it is assumed that use of the languages and predicates in which we may think, and in which we may refer consistently to that which we may think and talk of, does not imply that that to which predicates may be consistently used, may be *correctly* identified, and thus that a notion of *consistent use of predicates* does not require a notion of such correct determination. However, there are good reasons for ignoring this possibility since it would render void not only the notion of consistent use of predicates, but also *representations* and *determination of extension of predicates*, and thus *rules for applying predicates*. If so, it would seem that Fodor with his computational account of how natural languages may be acquired and used, has managed to solve in one stroke two crucial problems of language and use of language, i.e. the problem of *reference* and *truth*: for in his model both *truth* and *reference* boil down to *consistency of use* - which in its turn is guaranteed by innate biological structures. However, if this goes, then anything goes - and we might as well admit that philosophy of cognition and language is a discipline not worth taken seriously.

However, apart from the problems of assuming the existence of an innate language and the logical apparatus involved in its use, which I discussed in the previous sections, the conditions for learning a natural language and for understanding and applying its predicates as suggested in Fodor's computational model, present other problems worth mentioning. Thus, Fodor may have avoided the problem of an infinite regress by saying that "though the machine must have a compiler if it is to use the input/output language it doesn't also need a compiler for the machine language" - since the formulae in the machine language correspond directly to computationally relevant physical states and operations of the machine. However, by the same token, what is meant by *understanding* predicates about natural language cannot be a property or part of that language; that is, it cannot be a language in which predicates may be generated, known or used about the language itself. If so, it would seem that the predicates and propositions that users of a natural language may put forward about predicates and propositions in the language they talk, have no expressible extensions in LOT. Hence, this feature of *reflexivity* of natural language must clearly be an example of the *greater* expressive power of natural language than that of LOT. So, how is that learned?

15.4 Principles for description: Conclusion

For millennia is has been considered self-evident that the principles of formal logic are principles, the validity of which cannot be doubted or questioned. They are beyond truth and falsehood in the sense that it is impossible to doubt or question their validity - and still talk sense. Such doubt or questioning would require the possibility of going beyond the principles of formal logic, but the point is that we cannot analyse these principles, still less express doubt about or question their validity, without expressing this doubt and these questions in a language in which the principles of formal logic are valid. In other words, we cannot doubt or question these principles without contradicting ourselves.

In the same way the conviction has reigned that whatever we talk about outside the domain of formal logic, for example reality and our cognition and description of reality, may be talked about quite arbitrarily. Whether the point of view of one or other of the existing philosophical positions was championed, became eventually a matter of taste or conviction (or political power or religious compulsion), but it could certainly not rest on logical necessity. To such status only the principles and implications of formal logic could attain.

However, as Zinkernagel has argued in his, for contemporary philosophy tragically overlooked book, *Conditions for Description* (1962), this view does not hold. By shedding light on our use of elementary rules of everyday language, and by demonstrating the epistemological consequences of his findings on the logical relations of these rules, Zinkernagel has convincingly shown that we must assume the existence of principles beyond the domain of formal logic which pertain to the same necessity as the ones belonging to formal logic. In general he has shown how important it is for the development of a consistent theory of epistemology that such conditions for description exist.

That fundamental, logical conditions for language and use of language exist which *have to be taken for granted* means that it is not possible to ask what the status of these logical conditions is. We cannot ask, for example, whether such logical conditions and implications have an "a priori" or "a posteriori" status, or whether they are "analytic" or "synthetic", or the like.

Let me quote Zinkernagel on this vital point:

> "Epistemologically, this bifurcation [a priori versus a posteriori, my addition] of our knowledge does not present very cheerful prospects. Because of it, we are faced with the alternative of regarding epistemological statements as expressing

> *a priori* knowledge of our own words or concepts, regardless of the applicability of these words and concepts, or as statements about empirical matters, the validity of which may tomorrow be perfectly unwarrantable.
>
> It is possible to escape from this unpleasant dilemma, if, beside formal logic, there are other general conditions for description of experience. The knowledge of such conditions cannot be characterized as *a priori* knowledge, which concerns only our own words and concepts, nor experience; in fact, what we aim at is knowledge about conditions for describing experience. But neither can that kind of knowledge be characterized as knowledge of experience, which may prove invalid if our experience change, seeing that such knowledge is about our conditions for describing experience and, in consequence, about our conditions for giving well-defined descriptions of the change of experience. The distinction between *a priori* and *a posteriori* knowledge cannot be applied to statements about conditions for describing experience." (ibid., p. 43)

In the chapters of Part II, I have argued that the *Principle of the general correctness of linguistic description*, i.e. the presupposition that we can use language to talk correctly about reality, and, conversely, that reality exists as something which we may talk correctly about, must be a logical condition for description and use of language. I have done so by showing that we cannot talk about language and propositions put forward in language about reality independently of or without talking about reality - and conversely - that it does not make sense to talk about these conditions for describing reality independently of these conditions. Furthermore, I have attempted to show what are the logical implications of these conditions. However, the fact that conditions for language and use of language are fundamental, does not mean that they are conditions which could exist *before* use of language and concrete situations in which language users may find themselves in reality, i.e. that they are "transcendental" conditions, the origin of which we could account for or explain in terms of innate "structures" or "ideas" about necessary principles and the like. Neither could such "structures" or "ideas" be explained by means of empirical investigations. For, any account of the fundamental conditions for language and use of language to describe reality - or any other things on which we may carry out empirical investigations, including language - would necessarily have to be formulated in a language to which these conditions apply.

So, the reason why it is not meaningful to ask about the status of the logical conditions for description and use of language, is that knowledge about these conditions cannot be characterized as knowledge about language *as opposed to* knowledge about reality. Conversely, knowledge about these conditions cannot be characterized as knowledge about reality *as opposed to* knowledge about language.

Whilst acknowledging that there is nothing new in my calling attention to the necessity of assuming the existence of logical conditions for description and use of language, it has to be said that, apparently, this necessity still awaits proper recognition and understanding. And whilst acknowledging, furthermore, that there may be many more important consequences of the *Correctness principle* for language and knowledge than I have so far been able to pin down, I hope with this principle to have taken significantly further some of the points which only lie dormant in Zinkernagel's work. Not least, I hope to have made clear the crucially importance for both science and philosophy of the limitations *and* possibilities imposed by the conditions of a logical nature, on which our use of language to describe reality are necessarily based. These conditions, as well as the limitations and possibilities they impose, I have argued, are crucially important for philosophers and scientists to be aware of and constantly to investigate - in their efforts to determine what kind of questions may be asked and what kind of accounts may be given about reality and the language used to describe reality. This applies whether our accounts are about matters dealt with within the Natural Sciences, or whether they are about matters of concern to Psychology, or to accounts of phenomena of a cultural, artistic, religious or fictional kind - or just to normal everyday situations. In particular, I have tried to show what it means and implies that any description of reality is limited, and to argue what possibilities follows from the fact that whilst limited, any description of things in reality put forward in any particular situation is necessarily related to other descriptions of the same things and reality put forward in other situations. In Chapter 16 of Part III, I shall further expand on the consequences of this for the possibility of having consistent notions of 'identity' and of 'the same thing', and hence of 'same' and of 'thing'.

Since Galileo the view has been endorsed that for a research area to acquire the status of a scientific discipline, certain requirements must be met. These requirements include definition and delimitation of the range of features and properties of the phenomena or subject matter to be investigated, as well as basic assumptions about their nature. To this must be added development of methods and concepts by which unambiguously and adequately to account for the phenomena being investigated. These are basic requirements to be met by any discipline which claims that its inquiries, as well as the results emerging from them, make a difference vis à vis common sense and superstition. It has been the aim of the various chapters of Part II to show that *how* and *in what way* a discipline defines and determines its subject matter and delimits its area of research from those of others studying *radically* different phenomena, is just as essential to secure this difference. And, not the least, it has been the aim to show that whether or not a science succeeds in developing *consistent* theories about its subject matter, crucially depends on whether or not its definitions and delimitation of this matter accord with the general logical conditions for describing reality and whatever else we use language to describe.

PART III

IDENTITY

Chapter 16

Identity and identification – same and different

16.1 Introduction

In speech act theories from Austin through Searle to Habermas and Grice there has been a general consensus about the conditions which constitute a dialogue, and the situation in which verbal communication between people takes place. It is agreed that a dialogue is a social activity which, formally speaking, presupposes the presence and acting of at least two persons in the roles of speaker and listener; a "something" which is the object being talked about; and verbal utterances being "sent" and "received". In the type of speech acts with which this text so far has been mainly (if not entirely) concerned, i.e. statements and descriptions about things in reality, it is fundamental that the speakers have identified not only the things being talked about, but also themselves as persons belonging within the same reality.

Furthermore, for any description or statement put forward to be meaningful for both speaker and listener, they must know the implications of the statements being used. Or, if they do not, it must somehow be possible for them to discuss and reach agreement as to what the implications may be. Likewise, they must know in what situations and about what the statements are correctly applied. The latter conditions may become obvious if we consider the situation in which two speakers are talking, say, about Egyptian pyramids. If one of them believes that pyramids are cubic and 6 feet tall, and that the term 'pyramids' implies descriptions to that effect, while to the other the term 'pyramids' has quite different implications, no further sensible discussion among them about Egyptian pyramids would seem possible. The same would apply if they did not agree that the term 'pyramids' is correctly used to refer to some particular objects to be found at some particular places in the Egyptian dessert.

This set of conditions would seem a reasonable tentative basis on which to carry out further analysis and characterization of speech acts. In particular, it seems reasonable to presuppose that we have to know *what* we are talking about in order to talk about it, and thus that the "something" being talked about has been *identified*. But what does it mean to have identified something? What does it take and what must be implied in an identification of an object or event, which sufficiently determine the object being talked about?

In most speech act theories it is assumed that the identification of the object of conversation, necessary for the linguistic determination and description of it, must in the final analysis rely on some kind of non-verbal act of identifying the object (e.g. visually). Accordingly, the "something" being talked about belongs to a non-verbal context of action, which forms the basis of linguistic discourse. How this non-verbal identification is brought about, and how more precisely to characterize the relation between the non-verbal determination of the object and a verbal description of it, is of no particular concern to speech act theories themselves. - This would seem a reasonable stance, especially in view of the argument that no *linguistic* description, no matter how comprehensive, would ever amount to an identification which could *exhaustively* determine the object or event being talked about and that, in case of doubt, only recourse to some non-verbal exhibition of the object or event in question would resolve the problem of what is being talked about.

Now, in other quarters of philosophy which deal with epistemological problems of language and use of language, the "problem of identity" or of "identifying" an object is not taken so lightly. Among the problems being debated are problems such as "what is it that makes a thing a *particular* thing", or "what is it that identifies the thing as the *same* thing, and thus makes it preserve its identity or "sameness" across different situations in which the thing may be described differently?" Indeed, how can we be sure that we are in fact talking about and referring to the same objects or events in situations in which they may be referred to and described differently? These are questions which seem to be obvious for any theory of language and cognition, and on the answers to which any theory of language and cognition would seem to hinge.

An extended presentation will not be given here of how these problems have been dealt with in the philosophical literature, but only an outline of some recent positions and solutions relevant to the analysis of *the problem of identity*, which will be subsequently suggested.

16.2 Recent positions on the problem of identity and reference

I shall start with what Wittgenstein had to say about the issue. In his discussion of how we use words to refer to objects, Wittgenstein (1958a) criticized the assumption that for a general term (say, 'gold', 'dog', or 'book') to refer or apply to an object, the object must posses the proper *essence* or defining properties. The view that there must be defining properties for a general term goes back to Socrates, who maintained that one could gain understanding of terms only by discovering the essential property that would make something an instance of the term in question.

This view, held by many philosophers ever since, is challenged by Wittgenstein, who points out that for many important terms in language we cannot specify defining or essential properties. He writes:

> "The idea that in order to get clear about the meaning of a general term one had to find the common element in all its application, has shackled philosophical investigation; for it has not only led to no result, but also made the philosopher dismiss as irrelevant the concrete cases, which alone could have helped him to understand the usage of the general term. When Socrates asks the question 'what is knowledge?' he does not even regard it as a preliminary answer to enumerate cases of knowledge." (Wittgenstein, 1958a, pp. 19-20)

Wittgenstein maintains that it is *not* because of the inadequacy of language that we cannot specify defining or essential properties; rather, the point is that use of language *does not require that things have essence*. He uses the example of the term *game* to illustrate his point. Thus, there are a variety of quite different activities for which the term game is used, i.e. for ball-games, board-games, card-games, nursery school games, Olympic games etc.; some of these games involve competition, some require skills, others have an element of chance, still others are amusing etc. However, there is no defining property nor "essence" of games being shared by all and only games, but merely a variety of overlapping similarities between different games. To describe his alternative view of what may group things into different kinds, i.e. identify them within a particular category, Wittgenstein (1958b) introduced the notion of "family resemblance". There are, he says no necessary and sufficient conditions that determine membership in a category, but, just like members of a human family, they may resemble each other without there being one or more characteristics shared by all.

Kripke (1971, 1972) goes even further and denies that there are any sets of properties that determine the meaning of terms for such categories or, indeed, that such terms have any meaning. He proposes the alternative and much debated view that *proper names* referring to individuals, and *common nouns* referring to "natural kinds" like 'carbon', 'gold' or 'apple', only have referents. They are so called *rigid designators* which are used to pick out things, but have no meaning which refers to the properties of the things that they pick out. In particular, names and nouns for things are not *descriptions* of things, nor do they imply that something has been predicated about the things to which they refer. On the contrary, terms apply *directly* to the object, and they do so by a connection which was fixed by the initial naming of the object. How Kripke arrives at this view is a rather long story, which I shall try to cut short.

The story originates in an attempt to sort out the meaning and truth of sentences which contain *modal* words such as "necessarily, "must", "possibly", or "may". An example is, "It is possible that Nixon might not have been president". Now, this sentence is perfectly understandable. Most ordinary people know from experience that things may not always turn out as expected. All kinds of things may happen - some quite accidentally, others not so accidentally - which may change the course of events. Nixon, for example, might not have gained enough votes to win the election; he might have been assassinated before his inauguration; or events and choices he made in his earlier life might have steered him clear of a political career, etc. In view of such contexts of possible events and choices, the sentence both has a meaning, which may be understood, and it may be a sentence which is perfectly true.

However, as pointed out by Kripke and other logicians, because the sentence is a *counter factual* sentence, i.e. it asks us to imagine that things might have been different, it cannot be determined what precisely the sentence asserts by determining whether Nixon was in fact elected President. Nor may the truth of *counter factuals* be determined by reference to *actual* facts of reality. In order to represent what such statements are affirming, Kripke introduced the notion of a model, consisting of a set of objects called *possible worlds*. It is a notion which invites us to think of alternative universes that are defined in terms of specific changes from this universe, and which help us consider how other things would be different under these situations.[1] In an actual case we may choose a world in which Nixon was not born - and then fill in the rest of the scenario for that world. The point is that granted we manage to imagine the right kind of possible worlds, the meaning of the modal sentence about Nixon may be explicated, and so may the conditions in which a modal statement be true or false.

[1] Again, this is a strategy which does not differ much from the strategy adopted by ordinary people in handling their ordinary businesses of life (cf. the comments above). In order to cope with the demands of society and to plan one's activities, it is often necessary to consider several different possible outcomes of events, as well as of ones own actions, both prospectively and retrospectively.

By using this notion of *possible worlds* it can be shown, according to Kripke, that it is true that an object necessarily has a *particular* property only if it is the case that it has the property *in every* possible world in which the object exists. Thus, "Nixon was necessarily a lawyer" is true if, in every world in which Nixon exists, he is also a lawyer. However, if there is a possible world in which he exists and is not a lawyer, the statement is false.

This is in broad outline how Kripke arrives at the view that the names and nouns we use to pick out referents are *not* equivalent to any description of them. The view rests on the argument that we can envisage a possible world in which a person or a thing would not have the properties by which they are identified in the actual world. For example, we may pick out Richard Nixon as the person who was the 37th President of the United States; or as the attorney who prepared the court cases for the MacCarthy senate committee on "un-American activities"; or as the first Western head of state to visit communist China. But we can also envisage the possibility that he would never have been elected President; that the MacCarthy committee never existed; or that Richard Nixon's trip to China came to nothing due to cold-war complications. So, Kripke claims, the name is not identical with any description of the individual being referred to by the name (in Kripke's terminology, such descriptions are *non-rigid*); indeed, the name may be used to pick out a person, irrespective of what properties that person might have had in the possible world under consideration.

The argument for this view about proper names is much the same when it comes to common nouns. Because an object might not have the property we normally associate with the noun (e.g., gold might not be yellow in some possible world), the property cannot determine its reference; nor does the noun refer to that property. However, since both proper names and common nouns may in fact be used to refer to the *same* individual or object respectively, they may be conceived of as *rigid designators* which serve the purpose in this world as well as in other possible worlds to convey their referents to someone else.

But how, more precisely, does this fixation or linking of nouns or names to objects and individuals come about? Kripke and other proponents of this modal approach advocate a *causal* theory according to which names and nouns get linked to their referents through a causal chain. For example, at a baptismal ceremony a name might have been assigned to a person. All subsequent use of that name for that person is traced back to the original naming. This means, according to Kripke, that an individual in any possible world who has the same parents as, say, the actual Richard Nixon, i.e. comes from the same fertilized egg, is Richard Nixon, even if, in other possible worlds, he has a different life history. Granted that a successful "naming ceremony" was performed to begin with, and that the right sort of continuity in the later use of the name exists, people may subsequently use the name 'Richard Nixon' to refer to the individual to whom it was given at that ceremony. Similarly, when someone first encounters an "instance of a natural kind", such as a piece of gold, they might assign the name "gold" to that kind. Subsequent use of the name for "substances" of that kind will then be tied to it through a causal chain. However, as it applies to a name, the fact that the reference of a noun has been fixed does not mean that the properties of the instance to which it refers figures as part of the meaning of the noun. The noun does not *mean* any of the properties of the instance to which it actually refers, just as a name fixed to an individual does not *mean* any part of the life history of that individual.

Putnam, a proponent of this causal theory of reference, contends that the question of the *meaning* of a term only arises in the extremely rare cases where terms *are* linguistically associated with *necessary and sufficient conditions* in the way that, say, 'bachelor' is. He writes,

> "If you know that 'Quine' is a name and I know that 'Quine' is a name and, in addition, we both refer to the same person when we use the word [...] then the question of whether 'Quine' has the same meaning in my idiolect and in yours does not arise. [And] once the term 'electricity' has been introduced into someone's vocabulary [...] whether by an introducing event, or by his learning the word from someone who learned it by an introducing event, [...] the referent in that person's idiolect is also fixed, even if no knowledge that that person has fixed it. And once the referent is fixed, one can use the word to formulate any number of theories about that referent, [...] without the word's being in any sense a different word in different situations. Thus, [...] it does not follow at all that *meaning* [e.g. of the term 'electricity'] depends on the theory you accept." (Putnam, 1975, pp. 201 - 202).

What to Putnam seems right about Kripke's account is first of all that the knowledge which an individual user of a language has, need not at all fix the reference of the proper names in that individual's idiolect; for, the reference is fixed by the fact that that individual is causally linked to other individuals who were in a position to pick out the bearer of the name. This means, secondly, that the use of proper names is *collective*. Anyone who uses a proper name to refer, is thus a member of a collective which had 'contact' with the bearer of the name.

The views and points so far presented raise a number of questions. First, as Putnam himself concedes, the question of the meaning of terms *can* arise. If, for example, a person uses the word 'Quine' to pick out someone other than the bearer of the name, i.e. *Quine*, then there would be a difference in meaning in that persons idiolect and the meaning of the term in the language in which its reference has been "collectively" fixed. But how, we may well ask, could we ever begin to determine *differences* in the meaning of a term in the idiolect of a speaker and that which have been collectively fixed, without presupposing at least *some* (shared) knowledge about the things and their properties being referred to by the term - and without presupposing, therefore, that terms not only refer to, but also imply or *mean* those properties? Indeed, how could we talk sensibly about *incorrect* uses of terms of a "collective" language without this presupposition?

Secondly, according to Putnam's treatment of the electricity example, it does not seem relevant to theorists in this tradition that the identification of other, *subsequent* instances of a natural kind, will have to rely on some identification procedures or other. However, unless such procedures are thought to depend entirely on non-linguistic identification in which *nothing* is being predicated about things, a procedure in which a thing is identified as being an instance of a particular, previously named natural kind, must necessarily at some stage contain terms, which not only refer to the "substance", but also to some of its properties. If not, descriptions involved in the identification procedure such as, "substance A is X, Y, and Z", would be descriptions in which terms like A as well as the X's, Y's and Z's only referred, but had no meaning or implications - in which case nothing is being predicated about A. So, what kind of verbal identification of A would that amount to?

But these are not the only problems. When subsequent to a "naming ceremony", let us say forty years hence, we use the name 'Richard Nixon', we do not, of course, use it to refer to the *baby* to whom the name was fixed at that ceremony, but to a person who has undergone a tremendous range of radical changes - physically as well as psychologically and socially. Indeed, *being* that person forty years hence, physically, psychologically and socially, *relies* on a range and history of such changes and the conditions which made them possible. To hold, as Kripke does, that for a name to pick out a person subsequent to an appropriate "naming ceremony" requires that *the right kind of continuity in the later use of the name exists*, only makes sense if by *continuity* is implied reference precisely to a particular set of changes and historical events. It may well be the case, indeed it *is* the case, that depending on the actual situations in which we talk about e.g. Richard Nixon, different descriptions and determinations of the person to whom the name 'Richard Nixon' is fixed will be implied. Thus, we may talk about Richard Nixon as the 37th president of the United States, or as the first Western head of state to have visited communist China. However, for the name 'Richard Nixon' to pick out a particular person, the *same* person, in different situations or at different times necessarily requires that 'Richard Nixon' implies and means the person who lived through this set of different historical events. To repeat, this, if anything, is what must be entailed in the "continuity" of the use of the name 'Richard Nixon'.

The same applies to nouns, i.e. terms we use to refer to natural kinds. Although a term was originally fixed to a natural kind in a particular situation in which we had access to, and thus knowledge about, a particular, *limited* set of its properties, we are still able to use the term to refer to that same kind in other or new situations, in which we have access to different sets of its properties - or in which its properties have undergone some or other changes. In other words, we may still be able to use the term to pick out the natural kind in questions as *the same*, which in other situations is presented differently and by different properties. But to be able so to do, requires that the same term may refer to and mean *different* properties of the same thing in different situations, and therefore, that being in a situation and having identified a thing in that situation implies knowing not only what is the case with the thing in that situation, but also what may be the case about it in other situations. Going a step further, to have identified a thing in a situation implies knowing that there may be something else about the *same* thing or situation, which is (as yet) unknown, but which we may come to know about - in the present or in other situations.

With these considerations in mind, it is possible to point out more clearly a series of problems entailed in the employment of counter factuals in order to determine what properties an object may or may not necessarily have. There is the problem, firstly, of how we determine which entity in another possible world is Richard Nixon, or a piece of gold, i.e. how the "transworld identity" of objects appearing in different possible worlds are preserved. To Kripke, however, such an objection is fundamentally mistaken. For, possible worlds are *stipulated* worlds, not discovered. We stipulate which *individuals* or *objects* exist in the possible world and what properties they have. Hence, we never need to raise questions as to which individual or object corresponds to an individual or object in the world. If we take the case of *gold*, for example, we decide whether or not it exists in the possible world we are imagining, and if it does, then we attribute to it all its *essential* properties (which, in Kripke's view, are typically such properties that are not used to identify gold under normal everyday conditions, but rather its physical properties, such as its atomic number), and whatever other properties we ascribe to gold in the possible world.

Although this proposal may get Kripke off the "transworld identity" hook, it certainly puts him on another and no less uncomfortable one. The view proposed positions him in *Essentialism*, according to which there are some *determinable* properties, which are essential to objects in the sense that if the objects lack those properties, they would not be the *same* entities. Such an assumption, as pointed out by Wittgenstein, won't work. Apart from the reasons already pointed out by Wittgenstein, and the ones which will be given in the following sections, here are some of the most obvious problems entailed in Kripke's view, a view, as we have seen, shared by Putnam.

First, to identify the essential properties of an object, Kripke relies heavily on his *intuition* about what makes an object the object it is. For example, in the case of human beings, Kripke takes their *origin* to constitute their essential property; that is, although Nixon might have chosen a different career - Nixon might have become a sumo wrestler - he could not have been born of different parents. In the case of natural kinds such as gold or water, it is Putnam's intuition that it is their physical or chemical composition as "unpacked" by current physics which is essential - since it is the atomic composition that determine the law-like behaviour of a substance. To both Kripke and Putnam, "things which are given existentially and not by criteria help to fix reference" (Putnam, 1983, p. 73). What makes something a lemon, for example, is having the same *nature* (e.g. the same DNA), and not fulfilling the same set of criteria (yellow colour, thick peel, tart taste,...) laid down in advance. Accordingly, a term refers to something if "it stands in the same kind of relation (i.e. causal continuity in the case of proper names; sameness of 'nature' in the case of natural kind terms) to these existentially given things". Although the meaning of the term 'lemon' in the idiolect of different people may differ, the term only has *one* referent, namely the "stuff" of which the lemon is objectively composed; what that is, is established by science.

To Kripke the essence of things thus established are properties which things have in any possible world and, consequently, the truth of statements about such properties is *necessary* - even though we cannot know it *a priori*, but come to learn it empirically. (Conversely, the truth of statements about the properties of things, which they do not have in all possible worlds, is not necessary, nor are they necessarily about the same thing). Water, for example, is H_2O in any world in which it exists, although it might differ in other properties from water there, i.e. may not be a wet and tasteless liquid - or whatever other meaning people have in mind when they describe water outside scientific conditions. Thus, once we discover the composition of water in the actual world, we also discover how a substance must be composed to *be* water, and thus discover the essence or "existentially given nature" of water. This suggests, says Putnam, "that the old idea that science discovers necessary truth, that science discovers the essence of things, is, in an important sense, right not wrong" (Putnam, 1983. p. 55).

The problem with this modal approach to determine the reference of terms, however, is that the intuitions of Kripke and Putnam about what constitutes the essential properties of objects or individuals, are not necessarily shared by everyone, indeed they may be completely unknown to most speakers. To some it may appear that the physical appearance of Nixon is crucial to someone being Nixon, to others that he was a politician, or a lawyer. And to some it is a crucial property of water that it is a colourless liquid, which can be used to make tea, to wash clothes, or which turns to ice at a particular temperature.

This means, secondly, that in stipulating a "possible world", *any* of the qualities of an object or individual could have been chosen to identify the object or individual. If so, the notion of possible worlds and the whole business of using and constructing counter factuals for determining what are the essential properties of objects and individuals, rests on a set of totally arbitrary choices, and thus it cannot be used as a rigorous vehicle to determine what constitutes the identity of objects and individuals. But nor may it be used to substantiate the claim put forward by Kripke and Putnam that names and nouns have no meaning. Rather, so to claim amounts to avoiding the issue of how and in what sense names and nouns may refer to particular things at all, and thus may be used to pick out and 'mean' those things - as opposed to others. And to claim this problem to be solved by suggesting that our identification of things relies on the essential properties which things objectively "have" (whether known or unknown to the ordinary speaker) and, consequently, that linguistic terms are used to pick out things having *those* properties, is similarly to turn a blind eye to the impossibility of accounting for our everyday knowledge of things and the meaning of nouns and names in our everyday language in terms of such properties - whatever their nature.

So, although Kripke and Putnam might well have intended from the outset to radicalize the points put forward by Wittgenstein and to propose a better, more consistent alternative avoiding the problems of the traditional view of things having defining or essential properties, their so called "new theory of reference" appears in the end to re-instate, in a new *scientistic* guise, this very same traditional view and its problems. Luckily, as Wittgenstein pointed out, neither identifying things in the world, nor using language to talk consistently and unambiguously about them, relies on things having defining or essential properties, nor on terms referring to such properties.

What I find particular confusing in the current discussion of how things we talk about are identified, is that "the identity of a thing", i.e. that which make a thing a *particular* thing and the *same* in any circumstances, is treated as if it is something which "resides" in the thing - *as opposed to a notion belonging to an account of how people carry out acts of identifying things and events in particular situations*. So, from this point of view I shall suggest the following analysis of what identifying things implies - as well as an alternative solution to the problem of the identity of things. Let me stress that this analysis is a conceptual analysis, not to be confused with a phenomenological analysis.

16.3 To identify a thing as the same: an alternative view

Whenever we experience or observe a thing, i.e. an object or event, the thing is experienced or observed as a *particular* thing. We rarely, if ever, experience or observe a thing in material reality as just a "something", a mere "blob", a "substance" with no meaning, function, determinate shape or extension. Even when confronted with a thing we have never before encountered, we normally experience it as something with a particular shape, colour, or texture; as a living or a dead thing, occupying some determinable amount of space at a particular location relative to other things and ourselves. In this sense, a "thing" is always experienced or observed as a thing about which something has been *predicated*, a thing about which something is the case and something else is not the case, and about which, therefore we have knowledge - however rudimentary that knowledge may be, as in the case when we encounter a thing for the first time. And in this sense a thing identified is a thing *described*, and what applies to putting forward descriptions or assertions of a thing applies to identifying a thing:

Just as we may describe a thing differently in different situations, so identifying a thing in one situation may differ from identifying the same thing in a different situation. And just as in describing things, a whole series of possible ways of identifying things is at our disposal. For example the bottle on the table in front of me I may observe or identify as the bottle of Port which I received as a present from an acquaintance; or I may identify it as the *transparent* bottle (if there are several bottles and the others are not transparent), or the "little" bottle, etc. If, by chance I am a physicist carrying out experiments on the bottle in a laboratory I may describe and thus identify it as a particular structure of molecules. What is common to these different identifications and descriptions of the thing - and almost infinitely many others which could have been suggested - is that they may be considered as "choices" or "acts" of the person identifying the thing with regard to, indeed which *presupposes* different sets of possible descriptive or categorical systems and means of observing and acting with the thing - irrespective of whether the identification is expressed verbally or not.

These examples show, first, that a given thing or event may be identified or described in *different* ways; how, in a particular situation, a thing is identified will - just as is the case for describing the thing - depend on our aims, purposes, interests, as well as conditions of observation and action in that situation. Second, that an identification or description is always made within concrete *situations* or *contexts*, i.e. situations permitting particular actions and observations. Thirdly, the identification, its meaning or implications, will be determined by a *limited set of predications*, i.e. those which are relevant to the situations or context of action and observation in question. (This issue will be further developed in the section: Identification, reference and truth.)

However, having identified a thing also implies having distinguished it from other things. The bottle on my table, for example, is a separate thing which is not one with the table. The table, in its turn, is a separate thing, which is not one with the floor on which it stands, etc. Again, how things are separated or distinguished from other individual things depends on the situation in which the distinction is made, i.e. is not well-defined independently of the possibilities of action and observation, descriptive or categorical systems, etc., which are not only *relevant* to, but which define the situation in question.

Now, it rarely if ever happens that, in a concrete situation, a thing is identified only by its membership of some particular category. The thing has inevitably at the same time also been identified as *this* thing, which is *different* from other members of the category, and as the *same* thing which in different situations may be observed, identified or described in different ways, as in the examples above. Let me for a moment tentatively suggest a distinction between what may be called an *external identification* and an *internal identification.* According to this distinction, an external identification refers to the fact that a thing is identified by its relations to other things in the situation, and according to the possibilities of observation, action and description available in that situation. An *internal identification*, on the other hand, refers to the fact that the thing being identified is the *same* as that which in other situations may be identified and observed in these and those other ways.

The important point of the argument I shall now attempt, is that for any concept of identity as the basis for identifying a thing, these two aspects of identification are always present. If a thing has been identified according to the one, it has necessarily also been identified according to the other. Thus, to identify a thing as a *particular* thing implies that it is *this* thing, which I now look at, and which, in the next moment, I touch, walk around, think of etc. In other words, it is *this* thing which in the current situations is related in these particular ways to other things, and towards which I have these particular opportunities of action, observation and description, the *same* thing towards which in *other* situations I may have these other possibilities of action, observation and description, and which in those situations is related in these other ways to other things. This, I contest, is what it means to have identified a thing.

Let me give a very simple example, which may clarify what I take to be involved in having identified a thing as a particular thing (and thus to have identified it both "externally" and "internally"). Please note, the example is to be understood as an example of principles, and *not* as an attempt to account for how identification of things actually develops.

An object, e.g. a "cup", is identified as a particular thing by a person, say a child, only when the child is able to appreciate that the cup is a thing which in different situations may have different functions, and a thing one may observe and towards which one may act in different ways. That is, when it is part of the child's cognition that the thing he is now knocking against the table, is the same as that from which a moment ago he was drinking, it only *then* makes sense to say that the child has identified the thing as a particular object. This could also be expressed by saying that if the thing is identified with regard to *one* situation only (i.e. is only identified *externally*), nothing like an identification of *this particular object* could be said to have taken place. If, furthermore, identifying a thing as a particular object were just a matter of identifying the object as belonging to some category or other, the thing could not be said to have been identified. In that case there would, most probably, be as many objects as there are ways of observing and categorizing the object and its various properties, or ways of acting towards or manipulating the object. And in the example just mentioned, the child's cognition of the cup would not amount to an identification of the cup; rather, what would exist for the child would be some sensation in the hand, a visual impression of some shape and size, a sound etc. And if the object is only a cup when the child uses the cup to drink from, it is questionable whether the child would be able to distinguish the "thing" and the act of "drinking" from "the cup".

Thus, an identification of a thing as a particular thing, i.e. *this* thing, the *same* thing, inevitably implies and presupposes the notion of: '*different descriptions of the same*'. It is not because we can identify and cognize "something" about the thing, which is the *same* throughout different situations that we are able to identify the thing as the same thing in different situations in which it may be described differently. It is the other way round. The thing is identified as a particular thing, the same thing, *in virtue* of the different descriptions of it. It simply would not make sense to talk about the thing being the *same* thing, and we would not have, indeed would not need the notion of *the same thing*, unless the thing in different situations appeared and could be described differently.

This applies when the thing is identified "across" different situations (as a bottle, a present from my acquaintance, a bunch of molecules structured in a particular way etc.) as well as in the quite elementary situation in which I identify the thing as the same thing which I *look* at and *hold in my hand,* and which in the next moment, e.g. when I move relative to the object, looks in this or that different way. Thus, to have identified the cup on the table in front of me implies that the cup seen from this particular perspective looks this way, while seen from a different perspective will look in this or that other way. (For example, from one perspective I cannot see the handle, from another I can see that the cup is hollow, etc.)

To have identified a thing, then, may be characterized by saying that, *logically*, one has committed oneself to describing and observing it in these and those possible different ways, in which the thing may present itself to the observer in the current as well as in different situations. Put differently, identifying a thing as a particular thing, implies that the identification of the thing in the actual situations is but *one* of the possible ways in which the thing may be identified. It is *this* fact about identification of a thing, which ensures the possibility of having to do with "the same", or "the same thing". That is, the notion of "the same", or "the same thing" only exists *in virtue* of the notion of "different" or of "different descriptions of the same thing".

It is for this reason that it is wrong to ask: "what (particular property, feature, trait or the like) is it that makes the thing "preserve" its identity throughout different situations". This reason applies whether the identification takes place in the kind of "stipulated" worlds, which philosophers like to imagine, or whether it takes place in the kind of "discovered" world, in which we observe and carry out investigations or acts on objects in normal every day as well as in scientific situations. Contrary to how the "new" traditional view has it (i.e. the Kripke-Putnam view), the "meaning" of an identification of a thing is not only determined by its relations to other descriptions or predicates within any *particular* descriptive or categorical systems (say, scientific), nor by any particular situation defined by such systems - but also by its relation to descriptions and predicates about it in *other* situations. Indeed, it only makes sense to say, for example, that the chemical identification and definition H_2O applies to the *water* in the cup on the table, *because* or *in so far* it is presupposed that the chemical determination of the "stuff" in the cup is *logically* related to our everyday identification and description of it (cf. Chapter 14).

Thus, it is not due to the fact that we are in a position of being able to deal with "something" which is the *same* about the thing across different situations or worlds, which is the condition for our dealing with different descriptions as different descriptions of the same thing or event - as is assumed in traditional theories of identity as the basis for identification. For, if by *the same* we were referring to some determinable unchangeable "essence", something which the thing really, "physically" or objectively were; and if this "sameness of nature" were some kind of observable "existentially given" which ensured both that the thing in different situations preserved its identity, and that terms for it *thus* have a determinate reference, then it is difficult to see how it could be maintained that there could be different versions or descriptions of "the same" - let alone different *correct* descriptions or versions of it.

In the alternative notion of identity suggested here, there is no reference to an underlying "essence", nor to some unchangeable properties of a "something", which somehow ensures that it is the same which, nevertheless, we describe and identify differently in different situations. On the contrary. This alternative notion implies that the identification of a thing being *this* thing, is part of the cognition that the thing in other and different situations is necessarily described differently and that, by identifying a thing as a particular thing, *we have inevitably, logically committed ourselves to other and different descriptions or identifications of the thing in other situations.* I believe this to be a matter of principle for identification. What the principle implies, I think, can best be captured in these expressions:

To have identified a thing, and thus to have described or predicated something about it, implies that *more* can be said about the same thing. Or even stronger: *We cannot say anything about anything without being able to say more about the same.* I call this *The Principle of Identity.*

16.4 Problems in traditional views on the identity of things

There is nothing new in what I have just said about the identity of things and about how things are identified. Indeed, it is something we all take for granted in every one of our acts with things in reality. It comes so naturally and automatically that we barely notice that this is how we identify things as being *particular* things, how things appear identified as the *same* things. Perhaps it is the immense truism of the matter that has made it so difficult to see. It is so fundamental to our knowledge of things that they present themselves to us in different ways in different situations and depending on the circumstances under which we observe them, how we act with or relate to the things, what we may do to them, etc. And it is so fundamental to our identification and description of things, indeed it is part of our notion of *correct* identification of things, that in different situations and under different conditions we describe and identify the same things in different ways. And it is so fundamental that this very notion of "different descriptions of the thing" is what ensures and makes possible the notion of "the same thing".

It is no wonder, then, that philosophers in their attempts to determine the essential properties or *essence* of things (by whatever means or criteria), i.e. the properties of things which unambiguously identify them, have not been able to come up with descriptions which sufficiently, let alone exhaustively capture the identity of the things; and no wonder that all accounts of the identity of things, which have been suggested along such lines, are deemed *indeterminate*. And it is no wonder that it may be argued that all such attempts are based on *arbitrary* choices of how things may be described or identified. Now, how a thing is identified in a particular situation is *not* of course arbitrary; what *is* arbitrary, however, is the choice of the situation (or "stipulated world") in which the identification and description of the thing is carried out. This applies equally when philosophers discuss the descriptions and identity of things in different, stipulated possible worlds, as in our ordinary everyday encounters with things in reality in different situations.

Rather, I think, we may well ask: Why give a thing a name, or why have particular nouns to refer to things, if it were not for the fact that the same thing in different situations is a thing which look differently, which may have different functions, properties and the like? A tomato, for example, is a "thing" which grows on a particular plant; but it is also a thing - the same thing - which may lie on the ground or on this table; and it is a thing which may be sliced up and put into a sandwich, which we may eat; or it is a thing from which we can make a sauce with a particular taste which goes very well with pasta. Or it is a thing we may use as ammunition to throw at unpopular politicians, etc. If, say, a thing was only identified as a tomato when it hangs on a particular plant - we quite simply would not need a name for it. Why have a "name" for it at all, unless or if it was not for the case that it helps us to pick out the thing that may appear to us so completely different in different situations, and have so many different qualities and functions?

According to the classical and still prevailing (though fallacious) view, the identity of things, i.e. that which makes things into the things they "really" are, is constituted by their unchangeable essential properties. Or, as essential properties of things are only observed in particular situation (i.e. scientific situations), the identity of things is a "sameness of nature" which is observed in those situations as some kind of "existential given". Indeed, according to this view (cf. Putnam op.sit), the truth of other or alternative descriptions of the things is not *necessary*, nor may they necessarily be about the *same* things. However, as I shall argue next, if our knowledge of the identity of things depended on some unchangeable *essence* existing "underneath" our various and alternative description of them, then descriptions of things, which deviated from the description of their essential properties, could just as well be descriptions *either* of something else, *or* be descriptions of the things which were incorrect, *or* be descriptions devoid of meaning. This is, as far as I can see, the highly unfortunate consequence of the "new" traditional view suggested by Kripke and Putnam.

The nearest I have come in my efforts to express this "new" traditional view, is by using this example: Different or alternative descriptions and identifications of a thing (except for the description of the "essence" of it) may be likened to "masks" superimposed over a "real" face; these descriptions and identifications, then, are really only descriptions of the masks by which the face is covered. Nevertheless, it makes good sense to say, so it is maintained, that the "mask-descriptions" *refer* to the (hidden) face underneath the mask. As, however, they are not descriptions of the face itself (i.e. the essential properties of the object, or their existentially given nature), their implications or meaning cannot be about the face itself; hence, such descriptions cannot contain any meaning about, nor be correct descriptions of the face (i.e. the object). As masks they can only refer to the fact that a face is hidden underneath it. - This being so, it is not difficult to understand why it is maintained that terms we use to pick out things have no meaning which refers to the properties of things. In the section which follows I shall further elaborate the consequences of these point and assumptions.

16.5 Identification, reference and truth

The fact that the so-called essential (i.e. physical) properties of things, are properties which exist throughout all situations (or all possible or imagined worlds) is in the prevailing traditional view taken to imply that the *descriptions* and *determinations* of these properties are likewise independent on the situation (or possible worlds) in which they are put forward. So to assume is to assume in effect that such descriptions and determinations are *context-free*. To emphasise what it means to assume that such determinations of things and their properties are context-free, we may call them "in-vacuo-determinations". Now, it is important to stress that although things and their properties do indeed exist independently of the situations and circumstances in which language users identify and describe them, *descriptions* and *identifications* of these things and properties do not exist independently of language users and the situations and context in which they describe and identify them. As already argued, a description of a thing will always be a description of the thing put forward within a particular context or situation, in which the meaning or implication of the description is defined by a set of well-defined consequences in the form of other correct descriptions of the thing; (in an everyday situation, for example, 'a cup' is "a solid and hollow object of a particular shape", "an object out of which one may drink tea'; in a scientific situation it is "an object made up of molecules of a particular structure"); contexts and situations, therefore, in which it makes sense to determine the meaning, truth or correctness of these descriptions. This applies as well to stipulated "in-vacuo-determinations", regardless of which examples of descriptions and contexts we might choose.

However, let us for a moment assume the traditional view that what makes a thing a particular thing, the *same* thing, is that things have an "essence", "an existentially given nature". And let us assume that it had been possible to determine the so called existentially given nature or properties of a thing, and thus a description of what the thing "really" and objectively is throughout all situations, contexts or possible worlds. Let us further assume that such a description exists of, say, *an apple*, i.e. a description which, in Putnam's words, refers to apples in virtue of "standing in the right relation to the existentially given nature of apples". It would of course - for the sake of argument - be unimportant what description and inside how broad or wide a "logical space" we choose, for example whether we chose a description of an apple in terms of its DNA-structure, or any of our ordinary everyday descriptions of apples. So, I choose a description which goes like this: "An apple is a thing which grows (or hangs) on a tree". Let me take one more example: "Water is water as it exists at room temperature".

Now, if these descriptions of *an apple* and *water* were descriptions of what *an apple* and *water* truly, objectively or "existentially" were, and which made an apple or water respectively the "same" things, then it is hard to see what could be meant by statements like e.g. "water boils at 100°C and freezes at zero", or "the apple is lying on the ground". Indeed, it is hard to see how it could be maintained that those descriptions were about *an apple* or *water* respectively.

Likewise, if the identity of things for all purposes and under all conditions relied on such fixed determinations of them, it is questionable whether it would make sense to say that by manipulating things and exposing them to the influence of various forces, or the like, we may observe what happens to them and find out more about the things and their properties. It would not make sense, for the simple reason that it could not be maintained that it were the very *same* things, which had undergone these changes - for they would no longer *be* the same things. Indeed, we would be cut off from attributing the observed changes or reactions to our manipulations of an object to the object in question, or saying that given such and such circumstances, which are different from the original, it will be correct to describe a thing in this different way. For, again, the descriptions of what exists during those other and different circumstances would logically, by this traditional determination of the identity of things, have to be either descriptions of other things, or descriptions devoid of meaning.

Even worse, however, if we are prevented from saying that it is *this* thing, the *same* thing, to which the changing circumstances pertain and, therefore, prevented from saying that the water which we have heated to boiling point is the same water that some time ago had room temperature, then we would be cut off from determining whether the description "Water boils at 100° C" is a correct or an incorrect, a true or a false statement. Indeed, if our knowledge about reality and its objects did depend on determinations of objects and their "real", "essential" or "existentially given" identity, there could not possibly be any notion of falsehood of determinations - nor therefore of truth. Not even the description "The apple hangs on the tree" would be a true or correct description. For there could be no *well defined notions of false or incorrect description of it*. The description "The apple is lying on the ground" would not be false; rather, it would be a description, which was indistinguishable from either a description devoid of meaning - or a description of something entirely different from that which was originally described.

Well then, the consequences of this traditional view about identity and identifying a thing as a particular thing, the *same* thing, is that it would not be possible to talk about descriptions which differ from the original as being alternative descriptions of the same thing, nor therefore to say that such alternative descriptions are descriptions of the thing as it may be *correctly* described under other conditions of observation. For, according to the view that the truth of such alternative descriptions of the thing is not necessary, nor necessarily about the thing (cf. Putnam op.cit), they would be descriptions which could not be distinguished from incorrect descriptions of the things and descriptions of some quite other things, or from descriptions devoid of meaning. Indeed, no criteria would exist on which to base such a distinction, nor therefore a notion of such a distinction.

It is, however, a matter of decisive importance for our cognition of things that such a notion exists and, thus, that we can talk in well defined ways about *incorrect* statements and descriptions of things, i.e. that something is or is *not* be the case about the things. This is just as decisive as the fact that something is *true* or *correct* about them. And it is important to any well defined notion of the identity of things that we are able to determine *when* and *that* a particular description of a thing is correct or incorrect - or a description devoid of meaning. And it is important that we are able to determine *when* and *that* a description is a description of a particular thing as opposed to being a description of some other thing, and *when* and *that* a description of the thing is a false description, as opposed to a description of it which would be correct in some other and different situation. Without the possibility of a distinction between these cases, there could be no well defined notions of false or incorrect description of things, and thus *no well defined notion of falsehood*. But if there could be no well defined notion of falsehood of statements about things, there could be no notion of truth of such statements, and *therefore* no notion of statements about things, let alone statements about the *same* things.

In contrast, the alternative view of the identity and identification of things I have argued for in this chapter, seems to offer much more promising - and consistent - prospects. According to this view, to identify a thing as a particular thing implies, firstly, that it is a thing which is correctly identified and described in this situation in this particular way, *and* which is correctly described in other situations in these other ways. To have identified a thing on this view, implies that, *logically*, we have committed ourselves to these other or alternative descriptions in other and alternative situations as well. Thus, the very notion of "the same" or "the same thing" only exists in virtue of the notion of "different" or "different descriptions of the thing". A consequence of *The Principle of Identity* is that it is not only possible to determine the correctness of a description relative to some arbitrarily chosen situation and logical space of descriptions and concepts, but also that any situation in which a thing is described and identified is but *one* of the various different situations and logical spaces in which the identity of the thing may be determined, and to which we have committed ourselves in the very act of identifying it. This condition granted, it is possible to talk about alternative - even new and as yet unknown - descriptions of a thing as alternative correct descriptions of the *same* thing.

Secondly, all *other* descriptions of a thing, except from some particular description of it, are not just descriptions of other things. And all other descriptions of a thing, except for a *correct* description of it put forward in some particular situation, are not necessarily descriptions of other things. For it so happens that *incorrect* descriptions of the same thing exist. But to assert "this is a false or incorrect description of a thing", only makes sense on the condition that something may not only be *true*, but also be *false* about the same thing.

Hence, if *being* a thing, i.e. this particular thing, implies that something is the case or true about it - and that something else is false about the *same* thing - then to have identified a thing not only implies knowledge of what is true or the case about it, but also of what is false or not the case about the same thing. Otherwise, it would not be possible to distinguish between plainly *false* descriptions of a particular thing, and *misidentifications* of the thing, i.e. cases in which a thing has been falsely identified and described as some other thing - of which the description would be true. In other words, such misidentifications would go quite unnoticed (cf. Chapter 12).

The dependence of falsehoods on truth, *and* vice versa, becomes obvious when we consider that saying "this is true about p" means that it would be false to deny it. Aristotle even made a logical principle out of it; he called it *the principle of contradiction*, and it states that A and non-A cannot both be true. This is what true means and what it means that a belief or proposition is true about something. And this is what it takes, i.e. *a principle of contradiction,* to have a concept of 'truth'. To say that something is true about a thing only makes sense provided that something else would be false about it - including the contradiction of what is true. Without falsehood, 'true' would not mean a thing, would not exist in any well defined way. Hence, we cannot have true beliefs about anything independently or without something else being false about the same thing.

I wonder if we have even started considering the decisive epistemological conditions for language and use of language, if we have not made up our minds about what it implies that we may put forward statements about things which are *incorrect* or *false*. That is, if we have not understood that to maintain that something is the case or correct about something inevitably implies that something else is *not* the case or incorrect about it. Meanwhile, I shall contend that consistent and well defined uses of the concept of truth necessitate a notion of identity as the basis for identification of the kind suggested in this chapter. Conversely, the notion of the identity of things requires the existence of the principle of contradiction.

16.6 Answers to objections to the analysis presented: Conclusion

It may be objected that important points were missed or left out in my discussion of the notion of essential properties; in particular, objections may be made to my choice of examples of descriptions of such properties. It could be argued, as Putnam does, that it has always been the aim of the sciences to provide rigorous definitions of the objective essence of things - the definition of water as H_2O within chemistry is a case in point - and that such definitions have had significant *explicative* relevance for other areas of description. For this reason, it is simply nonsense to say that we cannot make alternative correct assertions about the same things once "essential" descriptions of the things have been obtained. If, as an example of an essential description of water I had chosen the chemical description H_2O, instead of "water is water as it exists at room temperature", this would have been immediately clear.

Now, had I chosen the description of water which constitutes the chemical definition of water, I would certainly have had to find other examples of descriptions that would not go along with the original, stipulated description of water because, within a chemical definition of water, it makes good sense to say that water turns to ice at 0° C and boils at 100° C. Very well, but I see no reasons to maintain that a chemical description (or series of descriptions involved in the description of H_2O), in any particular way identifies water -

any more than does other descriptions of water, for example our everyday description of water being "a wet and tasteless liquid". And I see no reason to maintain that once scientific definitions have been reached, we have reached descriptions of what *water* essentially, objectively is. More importantly, if such scientific determinations may have explicative and definitional relevance to other areas of description, it is precisely because they - just as any other description - cannot be "final" or "essential" in the sense: the description of what water really and objectively is. So to maintain would be inconsistent. For, if water were not something which could *also* and objectively be a "wet and tasteless liquid" - or however else water and its properties may be identified and described in normal everyday situations - the chemical description of water could not have any explicative relevance for these other non-scientific descriptions of water and its properties. Indeed, a chemical descriptions of water only has explicative relevance to other areas of description because *more* can be said about water in these other non-scientific areas. And to the extent that the chemical description H_2O may be said to be *applicable* to *water*, those other and non-scientific descriptions are necessarily descriptions on which the chemical descriptions of water *rely*. In general, to the extent that scientific descriptions and laws are applicable to the behaviour of the objects and things we encounter in our everyday world, and described in terms of 'lemons' and 'gold', 'chairs' and 'tables', 'rocks' and 'trees', and so on, scientific descriptions rely on such everyday descriptions and determinations of these objects and things. As pointed out earlier, in this sense scientific descriptions rely on an *abstraction* in that they rely on non-scientific descriptions of the things. Non-scientific descriptions, which must be assumed to be about things and objects which exist just as objectively as do their scientifically determined components; descriptions, therefore, which are as objective as scientific descriptions of the same things. To say otherwise would be to say that scientific descriptions could have explicative implications for things which did not exist *objectively*, and of which true description could not be made.

Hence, for scientific descriptions and laws to have explicative relevance for our everyday notions and knowledge of things requires not only that the same things may be described differently in different situations, but also that *logical* relations exist between such different descriptions of the same things. We cannot, therefore, do away with these other non-scientific descriptions, or for that matter with any other objective description on which scientific descriptions and determinations of the things relies. Things would certainly be much simpler if we could; but things - and in particular our cognition and description of them - just are not that simple.

It might also be objected, that most if not all the problems of identity as the basis for identification would disappear the moment *numerical identity* was employed, i.e. when the identity of things were described in terms of exact space-time determinations. But however true it may be that an identification of a thing as a particular thing will involve a determination, explicitly or implicitly, of where the thing is located and when, a space-time determination is not sufficient to identify *what* the thing is. A thing thus identified could be an apple or any other object for that matter. And without some further specification of what the thing is, it could not be said that it was the *same* thing, if the thing was moved - or moved itself - to some other place. This would require *in addition* to the space-time coordinates of the thing, that some other determination of the thing be given - however rudimentary. Therefore, a space-time description will not be sufficient to determine the identity of a thing. Nor would it, as I have tried to argue in this chapter, be sufficient for an identification of the thing as being some particular thing to be made.

To conclude, according to *The Principle of Identity* suggested in this chapter, to be in a situations and to have identified the things in that situation implies that *more* can be said about both the things and situation - in the actual as well as in other situations. Indeed, it is to know that there is probably a lot about the things and situation which we do not know - but which we may come to know about by carrying out further investigations on them - in the actual situation or in others. However, such new knowledge and descriptions are necessarily, i.e. *logically*, related to the knowledge and descriptions which formed the point of departure for our further investigations of the things. Granted these conditions, it is possible to account consistently for the continuity in our acquisition of new knowledge and descriptions of things in the world, and thus to have consistent notions of the continuity of the identity of things that this development in our knowledge and descriptions concern.

PART IV

PERSONS

Introduction

In the chapters of Part IV an attempt is made to determine some elementary conditions for someone to be a person who can act, communicate and form communities with other persons. Needless to say, such an attempt requires that at least some elementary specification be made of what a person is – just as similar determinations of the conditions for persons to have knowledge and a language to talk correctly about reality require some elementary specification of what reality is. And yet, even after having been thus specified, both questions as to what a person is and what reality is have a ring of the unknown, and in the case of a person even of the unknowable. Now, we certainly have a lot of experience and knowledge of being persons, finding ourselves in material reality with other persons with whom we may act, communicate and form communities. However, if it is true that reality is inexhaustible, as argued in Chapter 14, and our knowledge of reality, therefore, can only be *partial*, then reality will always be something about which *more* and different things may be known and said, and thus about which something is unknown to us - but not unknowable. And if it is true that knowledge about reality can only be the knowledge of persons, who know there is something that they do not know about the situations in which they find themselves in reality, but may come to know, then, by implication, to be a person is to be someone about whom much more may come to be known about oneself.

In this sense, to be a person is to be someone in a constant and never ending process of *becoming* - each of us in our own and distinctly unique way. A person, as Kierkegaard puts it, is "that unique individual", constantly in the making. Therefore, there is something about us, which is unknown - even to ourselves. Equally significantly, there is something about us, which is forever and fundamentally unknowable to others. Thus, I shall never be able to know what it is like to be the person you are *in the way you know it*, or be able to know your pains, or know your perceptions and thoughts of reality and other persons *in the way you do*, because I cannot see with your eyes, feel your pains, think your thoughts, or have your experience of the world and other persons.

To this uniqueness of our personal experience must be added the differences in our conception of ourselves, others and the world in which we find ourselves, for example due to differences in our upbringing and training, or differences in socio-economical and cultural background. However, differences are perhaps most startlingly seen in the different "stories" told by different members of the same families about themselves, their families and the world around them, and in the differences in how they think they are seen by other members of their families compared with what these other members actually think about them (see e.g. Bruner, 1990). Bearing in mind this uniqueness and differences between persons, it is not difficult to understand why it has seemed reasonable to assume that the persons we are, and the reality we live in must somehow be the product of how each of us experiences and conceives of ourselves and reality - and also to assume that each of us lives or exists in worlds which are not only *personal*, but even *private*.

And yet, in spite of these differences in our personal experiences and conceptions of ourselves and the world around us, to be a person is something fundamentally social. Indeed, no-one can be a person, and thus someone who may realize that he or she is uniquely different from other persons *without* other persons from whom he or she may differ. Nor may one be different form others without having possibilities of determining *how* one differs from others - whether such differences concern one's notions and experiences of things or states in material reality, or one's inner feelings, thoughts, beliefs or emotions. Furthermore, as great poets and philosophers have long since known - and as has now been overwhelmingly confirmed by our colleagues working with clients in departments of clinical psychology - no-one can be or *become* a person without being the object of the *desire* of other persons to receive and see him or her as a person, and to understand the way he or she is a person in a world which in so many ways is unique to every one of us. Nor can one be or become a person without desiring to receive and see others as persons.

In previous chapters I have only discussed the conditions for knowledge and description of persons of things and states in physical material reality. According to these discussions (cf. especially Chapters 5, 6 and 14), it must be a condition for determining differences in the perceptions and knowledge of persons about such states and things that they are persons in a public world, of which they share a considerable amount of knowledge, and a language in which true statements may be put forward, including true statements about the things and states that their *differences* in knowledge and perception concern. But what about the knowledge we have and the descriptions we put forward about things which are not publicly observable, such as our so called "internal states", our thoughts, feelings, emotions, religious beliefs, pains, or taste? Can it be maintained that knowledge and descriptions of such inner states have the same status as the knowledge and descriptions of physical objects in material reality? And can it be maintained that the conditions for knowledge and description of such non-publicly observable mental states and events are the same as the conditions for knowledge and description of publicly observable material states and events? Indeed, can we be sure that different people are using language in the same way when describing their internal states, or be sure that the descriptions and assertions put forward about such states and other non-publicly observable things, have the same implications for different persons? Or do we here find ourselves in a "realm" of experience, knowledge and use of language which is not only *personal* but even *private*? And does it really make any sense to say that thoughts, emotions and feelings, which are so different from physical or material things, may be known in the same way as we know physical or material things, or to say that we are using language in the same way when describing non-physical things as we do when describing physical and material things?

These are some of the questions which have been extensively debated within philosophy for centuries and, more recently, within psychology. However, since the investigations of the status of internal states have relied on the same bifurcation between the "mental" and the "physical", on which rest similar investigations of the status of our knowledge and description of physical reality, it is not surprising that the discussion about the "mental" or "internal" has been beset with the same confusions as the corresponding discussion about the "physical" - as seen in preceding chapters. It is the intention of Chapters 18 and 19 and 20 to clear up some of this confusion. In particular, it is my intention to show that our knowledge and descriptions of the non-physical, the "mental" or "internal", do not present *epistemological* problems which are fundamentally different from the problems concerning our knowledge and descriptions of physical or material reality and its objects. Thus, I shall argue that, just as we have to assume that to be a person and a language user is to be someone, who has knowledge of and may describe and communicate with others about physical reality, we have to assume that to be a person is to be someone, who has knowledge of, and who may describe and communicate with others about his internal states. Conversely, just as physical objects and events exist as something which persons may have knowledge of and describe correctly, internal states exist as something about which persons may have knowledge and describe correctly.

The main line of the argument is that, despite the fact that every one of us only has direct access to and may experience our own thoughts, emotions, feelings and other internal states, we both can and do indeed communicate about such internal states. It even seems to be widely assumed that it makes good sense to discuss *whether* or *how* the internal states of persons may or may not differ - and thus to assume that we do indeed have a language in which it is possible to discuss, determine and talk sensibly about such states and differences. In particular, I shall argue that a person who may reflect on, be aware of, and communicate about his internal states, is a person who at the same time finds him or herself in a situation in physical, material reality, i.e. a person who at the same time knows of and is aware of him or herself as being part of physical material reality. Indeed, logically, we cannot talk about 'internal states' of a person without talking about 'persons' with 'bodies' having this knowledge, nor talk about persons, their bodies and other things being part of physical reality, without talking about and referring to internal states of persons. But neither can we talk about persons without talking about 'other persons'. The argument for the existence of a logical relation between such concepts is not an argument to the effect that there is no difference between internal states and physical objects - there are indeed many differences. Rather, the purpose of the argument is to show that we cannot talk about the 'inner' or 'mental' of a person in isolation from, nor independently of, the 'outer' or 'physical' of a person, and hence that the 'inner of a person' is very much part of the 'outer of the person' - and vice versa. Indeed, we cannot talk correctly about our knowledge of physical reality without being able to talk correctly about our internal states - and vice versa. Acknowledging then that there are significant differences between internal states and physical things, it is my aim to show that some of the arguments put forward - in particular by Wittgenstein - about the differences in status of our knowledge and descriptions of internal states versus physical things, rest on false assumptions based on the Cartesian bifurcation of "mind" and "matter".

However, first, in Chapter 17, I shall once more take up this bifurcation and how it has been radicalised by present day idealist philosophical positions, notably *Deconstructivism*, *Anti-Realism* or *Irrealism*, which deny the independent existence of anything outside the mind (or texts) of persons - thereby inevitably denying the existence of persons. Well, bearing in mind the interelatedness between the knowledge of persons and the reality which this knowledge concerns, and also that knowledge and description of reality can only be the knowledge and description of persons finding themselves in particular situations, and having particular possibilities of observation, action, interests, intentions, purposes etc., it is understandable why it has seemed reasonable to assume that reality is "what we make of it". And bearing in mind as well the significant differences in the purposes, interests and possibilities of observing and describing reality available to persons living under different historical, cultural and socio-economical conditions, and thus the significant differences in their knowledge and description of reality due to these conditions, it is understandable why it has been so obvious to assume that reality is as it is *in virtue* of how it is being conceived of, or even *constructed* by persons. And bearing in mind, lastly, that how persons in different communities and societies act in, acquire and describe their knowledge of reality to a large extent depends on conventions, it is equally understandable why it has been assumed that what exists - or does not exist - in reality is a matter of convention and agreement among persons.

Granted these assumptions, it would be true that what we say about reality and ourselves, is just as *fictitious* as the stories we construct in literary texts, for example in novels, myths and adventures. Hence, when giving accounts of ourselves, reality and what exists in reality - whether in scientific or in ordinary everyday terms - we are "narrating stories", just as and in the same way as we fabricate adventures. If so, there is no need of, let alone any possibility for distinguishing *imagined worlds* from *actual worlds*, nor for distinguishing *actual minds* from *imagined minds*. Consequently, there would be no possibility for distinguishing between *inventing stories* about reality and our minds, and *observing facts* of reality and our minds.

Now, before committing ourselves to such assumptions and consequences, we would do well to consider what it is that makes it possible to talk about and determine differences between persons, be they the individual differences which exist between a person and any other person, or the differences between persons due to their different historical, socio-economical or cultural, political backgrounds. And consider what it is which, despite all these differences, makes it possible for persons together to determine that and how they differ, but also what makes it possible for them together to determine the situations in which they find themselves, and in these situations to develop conventions and rules for how to act, communicate and perceive themselves and reality. That is, consider what is the *basis* on which conventions and rules for co-action, perception and communication may be negotiated and agreed among people being thus unique and different. Conversely, we might well consider that if reality and we ourselves are nothing but products or fabrications negotiated between our inventive minds, then nothing would prevent us from constructing, negotiating or conventionalising ourselves, other persons and reality out of existence. Admittedly, situations do exist in which we are only able with considerable difficulties to determine what is fictitious and what is real. But despite such occasional difficulties, it still makes sense - even to proponents of the above assumptions - to preserve such terms as "fiction", "reality" and "facts"; indeed, they use those terms all the time. In so doing, they necessarily concede that, generally, it is possible to determine and distinguish what these terms refer to, and also that these terms have correct implications and uses.

Whilst acknowledging that we cannot talk about reality independently of or without at the same time *describing* reality, nor talk about descriptions of reality without referring to the interests, possibilities of action, observations and descriptions of persons in different situations, proponents of *Epistemological Idealism* ignore the *asymmetry* which necessarily exists between the knowledge and description of persons and the reality which this knowledge and description concern. However (as argued in Part II), to ignore this asymmetry is to ignore that we cannot talk of knowledge *about*, nor of descriptions *referring to* and being *true* or *false* of anything without assuming, logically, that *what* we describe and know of exists independently of our knowledge and descriptions of it. Conversely, the consequences of assuming that things may only exist in situations in which we observe and describe them, leaves two options open regarding the status of things when "outside" such situations. *Either* they cease to exist - which amounts to assuming that things do not exist independently of what we do *not* know or say about them. *Or*, we cannot be sure whether they still exist as the same kind of things; they may well exist as something radically different or, at any rate, *in principle* unknown to us. This amounts to assuming that although things may exist independently of the situations in which we observe and describe them, things existing independently of us and such situations, are not the same as those we know and describe in these situations - in which case we would be hard pressed to maintain that our assumptions in fact concern the *same* things.

In the chapters which follow it is my aim to determine what - in spite of the overwhelming uncertainties - we do know about persons, and what we *can* say and necessarily *have* to say about someone who may function as a person, which may form the basis for our further inquiries into the social existence, communication and knowledge of persons about themselves, reality, and other persons. As in previous chapters, the discussion will mainly concern problems of an epistemological kind - and only touch on the psychological ones. I say that somewhat apologetically. For, if we do not appreciate the profound importance of the *psychological* conditions for people to develop different conceptions about reality, themselves and other people, there is little hope that we shall have anything of significance to say about the epistemological conditions for talking about individual differences, and for persons to communicate their differences among themselves. And we shall probably have very little to say about the epistemological conditions for people to understand each other without recognizing how crucially important it is for people *psychologically* that they do understand each other. An account of the epistemological conditions for people to act as persons in communities with other persons will probably be of limited interest, if not based on at least some understanding of what "drives" human beings to act and become persons; persons that is, who are part of social and cultural communities.

There are good reasons for apology if one does not address the conditions of a psychological nature in a discussion of the epistemological implications of people to be persons. Indeed, there would be no epistemological problems to address if there were no *desires* of people to become persons. It may well be, as I shall argue in Chapter 20, that, logically, no one can be a person, and thus be someone who may differ from other persons, without other persons. And it may well be that to be a person who is different from other persons both presupposes and requires an inter-subjectivity of language and cognition, which is such that one's description of what one experiences and knows, would also be true for other persons, could they know and experience what oneself does. And it may well be, therefore, that no one can have a notion of 'truth' without a notion of 'other persons'. However, what in the final analysis makes people understand each other is that they *want* to understand each other and *want* to make themselves understood. In this respect, any additional question about the *formal conditions* which makes this understanding possible, must come second.

So, although I do not address at great length the psychological conditions for people to be persons, this does not mean that I do not recognize the importance of these conditions for an account of the formal, epistemological conditions for people to function as persons. Indeed, in my discussion of these formal conditions I hope to make it clear that if we ignore the social and psychological conditions for people to be persons, we risk developing epistemological assumptions about persons which are at best totally inadequate. Needless to say, if such inadequate epistemological assumptions are accepted as basic assumption for psychological theories, these theories will be just as inadequate.

It is with this danger in mind that the following analysis of the epistemological conditions for being persons will be carried out. In view of the crucial importance of appropriate epistemological assumptions about persons for determining what are relevant issues to pursue, and inquiries to carry out within psychology concerning the ways in which persons actually function as persons, such an analysis needs no further justification or apology.

Chapter 17

Some consequences of epistemological idealism

17.1 Introduction

Descartes concluded his systematic doubt about the existence of everything he could think of, including himself, by asserting: I think therefore I am (Descartes, 1637/1991). Descartes could not doubt that he doubted, and thus could not doubt that he existed as someone who could doubt. But nor could he doubt that doubting is an act carried out by someone (regardless of *what* made him doubt). So, he knew something about himself, namely that he *existed* as someone, *an agent*, who could doubt.

However, he could not be an agent who could doubt, without knowing *what* he had doubts about and, therefore, there had to be *something* he knew about that which he was doubting. Thus, a person carrying out the act of doubting, no matter what the object of his doubts may be, or what he may doubt about it, must know at least *something* about the object of his doubt. In particular, an agent who may *express* doubts about things in reality, cannot doubt that he may *determine* and thus *refer* to such things, nor doubt that he is an agent who necessarily exists in the very same reality about which he may express doubts. Thus, it would not make sense to say that we are agents, who may determine and refer to the things in reality which our doubts concern, and about which we therefore have some knowledge, but we ourselves do not exist in the reality of which we have knowledge and may express doubts. Conversely, it would not make sense to say that we ourselves and our thoughts or doubts are *all* that exists in reality, but the things and events which we can think of and express doubts about, do not exist in reality. For, take away *either* the person, who doubts and knows that he doubts, and who may determine and know something about that which his doubts concern; or take away the possibilities that things he may doubt about exist as things about which he may doubt and thus have some knowledge - and the doubt will cease to exist. So, Descartes discovered - and we after him - that no doubt can be exhaustive; our doubts, just like our knowledge, can only be *partial,* i.e. incomplete. We cannot doubt ourselves out of existence; nor can we doubt out of existence any other things in reality about which we can express doubts.

But neither do things in reality that our doubts concern only exist *in virtue* of being objects of our doubts. For, as just argued, things of which we can have any well defined doubts, are necessarily things to which we may refer and have some knowledge about. But to say that things of which we may have doubts are things about which we have some knowledge and to which we may refer, inevitably implies that they are things, which are different from and which exist independently both of our knowledge and doubts about them. This applies whether the doubt concerns the properties of things in reality - or whether the doubt concerns ourselves.

Now, according to the assumptions of various versions of *Epistemological Idealism* (such as Constructivism, Relativism, Subjectivism, Conventionalism, Anti-realism), things do not exist independently of the descriptions and assertions we put forward about them; things only exist in virtue of our descriptions of them, and in virtue of the knowledge we have about them. We, i.e. persons and language users, literally, so it is claimed, "construct" or "make" things with words. However, these claims and assumptions are so much easier for its critics to refute than for its proponents to defend. Any attempt seriously to discuss what it means and what follows from these claims and assumptions leaves one tied up in a web of obscurities, absurdities and contradictions. Does it follow, for example, that before people inhabited and acquired knowledge about the world, nothing existed? And does this mean that, for example, contemporary cosmology is a mere product of description or statements somehow originating in the minds of contemporary cosmologists? That is, *either*, what existed before we entered the world is something which relies on the minds of cosmologists; *or* the physical evolution of the universe is something which has been made to exist by some sort of backward mental causation.

Few proponents of idealism, I think, would seriously attempt to defend such implications. Rather, they would maintain, as did Kant, that because the notion we have of reality is dependent on how we describe reality, it is not possible to say what reality is apart from our descriptions of it. But, so they continue, what exists "out there" in reality, and thus what Reality is in itself *beyond* the descriptions of persons, may not be the same as the reality, which exists according to our descriptions of it. Well, it does not really matter whether Idealism implies one or the other; the result, as far as I can see, is the same. In either case, reality and things in reality as something which we may talk sensibly about are being either "created" or "talked" out of existence.

Matters become even worse, if not downright mind boggling, if one attempts to address the issue about the status of *persons*, and what, according to Idealism, determines the existence of persons necessarily being part of the reality they themselves create. For, if reality only exists in virtue of being created by persons, and if persons themselves are part of this reality, then persons, in the process of creating (the rest of) reality, must at the same time create themselves into existence. Alternatively, if what is implied in Epistemological Idealism is that the reality which "really" exists, is not the same as the reality which exists according to our descriptions of it, then it must be similarly implied that the persons which "really" exist, are not the same persons as the ones which exist according to their knowledge and descriptions of themselves, and so on. The point is that proponents of Idealism simply fail to realise that what we say about the existence of reality, and the possibility of having knowledge about *reality,* inevitably has consequences for what we may say about the existence of *persons*, and the knowledge of persons about themselves. That is, irrespective of how Epistemological Idealism chooses to do away with reality as something we may talk sensibly about, the consequences will inevitably be that persons, being part of such reality, are being done away with as well.

Present day versions of radical Epistemological Idealism, such as *Deconstructivism*, certainly cannot be accused of ignoring the problems which Idealism gives rise to, nor of failing to notice that creating or talking reality into existence à la *Subjective Idealism*, actually means creating or talking reality *out* of existence - and with it the subject who does the talking. Rather, this insight is the central tenet of Deconstructivism (Derrida, 1972). According to its proponents, Deconstructivism is the only possible alternative to emerge from the ashes of a fallacious view, which for too long has influenced our sciences and civilisation. This is the view of a *Classical Realism* which holds that a "ready made" reality exists which, depending on preference of terms or temperament, either *lends* itself to, or *enforces* upon us its own uniquely true description of itself. The consequences of the collapse of such realism are that *no* realism is possible, according to Deconstructivism. Without captiousness the very subject, or person, presumably being in a position of knowing the implications of his own declarations, is declared as "dead" as reality. There is no subject, nor any position from which a subject may talk about reality or himself, but only *subject positions*, i.e. positions of discourse, from which we ourselves as well as all other thing are produced in language. Although a notion of truth is granted necessary, "*for all practical purposes*", proponents of

Deconstructivism deem void any *general* concept of truth. For, what is *true* is just as dependent on and relative to situations, positions and purposes, as the concept of reality has proven itself to be dependent on and relative to ideology, theory and power. Of course, such radical relativization of *truth* leaves proponents of deconstructivism non-committal to just about everything they may talk about. Needless to say, it leaves equally relative the truth of any deconstruction carried out and assertions put forward in the name of deconstructivism.

The message of Deconstructivism has not until recently had nearly the same appeal to psychology - or, for that matter, to any other science struggling to grasp what goes on in reality - as it has had to the studies of literary "texts" and fiction. Few scientists would rush to greet an epistemology that attempts to annihilate the very subject of their inquiries, in the case of psychology, the person. However, radical versions of *Constructivism* and *Anti-realism*, having consequences which are indistinguishable from those of Deconstructivism, have recently been promoted as an epistemological position on which to base not only theories of art and fiction, but inquiries and theorizing within psychology as well (Bruner, 1985, Gergen 1986, Burr, 1995). In what follows I shall discuss these consequences. The point of departure will be the assumptions of these "isms" as taken to their logical conclusion by Nelson Goodman in his arguments for what he calls *Constructionalism*. In the course of this discussion it will hopefully be clear why Constructionalism - or any other version of Constructivism and Anti-realism - cannot possibly be an epistemological position on which to base psychology, let alone be a better or more tenable alternative to the position they are meant to replace, - i.e. Naturalism in its various guises (such as Physicalism, Computational Functionalism, classical or "Metaphysical" Realism). In the section that follows I shall begin by giving an outline of Goodman's epistemology.

17.2 Constructivism and the disappearance of reality and persons

To Goodman, no such thing as an "objective reality" exists; the world we talk about and inhabit is nothing but a product of our descriptions, perception, ordering and categorisation. Indeed, there is not just *one* "world", but a plurality of "worlds", as many *versions* of worlds as our creative minds permit us to produce. According to Goodman

> "The worldmaking mainly in question here is making not with hands but with minds, or rather with languages or other symbol systems. Yet when I say that worlds are made, I mean it literally. (Goodman, 1984, p. 42). [...] Any notion of a reality consisting of objects and events and kinds established independently of

discourse and unaffected by how they are described or otherwise presented must give way to recognition that these, too, are parts of the story. If we dismiss measures of referential distance as not matters of fact because they are discourse dependent, we shall have trouble finding features that *are* matters of fact." (ibid., p. 67)

Thus, because reality or "worlds", and the things which furnish any particular "world", are mere fabrications of our creative capacities, the worlds and things we experience do not exist independently of, but only *in virtue* of having been made by us. And whereas descriptions or world-versions exist independently of worlds, worlds do not exist independently of versions, i.e. *independently of versions which make worlds they fit.* In Goodman's own words:

> "a statement is true, and a description or representation right, for a world it fits. And a fictional version, verbal or pictorial, may if metaphorically construed fit and be right for the world. Rather than attempting to subsume descriptive and representational rightness under truth, we shall do better, I think, to subsume truth along with these under the general notion of rightness of fit. (To this a note is added which reads:) Readers of foregoing pages will be well aware that none of this implies either that any ready-made world lies waiting to be described or represented, or that wrong as well as right versions make worlds they fit." (Goodman, 1978, p. 132.)

It may be a bit difficult to understand how the "making" of a world is dependent on the creation of the "right" rather than "wrong" version of a world that fits it. It may perhaps help if one thinks of a version or description of a world as being some sort of *key*, and the world it fits as some kind of *key-hole*. Thus, we may create various versions of keys - but whether they fit a key hole (i.e. are the right sorts of version) does not depend on the fact that a key hole already exists into which these keys may fit; rather, a key is "right" if in the process of creating the key, a key hole emerges (bingo!) into which it fits.

Worlds then, to the extent that they are granted any existence at all, only come into existence *in virtue* of the creation of *versions* which gives rise to the making of worlds. If they do, the descriptions or versions are said to be "right" rather than true - for the problem of whether a particular version or description fits the world it creates, cannot be a matter of whether the descriptions refer to some objective facts, which exist independently of the descriptions, and about which the implications of the descriptions hold true. Indeed, because of the "coalescence" between worlds and their versions, there is really no way of distinguishing between, on the one hand, *descriptions* or *versions* of worlds and, on the other, the *worlds* they fit.

How Nelson Goodman arrives at this rather unusual view is a longer story, which is best understood - if at all - if we start at the beginning.

It begins (in Ways of Worldmaking, Goodman, 1978) by the observation - presumably dating back to the time when humans first became language users - that the world we live in may be described in a variety of different ways, which are equally true. Thus we have many different ordinary everyday descriptions, as well as several sorts of artistic and fictitious accounts, of ordinary as well as not so ordinary everyday objects and events; and we have several different accounts of "scientific objects", e.g. of "physical objects", described within physics as molecules, atoms, elementary particles etc. However, none of these different descriptions or world versions have any special claim to objectivity or truth. For, says Goodman, different descriptions or versions are not just different *true* and - on the face of it sometimes - *conflicting* true descriptions of *the same facts or world.* Although the two statements, "The sun always moves", and, "The sun never moves," may both be true, it does not make sense to say that they are both true of the *same* thing or "world" - however true within different frames of reference. Indeed, the very notion of "different descriptions being different descriptions of the same world" is to Goodman a void notion. Here is why:

> "We are inclined to regard the two strings of words not as complete statements with truth value of their own but as elliptical for some such statements as "Under frame of reference A, the sun always moves" and "Under frame of reference B, the sun never moves" - statements which may both be true of the same world. Frames of reference, though, seem to belong less to what is described than to systems of descriptions: and each of the two statements relates what is described to such a system. If I ask about the world, you can offer to tell me how it is under one or more frames of reference; but if I insist that you tell me how it is apart from all frames, what can you say. We are confined to ways of describing whatever is described. Our universe, so to speak, consists of the ways rather than of a world or of worlds." (ibid., p. 2.)

Thus, if someone asks "what *same* world or fact is being described in different ways", we can only answer by offering new "frames of reference" - but we cannot point to that which independently of (or "underneath") these frames of reference *is the same fact or world.*

However, it would seem that what applies to frames of reference must apply equally well to *world versions*: just as "frames of reference belong less to what is described than to systems of descriptions", so do Goodman's worlds-versions belong less to what is being described in those versions than to systems of description. So, what may be said of one world may be said of the many: we only have their versions. And to say that versions match or fit worlds "underneath" their versions, is just as empty as it is to say that descriptions belonging to different frames of reference are descriptions of the same facts or world. At best, such turns of phrase are mere matters of convenience.[1]

Now, the fact that worlds "un-described" or beneath *both* their "frames of references" *and* their version must be equally void, makes Goodman talk sometimes as if there were no worlds at all, and sometimes as if worlds and version are the same things. However, at other times he speaks as if many worlds actually exist. This would seem to make sense when at these *other* times discussions of the implication of the "rightness" of a version is on Goodman's agenda; in such discussions the existence of at least one world per version would seem required. For, what point could there be in maintaining that a version is "right", i.e. fits a world, if there were no world at all? Indeed, if we allow ourselves to think at one and the same time what Goodman may only think at different times, we may well ask: if worlds, just like "the world", are *nothing* independently or "bereft" of the versions which makes them, how, and in particular, *what* could any version possibly fail to fit, be right or wrong about? Indeed, how could notions like "right", "wrong" or "fit" fail to be just as empty as the Goodmanian world(s)?

Although this must be just as obvious to Goodman as to anyone else, he proceeds as if it were not. For there is, he says, yet another and more important reason for maintaining the existence of not only one, but several worlds: So long as constructing right versions are not all translatable or reducible into one (e.g. to physics, as physicalism would like it), there is a non-trivial reason for insisting that there must be many worlds (Goodman, 1978, p. 4). And since there are many true versions which are conflicting, there must be many worlds. The arguments leading Goodman to this conclusion are these: "Some truths conflict. The earth stands still, revolves about the sun, and runs many other courses all at the same time. Yet nothing moves while at rest" (Goodman, 1984, p. 31). However, in order to escape this contradiction "we (usually) seek refuge in simple-minded relativization: according to a geo-

[1] Thus, Goodman writes: "Yet doesn't a right version differ from a wrong one just in applying to the world, so that rightness itself depends upon and implies a world? We might better say that 'the world' depends upon rightness. We cannot test a version by comparing it with a world undescribed, undepicted, unperceived. [...] While we may speak of determining what versions are right as 'learning about the world', 'the world' supposedly being that which all right versions describe, all we learn about the world is contained in right versions of it; and while the underlying world, bereft of these, need not be denied to those who love it, it is perhaps on the whole a world well lost. For some purposes, we may want to define a relation that will so sort versions into clusters that each cluster constitutes a world; but for many purposes, right world-descriptions and world-depictions and world-perceptions, the ways-the-world-is, or just versions, can be treated as our world." (Ibid., p. 4)

centric system the earth stands still, while according to a heliocentric system it moves". (ibid., p. 30.)

But,

> "there is no solid comfort here. Merely that a given version says something does not make what it says true. [...] That the earth is at rest according to one system and moves according to another says nothing about how the earth behaves but only something about what these versions say. What must be added is that these versions are true (ibid., p. 30). [...] How, then, are we to accommodate conflicting truths [...]? Perhaps by treating these versions as true in different worlds. Versions not applying in the same world no longer conflict; contradiction is avoided by segregation." (ibid., p. 31)

Later, on the same page, there is no "perhaps" about it; Goodman simply contends that, "In any world there is only one Earth". And in order to press home his point, Goodman asserts that different worlds do not occupy the same space or time. Space-time, according to Goodman, is an ordering *within* a world, and space-times of different worlds are not embraced within some greater space-time (ibid., p. 31). This, in his view, ensures that there is no risk of collision between several different "earths going along different routes at the same time" (ibid. p. 31).

Well, by "segregating" Earth and the movements of Earth to different versions applying to different worlds, and occupying different space-times, Goodman has efficiently avoided a collisions between earths existing in those different worlds - as well as conflicts of the truth of versions and descriptions "assigned" to different worlds and things in those worlds. Indeed, no such collision or conflicts of truth may occur, because the things existing in different world are simply not *the same sorts of things*, neither can there be any "overlap" of identity of things belonging to different worlds. There can be no such thing as the same thing "traversing" different worlds, for such a notion would require the existence of a "bigger" world into which both versions of, say, the movements of Earth would fit - and it is exactly this notion and the existence of such a world that Goodman wants to get rid of:

> "By assigning conflicting versions to different worlds, we preclude composition of these totalities into one. Whatever we may mean by saying that the motion of the Earth, or of different earths, differ in different worlds, we rule out any more comprehensive whole comprised of these. For a totality cannot be partial; a world cannot be a piece of something bigger." (ibid., p. 32)

Goodman's epistemological Universe, then, is a universe which is inhabited by infinitely many things in just as many worlds - as many as there are potentially "conflicting" world versions. It is a universe - to take a different example - in which a version containing the true statement, "this thing is a tomato which grows on a plant", belongs to one world (let's call it the world of everyday objects), while the statement "this tomato is made up of a particular structure of molecules", is a true statement about something entirely different, belonging to a different world (let's call it the world of "physical" things). And as was the case with the term "Earth", the term "tomato" used in different world-versions is not assigned to the same thing. This means, furthermore, that descriptions of "physical things" are only applicable to things within the world of physics - and not to things "made" in other world or worlds.

Now, however much Goodman may have managed to explain away any "conflicts" of the truth of versions that may arise within his Constructivist epistemology, the solution he proposes seems to "achieve" much more than anybody would want; he has, as far as I can see, thrown out the baby with the bath water. For by assigning "Earth" and "tomatoes" in different versions to different things in different worlds, he has at the same time efficiently debarred himself of maintaining that things being made in one world *at the same* time may have the sort of properties of things belonging to other worlds. And if there can be no "preservation" of the properties of things when "traversing" different worlds and their peculiar space-times, he has debarred himself from saying that it is the same things to which *identical terms* in different versions refer, indeed that the same things may be dealt with in different worlds. There is, as far as I can see, no way that such an identity of things can be established in different worlds. If so, Goodman is at the same time denying himself that which would ensure the possibility of separating or distinguishing between worlds, i.e. from maintaining that "Worlds are distinguished by the conflict or irreconcilability of their versions" (Goodman, 1984, p.31). For, what conflicts of truth or irreconcilability between versions could possibly occur between different versions and their descriptions - if they were not about the *same* things?

The solution by "segregation" proposed by Goodman clearly leaves him with other and far worse problems than the ones he set out to solve; problems, which now need being sorted out. Here are some of them:

If identical terms used in different versions are not "assigned" to things which *at the same time* may belong to different worlds, what, then, *are* the things to which such identical terms are assigned? Indeed, how does it come about that in different world-versions the same term, e.g. "Earth" or "tomato", is used to refer to entirely different things being "made" in different worlds? Is it due to mere sloppiness on the part of language users, or is it due to some spurious property of our creative minds hitherto unnoticed - even by Goodman? Whatever it may be due to, the reader of Goodman's books will look in vain for any explanation of what, according to his

Constructivist epistemology, must be an utterly puzzling "inconsistency" in our use of terms.

Secondly, how is it that knowledge about things "made" in physics has proved to have *both* explanatory relevance *and* has come to be applicable to things "created" within the world of every day things? How is it, to choose a rather simple example, that a physicist may predict that if in his world of physics he manages to make atoms of hydrogen and oxygen "fuse" in a 2:1 ratio, people will say that the result of the fusion is *water*, i.e. the sort of colourless and tasteless liquid, we know so well from our everyday lives. And how is it, conversely, that the same physicist may predict that as a result of applying various procedures, water as we know so well from our everyday life will be decomposed into *hydrogen* and *oxygen*? How do we account for these facts - and in particular that these facts are logically connected - without assuming that water, at one and the same time and *in one and the same world*, is a colourless liquid made up of hydrogen and oxygen? That is, how can we account for these facts, without *both* assuming that the physical (hydrogen, oxygen) descriptions apply to what comes out of kitchen taps, *and* assuming that this sort of thing, called water, does not cease to be describable in terms of hydrogen and oxygen when used to make tea, or to wash dirty clothes?

It is beyond my imagination to figure out what other epistemological notions Goodman will now have to render void (i.e. other than the ones already deemed void, such as truth, reference, exist, and reality) in order to account for facts such as these. What is impossible to imagine is which other notions of 'things', and 'the same things', as well of our cognition of them, would remain meaningful in his epistemology, following an account which does not imply that under different circumstances we may know different things about the them, and thus that different description may be different *correct* descriptions of the *same* thing.

However, one thing is patently clear: Goodman is not in the least disturbed by the ambiguities, contradictions and absurdities of his epistemological position. On the contrary, he delights in contradictions, "quips and cracks", "puns and paradoxes", as he calls them. Neither "quips" nor "paradoxes" are considered symptoms of flaws in assumptions or arguments, nor are the inconsistencies in usage of terms they give rise to (among them the terms "true" and "right"). Indeed, they are "part and parcel of the philosophy presented and the worlds made" (Goodman, p. 44, 1984).

Well, it has crossed my mind that Goodman is a real *wag*, someone who loves to make fun of philosophical eggheads and the pseudo problems they create - and in which they eventually find themselves helplessly trapped. If so, he must be laughing himself silly over the many attempts of his colleagues to sort out just what kinds of problem are entailed in the sort of views he proposes - apparently with no other purpose than to contribute to the entertainment. This may well be; however, at the risk of joining the laughing stock of Goodman, I shall discuss other disastrous consequences of his constructionist assumptions and arguments. My sole reason is that in Goodman's version of *anti-realism* taken to its extreme, it becomes so obvious what is equally wrong and unacceptable about, at least on the face of it, far less radical constructivist theories as well.[2] Just such theories, from time to time, seem to attract the attention of psychological theorists - most often as an alternative to some rabid "scientific-physicalism" of the day, claiming to explain everything worth explaining about the human mind and how it works.

A good place to continue the discussion of the consequences of Goodman's philosophical views would be to go right to the most basic assumption in his epistemological building - the one which both necessitates and gives rise to all the others and their problems. Here goes. Since we can only tell how the world is "under one or more frames of references" and, moreover, since "frames of reference belong less to what is described than to systems of descriptions", we cannot tell how the world is apart from or "underneath" all descriptions and frames of reference (Goodman, 1978, p. 2). And because we cannot do that, all we have and all there is, are frames of reference and ways of talking. Indeed, "bereft of" or "underneath" our descriptions and world-versions, reality is *nothing*; it is a world without "things and events", a "world without kinds or order or motion or rest or pattern." (Goodman, 1978, p. 20.)[3]

Now, it is of course perfectly true, that *we cannot describe reality without describing reality*. However, this self-evident fact certainly does not warrant the conclusion that how reality is *apart* from the way it is described, is different from what is being asserted in our descriptions of it. That is, it does not warrant the conclusion that how reality is according to our descriptions of it, is not what it is apart from being described. There is absolutely no compelling reason for so assuming, nor for assuming that it makes sense at all to talk about how reality is *apart* from what is asserted in the statements and descriptions we may put forward about it. On the contrary, as argued often enough, there are compelling reasons to assume that language and statements put forward in language may refer to and be true of what they are

[2] Anyway, it does seem to require some kind of explanation why one is prepared seriously to discuss Goodman's propositions. So, this is mine.

[3] In a comment on this he writes in his 1984 book: "Indeed I have argued in *Ways of Worldmaking* and elsewhere that the forms and laws of our worlds do not lie there ready-made to be discovered but are imposed by world-versions we contrive - in the sciences, the arts, perception, and everyday practice. How the earth moves, whether a world is composed of particles or waves of phenomena, are matters determined not by passive observation but by painstaking fabrication." (Goodman, 1984, p. 21.)

about; without this assumption there *is* no language, nor any use of language in which to discuss anything.

Therefore, it is utter nonsense to maintain that because statements and descriptions about reality are put forward within "frames of reference", then descriptions and statements belong to "systems of descriptions" (i.e. to language) rather than to what is being described. If we did maintain this, we would be saying (once again) that we have a language and in this language we may put forward statements and descriptions - but statements and descriptions are not (necessarily) about anything at all. However, if we do say so, we are doing away with language - and at the same time assuming that we can use language so to do.

But nor does the fact that we cannot describe reality without describing reality warrant the further conclusion of Goodman's that what is described, does not exist *independently* of being described, but only *in virtue* of being described. In particular, it does not warrant the conclusion that "bereft" of our descriptions, ordering, conceptualization, perception etc., reality is *nothing*. If we did maintain this, we would be saying not only that what reality is, depends on the ways we describe it, but also that what reality is - or rather, is not - "underneath" our descriptions - is just as dependent on *not* being described, and so forth.

However, in exchange for the "ready-made" world of things and events available for description which we find in various versions of realism, we have in Goodman's *Irrealism* a world "underneath" our descriptions with "no kinds or order or properties or patterns", etc.[4] However, such determination of what reality is independently of or "bereft" of our descriptions or version of it, can itself, of course, only be a *version* of reality.[5] A version, moreover, which just as any other may be right or wrong.

Now, this version better be right - or the basic tenet of Goodman's epistemology is wrong. Even so, Goodman's position is no better off if it *is* right. For, granted that this version of how reality is "underneath" all other versions is right, it must follow from its rightness that all other versions must be wrong; indeed, it will quite simply cancel or make void all other versions. Such world-versions and their worlds, fitted as they are with "things and kinds and properties and patterns", would indeed be nothing but mere fictions. Even worse, what applies to such versions of worlds must equally

[4] From this at least the following lesson may be learned: Even within Goodman's epistemology it is simply not possible to maintain views about what reality is *when* being described, or *in virtue* of being described, without at the same time referring to what reality is un-described - although *what* it is, may only be determined negatively, i.e. as "lack of" what appears in the world-versions we create.

[5] - Unless, of course, the insight about the nature of reality independently of any description or version of it, is an insight which is gained from somewhere outside the realm of Goodman's worlds.

apply to the persons constructing these versions: We ourselves with our bodies, limbs, etc., being part of the worlds we create, are nothing but fictions. Beyond or "bereft of" our descriptions or versions of ourselves we are - nothing.

Now, at times Goodman seems to allow doubts to creep in; at such times he writes:

> But I have not said that we can make a steak or a chair or a world at will and as we like by making a version. Only if true does description make things; and making true versions can be hard work. But isn't that begging the question? Doesn't that amount to saying that versions can make only what is already there? [...] If versions can make neither what is nor what is not already there, that seems a closed case against their making anything at all. Yet I am not ready to give up; 'being already there' needs further examination, and finding what is already there may turn out to be very much a matter of making." (ibid., p. 35)

Such "further examination" is offered by way of an example in which Goodman imagines himself sitting in a cluttered waiting room - and being unaware of any stereo system. However, gradually he "makes out" two speakers built into a bookcase, a receiver and a turn-table in a corner cabinet, a remote control switch on the mantelpiece - i.e. he finds a system that was already there. This finding of what is already there, he says, involves a lot of "making" by way of distinguishing the several components from the surroundings, categorizing them by function, and uniting them into a single whole, all of which requires a complex conceptual equipment. Another visitor to the same waiting room, fresh from a lifetime in the deepest jungle, will not find any stereo system, for he has no conceptual means of "making" any stereo system in that room. However, he may "make" other things, such as food and fuel of the plants and books that Goodman has found - or rather: *made*. This means, according to Goodman, that the things which he and his fellow companion find in the waiting room are not, after all, things which *are there already*, i.e. independent of the different conceptual systems with which he and the other person are equipped - and he concludes his examination of "what is out there" by contending with the physicist J. R. Wheeler (1981) that:

> "The universe does not exist "out there" independent of us. We are inescapably involved in bringing about that which appears to be happening. We are not only observers. We are participators ... in making [the] past as well as the present and future." (ibid. p 36).

Well, it is arguably correct that it requires complex conceptual equipment to be able to perceive and correctly describe the objects in the waiting room as a stereo system, books and plants, and to describe them as being inflammable or edible. However, if we take this fact to imply that the things being perceived and described do not exist in the waiting room *independently* of being perceived and described, serious, but by now well known problems arise. How, for example, would Goodman - in the actual situation as well as in general - go about determining any *differences* in his own and his companion's description and perception of objects, if no objects existed? And what sense would there be in saying that such differences in perception and description of objects between himself and his companion may be determined, if these different descriptions did not concern the *same* objects, i.e. objects which they may together identify and determine as some particular objects, having particular determinable extension, and existing at some particular places relative to and separated from one another in a room in which they both find themselves? That is, objects which, subsequent to such preliminary determination, may be known to be things having exactly the properties being perceived and correctly described in the different descriptions of them? (Described, for example, as readable books *and* inflammable objects, or as decorative plants *and* edible vegetables). If there were no identity of objects existing independently of the different world versions or perceptions of Goodman's and his companion, he could no more talk about *differences in perception and descriptions* of objects in the waiting room, than he could talk in the earlier examples about *conflicts of truth* between different world versions.

And if this were not serious enough, we have according to Goodman's constructionist position, when applied to his own example, that books simply cease to be inflammable and plants to be edible things just in case Goodman should not observe and thus "make" these properties of the books and plants. And just in case the visitor from the jungle does not know anything about books, the objects on the bookshelf cease to be books. Neither is there any "ready-made" stereo system already there in the waiting room independently of the "complex conceptual equipment" of Goodman's, which allows him to perceive and describe, and thus to "make out" or "find" such a system. But how did it happen that he became equipped with that conceptual system in the first place? Could it possibly be that one may find concepts about stereo systems in the cognitive system of Goodman's, but not in that of his companion, because stereo systems already exist in Goodman's everyday reality - but not in the jungle? And how did it occur to Goodman - and how does he explain - that his conceptual system in the actual situation enabled him to "make" or construct a stereo system in the waiting room - as opposed to a microwave oven or a koala bear - without assuming that, indeed, the stereo system must have been there already?

Yet, Goodman is not prepared to give up or accept what is obvious, i.e. that the reason why we cannot make stereo systems or koala bears "at will or as we like" is that we can't make out or find, what is not already there.

Now, the problems facing Goodman, should he choose to play his constructionalist game to its logical conclusion, are of course not confined to any particular object that he manages to "make out" or "make" in some particular situation. Thus, if it applies to *some* of the objects in the waiting room that they were not already there "ready-made", but only gradually came into being as a consequence of Goodman's conceptually making them out, it must apply to *all* objects - as well as *persons* - in the waiting room. Hence, what applies to the waiting room and its objects must apply equally to Goodman *in virtue* of being a person with a body in the worlds of objects he has himself created: he himself was not already there "ready made", but was only gradually made out or into being in that situation - but how did he get there? And *who* or *what* "made out" or "found" Goodman there - and from *where* did he, she or it do so?

But why involve oneself in the paradoxes and contradictions of Goodman's philosophy, when things may be said so much simpler - without involving oneself in contradictions and paradoxes: How, in any situation we describe what exists "out there" independently of our descriptions, depends on the knowledge we have about things, which in its turn depends on our purposes, intentions, interests, and possibilities of action and observations. Furthermore, in different situations we may describe the same things differently, and such different descriptions may be equally true descriptions about the very same things existing in the very same reality - without being incompatible or in conflict with one another. Indeed, as argued in Chapter 16 (on Identity - Same and Different), we cannot even talk about descriptions of "the same thing" independently of or without being able to refer to *any* description of a thing as being *but one* among several different descriptions of the same thing. Thus, to identify and describe a thing as a particular thing implies asserting what the thing may be when viewed from the point of view or "frame of reference" available in the situation in which it is put forward, i.e. a thing which necessary exists independently of this and any other description put forward in any other situation

The fundamental mistake committed by Goodman (and similarly by Putnam in his defence of "Internal Realism", cf. Chapter 14), is to confuse the *act* or *process* of describing and knowing *with that* which is described and known.[6] But although, as pointed out by both Goodman and Putnam, our descriptions of reality always *take place* with regard to particular conditions

[6] A mistake, by the way, which they share with the various attempts within physicalism to explain how knowledge of reality comes about as a causal process between "the world" and our "mind-brain system".

of observation, action, and conceptual systems, and that the truth of statements about reality may only be determined with regard to such particular conditions, it is equally important to realize (cf. Chapter 14) that no description put forward in any situation exists without having logical relations to other descriptions put forward in other situations about the same things.

And why not say, moreover, that descriptions of things, which exist independently of being observed, described, and acted with or upon, are put forward by persons having knowledge of and a language in which correct statements and descriptions may be made of situations in which they find themselves, of the things they act with in those situations, and of themselves and other person? That is, say that to be a person and a language user in a particular language community, say, the English, is to know that there exist tables and chairs, stereo systems and houses with walls and waiting rooms, tomatoes and molecules, etc. - and it is to know infinitely many correct plain every day words and descriptions in English of these things, and to know how they may be correctly applied.

Let me conclude this discussion of Goodman with the problems that beset the notion of "truth" and "rightness" within his philosophical constructions. Goodman, seeing himself as an epistemologist, cannot, of course, ignore these problems; however, he believes there is a way in which he may get around them. Firstly, he points out, we have to realize that truth is something that applies only to linguistic versions of worlds, i.e. to world versions which consist of statements. However, there are other ways of creating world versions - products of the arts, paintings, musical compositions, dance etc. are cases in point - and such non-verbal versions may be right or wrong as may verbal versions. Therefore, in order to account for the problem of "truth" and "rightness" the philosophical discussion must include all kinds of versions and cases of truth and rightness.

However, secondly, there is according to Goodman no way that we may state the necessary and sufficient conditions for either rightness or truth in *any* area, nor does any *general* characterization of truth and rightness exists that applies to all areas, but only partial characterizations of certain kinds of rightness in different areas. What we - or rather philosophers - may do, then, is to try to sort out the *standards of truth and rightness* in each of those cases or areas, i.e. on the basis of what is already believed to be true or right in those cases or areas. This task, Goodman admits, may be far more difficult to achieve for some areas and cases than for others; in some areas we do not even have the beginning of an account of such standards of truth or rightness. However, this is the challenge to philosophers; it only shows that there is a lot of work for philosophers to do.

But how can Goodman possibly believe that any significant problems of philosophy, science and art may be addressed, among them the standards of "truth" and "rightness" within those areas, when at the same time he believes that we cannot be sure that we ourselves and quite ordinary things in material reality exist as they appear in our descriptions of them? And how can he possibly think that we may understand or make sense of the knowledge of science and the wisdom of arts and myths, if we cannot be sure that our knowledge and descriptions of ourselves and quite ordinary things are descriptions of what is "already" or "really" there?

What Goodman is saying, in effect, is that language which exists *independently* of what it is about may be used by philosophers to talk correctly about language and the conditions for using language, however, without presupposing that they or we may use language to say anything correct about that to which these conditions apply. But how is it possible to believe that progress could be made in sorting out any problem of objectivity or subjectivity, of truth or rightness of descriptions, of fiction or fact, if *nothing* exists independently of our description, and hence no true – or false – description of it, not even of ourselves and the situations in which we find ourselves in material reality - including the situations in which we try to sort out those very problems?

In summary, Goodman's Constructionalism represents a version of the language-reality bifurcation, according to which human beings happen to be the lucky possessors of a language, which they may use to create themselves and the rest of the world - just as it is said in the Gospel according to St. John: "In the beginning was the Word ....". However, in Goodman's gospel with a difference: that which is created with words by Mankind, as opposed to what is created in the words of God, is not really what is "out there". Things being described and, by implication, we ourselves and other persons, only have an "actual" existence *for* us - *in virtue* of being described, and thus being "made" to exist *by* us. Goodman, in his promotion of these and other propositions of his "Irrealism", goes to great lengths to ridicule realism and realists who believe in the existence of an independent reality, as well as other assumptions traditionally associated with realism. However, I for one, do not mind in the least being called a "realist" - not even a "naive" one - for I have logical reasons for so being. How, for example, could I talk sensibly about anything at all, including myself, other persons, and material reality, without presupposing that in the language I speak, I can talk *correctly* about such things, and in so doing without presupposing that the things I describe exist independently of and as something to which I may *refer*, and about which correct and true statements may be put forward? Or, how could I have a language and put forward statements about such things, or talk about how these things may be depicted and construed quite differently in the arts and in fiction, if there *were* nothing to which any statements in language referred? And how could I use language to communicate with other persons about these and other matters without assuming that other persons assume what I assume?

Furthermore, one can certainly be a realist without claiming that the knowledge we have about reality and ourselves is or has to be "absolute" or "exhaustive", i.e. that Realism necessarily imply *Substantialism.* Realism, as proposed in this book, is fully compatible with the assumption that our knowledge and descriptions of reality and other persons - as well as our doubts - can only be partial; indeed it a crucially important point of my arguments that epistemological consistency *relies* on this assumption. Such a realist position precludes Substantialism. Hence, Realism need not be wrong; what is wrong is the assumption that Substantialism and absolute knowledge of reality, having been defended by traditional versions of Realism, is an inevitable part of any form or shape of Realism. And what is wrong is the assumption that Realism inevitably has to be just another philosophical position which, like any other philosophical "-ism", represents an attempt to "bridge the gap" between language and reality, or between knowledge of reality and reality itself, which arises as a consequence of the fallacious bifurcations within traditional philosophy. However, once realism has been freed of this traditional assumption and bifurcation, I think there are compelling reason for most of us to see ourselves as realists.

So, Goodman may have it all, scientific as well as everyday descriptions and knowledge of reality, even descriptions of reality employing metaphorical use of language; and he may have reality as depicted and represented in arts and dance, as well as the stories created in fiction and myths. Such products of science, art and fiction may indeed affect the ways we look at ourselves and perceive our position in the world we live in; and they may help us understand the ways in which we try to make sense of it. However, as already argued in Chapter 11, our understanding and ability to create and stage *artistic* expressions and depictions, as well as stories of fiction and myths, necessarily rest on our knowledge and ability correctly to describe the situations in which we find ourselves in material reality with objects and other persons. Thus, the very peculiar effect and force of the arts, fictions and myths, and our ability to create and understand the *extraordinary* produced in the arts, fictions and myths, is firmly based on our intimate knowledge and correct descriptions of the *ordinary*.

* * * * * *

In the course of the discussion in this chapter, it has become apparent how curiously difficult it has been to see that what we say about reality and our possibility of describing and having knowledge about reality, immediately affects what we may say about ourselves as persons - and vice versa. And clear, furthermore, that we cannot raise any doubts or questions about the independent existence of *reality*, nor about the very possibility that we have true and objective knowledge about this reality - without simultaneously and automatically cutting ourselves off from talking sensibly about ourselves, i.e. about *persons* and the *cognition* and *language* of persons. And clear, conversely, that we cannot say whatever we please about *persons*, for example, question or doubt the possibility that persons have knowledge of and may use language to put forward true – and false – propositions about reality - or of whatever their cognition concerns. If in our investigations of the knowledge of persons we suspended this possibility, we would automatically abolish the very conditions for talking consistently or meaningfully not only about this knowledge, but also about anything that this knowledge concerns.

In the chapters to come, I shall argue that to be persons and language users, living in social groups, communities and societies, logically requires that persons *know* that they are persons, and *know* that they share a vast amount of knowledge and correct descriptions of the situations in which they find themselves. This assumption is not incompatible with the fact that different persons - apart from the general determination of the situation in which they find themselves - may describe the same situations, things and events, very differently indeed. How this may be and why, is in my opinion where the challenge lies, for philosophy, psychology, sociology, and anthropology alike.

Chapter 18

Wittgenstein's theories of language

18.1 Introduction

In previous chapters I have discussed the conditions for knowledge and description of persons concerning things and states in physical material reality, just as I have addressed the conditions for persons together to communicate about the differences in their knowledge and description of such things. Among the latter must be that they are persons in a public world of which – their differences in knowledge and description notwithstanding – they share a considerable amount of knowledge, and a language in which true statements may be put forward – including, necessarily, of the things that their differences in knowledge and description concern. Without presupposing this, there is no way in which persons could together determine what their differences in knowledge and description concerned, even less determine or discuss what these difference were.

But what about our knowledge and descriptions of things which are not part of a publicly shared and observed world, such as our "internal states", our thoughts, feelings, emotions, religious beliefs, pains or taste? Can it be maintained that knowledge and descriptions of our inner states have the same status as knowledge and descriptions of physical object in material reality? And can it be maintained that the conditions for persons to have knowledge of and to communicate with others about their inner states are the same as those, which apply to their knowledge and description of things in physical reality? Indeed, can we be sure that the descriptions and assertions put forward about such states and other non-publicly observable things, have the same implications for different persons? Or, do we here find ourselves in a realm of experience, knowledge and use of language which is not only *personal*, but even *private*? And does it really makes any sense to say that thoughts, emotions and feelings may be known in the same way as we may know of physical material things, or to say that when referring to such inner states we are using language in the same way as we do when describing things in a public physical world?

These questions as to the nature and status of our knowledge and descriptions of internal states have been the topic of extensive debate within philosophy for centuries and in this century, in particular by Wittgenstein. In connection with arguments for his views and assumptions about the nature of language, and notably for his view about the conditions on which language can be sensibly used, he proposes that what one knows and describes must be part of a *public* world, and also that the language one speaks is a language one shares with others *in virtue* of being about things being observable in such a public world. This means, so Wittgenstein maintains, that the knowledge and description of a person about his internal state cannot have the same status as knowledge and description of things in publicly observable physical reality. Indeed, so he contends, it cannot be maintained at all that a person has *knowledge* about states to which only he or she has access and may experience, nor be maintained that when describing such states, the person is using a language which he or she shares with others. In the chapters which follow I shall argue that neither claim follows from the general conditions for persons to have knowledge, to describe and communicate with other persons about what they know. Furthermore, I shall try to show that the confusions which haunt the discussion about the status and conditions for knowledge and descriptions of such states, are due to the same fallacious Cartesian bifurcation between the "mental" and the "physical", which has given rise to similar confusions about the status and conditions for our knowledge and description of things in physical reality – as seen in preceding chapters.

In the course of this discussion I shall argue that our knowledge and description of the non-physical, the 'mental' or 'internal', does not present *epistemological* problems which are fundamentally different from the problems concerning our knowledge and description of things in the public physical world. Indeed, so I shall argue, we cannot begin to talk correctly and sensibly about our knowledge of physical reality without being able to talk correctly and sensibly about our internal states, our thoughts, beliefs, perceptions, feelings, and so on. Nor can we talk sensibly about any differences between our knowledge and use of language concerning physical and internal states, let alone begin to address questions of how the internal states of different persons may or may not differ, unless we assume that we do have knowledge about and a language in which it is possible to determine, describe and talk sensibly about both such states and differences. That is, without assuming that internal states, just like physical things, exist as things about which we may have knowledge and put forward descriptions and propositions which are true.

It is not, of course, the intention of these arguments to show that there are no difference between physical things and internal states – there are indeed many differences. Rather, it is the intention to show that the arguments put forward by Wittgenstein as to the differences in status of our knowledge and description of physical things versus our *experiences* and *expressions* of internal states rest on a false bifurcation between 'mind' and 'matter'. This is a bifurcation which ignores that a person who may reflect on, experience and communicate about his internal states is a person who at the same time finds him or herself in a situation in physical material reality, and thus is a person who knows and is aware of him or herself as someone having a body, which is part of this reality. Indeed, it is a bifurcation which ignores that, logically, we cannot talk about persons, their bodies and other things being part of physical reality, without talking about and referring to internal states of persons. In contrast, the arguments being proposed in this text for the existence of a logical relation between the concepts we use to characterise a person, are arguments to the effect that we cannot talk about the 'inner' or 'mental' of a person in isolation from, nor independently of the 'outer' or 'physical' of a person; indeed they are arguments to the effect that the 'inner' of a person is very much part of the 'outer' of a person – and vice versa.

In this chapter I shall attempt to give an account of the main assumptions and arguments about language and its use as first presented by Wittgentein in *Tractatus Logico-Philosophicus*,[1] and later in *Philosophical Investig-- tions.*[2] Put very shortly, the main aim of *Tractatus* is two-fold: it is to plot the limits of language - and by this Wittgenstein means the limits of *factual* discourse, which makes sense. And it is to show that the majority of problems of philosophy of language and knowledge are posed in ways in which the logic of language is misunderstood, because they are posed in terms which do not clearly distinguish them from problems of factual discourse. However, since they fall outside factual discourse they are nonsensical problems to which no solution may be found, but, on the contrary, which need to be "dis-solved" by way of a critique of language, which fixes the limits of what it is possible to talk sensibly about. I shall discuss some of the problems of the theory of meaning and truth presented in *Tractatus*; for

[1] for short: *Tractatus*
[2] for short: *Investigations*

although his views about the status of the "mental" versus the status of the "physical" are mainly to be found in his latter work, i.e. in *Investigations*, it seems to me - as it did to Wittgenstein himself - that these later views may only be fully understood in view of assumptions and arguments presented in his earlier work. Indeed, the rejection of the central doctrines of *Tractatus* came to be the very cornerstone of his later philosophy. Furthermore, the differences notwithstanding, there are significant similarities of assumptions as well; similarities which reflect the fact that diametrical opposites very often spring from the same root.

18.2 Wittgenstein's language games[3]

It is well known that Wittgenstein throughout his philosophical life was acutely preoccupied with the question of what philosophy is, ought to be, and how it must be done. Inspired by Moore's and Russell's approaches to the studies of philosophical problems, he had come to appreciate the importance of the fact that all philosophical problems - in order to be problems which may be thought of - had to be formulated in language and, thus, that the problems of philosophy could only be solved if one had a proper understanding of what language is. Problems of what it is to be a human being, and of what ultimately exists; of what knowledge is and how we come to acquire it; of what we think, feel, or what it is to *mean* or *understand* something, are inevitably posed in language – and so are their solutions. Language, therefore, had to have a central position in the investigations of philosophical thoughts and ideas.

It is equally well known that the underlying theme running through Wittgenstein's works in both of his philosophical periods is that most philosophical problems (i.e. metaphysical and ontological) are the product of misuse of language. Since classical antiquity the philosophical debate has been based on the consensus that the matters of existence, knowledge, truth, value, etc., are deeply important conceptual and logical problems (as opposed to empirical problems), which require painstaking conceptual and logical investigations. In Wittgenstein's view, however, it is not the proper task of philosophy to engage in such matters, for they involve illusory problems which arise as a result of misunderstandings about language. What may appear to be genuine thoughts and problems are in reality pseudo-problems, which occur when, unwittingly, we go beyond the limits of language and

[3] My understanding - not so much of what Wittgenstein says, but *why* he says it - has been greatly improved by reading the eminently readable books on Wittgenstein by Pears (1971/1985), Ayer (1986), Hacker (1993), and Grayling (1998) - and so has my account of the development of Wittgenstein's doctrines and thoughts. I am indebted to all four authors on both scores.

what can sensibly be said in language. Once the structure and limits of sensible discourse has been clarified, such mistakes can be avoided. The outcome of such analysis of language and of the meaning of concepts is not new and better explanatory theories about metaphysical problems, as has been attempted over millennia since Aristotle and Plato. Rather, it is a *clarification* in which such problems have disappeared. In his first mayor work, Wittgenstein puts it this way.

> "Thus the aim of the book is to draw a limit to thought, or rather - not to thought, but to the expression of thoughts: for in order to be able to draw a limit to thought, we should have to find both sides of the limit thinkable (i.e. we should have to be able to think what cannot be thought). It will therefore only be in language that the limit can be drawn, and what lies on the other side of the limit will simply be nonsense." (*Tractatus*, p. 3.)

And in *Investigations*, the same point is put thus:

> "[Philosophical problems] are solved by looking into the workings of our language, and that in such a way as to make us recognize those workings: *in despite of* an urge to misunderstand them. [...] Philosophy is a battle against the bewitchment of our intelligence by means of language (*Investigations*, para. 109). [...] The results of philosophy are the uncovering of one or another piece of plain nonsense and of bumps that the understanding has got by running its head up against the limits of language. These bumps make us see the value of the discovery. (ibid., para. 119). [...] Where does our investigations get its importance from, since it seems only to destroy everything interesting, that is, all that is great and important? (As it were all the buildings, leaving behind only bits of stone and rubble.) What we are destroying is nothing but houses of cards and we are clearing up the grounds of language on which they stand." (ibid., para. 118.)

However, the methods and approaches adopted by Wittgenstein in order to uncover the structure and limits of language, which once and for all would liberate philosophy of "everything interesting", indeed his very conception of language - both of the language being analysed and the one being used for the analysis - are very different in *Tractatus* and in *Investigations*. Whilst it is the aim of *Tractatus* to show that any of the different forms of language, in which we may talk sensibly, has the same essential and logical form, it is the aim of *Investigations* to show that language may have many different forms, which, when used in different circumstances, make sense.

And while it is the aims of *Tractatus* to develop an account of the logical structure of language, founded on a widespread and long lasting traditional conception within philosophy of what language is, the change in the view of language from *Tractatus* to *Investigations* came to usher in a similar widespread abandonment of this traditional view. So first, a few words about this view.

According to the traditional view of language to which Wittgenstein subscribed when writing *Tractatus*, language consists of words, each of which gets its meaning from being the name of something. To learn a language is to learn what its words 'stand for' or 'name'. In contrast, a word which is not a name of something, does not mean anything, indeed it is not even a word, merely a 'sound'. David Hume, who like so may other philosophers based his philosophy on this view, expresses in this way its consequences for philosophical enquiry.

> "When we entertain, therefore, any suspicion that a philosophical term is employed without any meaning or idea (as is but too frequent), we need but enquire, from what impression is that supposed idea derived? And if it be impossible to assign any, this will serve to confirm our suspicion." (Hume, 1966, p. 22)

Well, not all words are names - the logical constants such as 'and', 'or', 'not', 'if' and 'is' are cases in point. However, they do not present any problems for the naming theory of language. According to this theory, words in a proposition, say, 'the cat on the mat has four legs', refer to or name a *subject* (the cat on the mat) about which some properties are *predicated* (has four legs) by using words which name or stand for these properties. However, there were other problem which did threaten the naming theory, and which required serious analysis to overcome. Take for example the expression, 'The round square does not exist'. Can it sensibly be maintained that 'The round square' is the subject – and thus that it refers to or names the round square? Or rather, how could this be maintained when, at the same time, its existence is denied? One cannot sensibly name something which does not exist. Only two options seemed possible:

Either to maintain, as did the German philosopher Meinong, that since the expression, 'The round square', is in fact an expression which refers to and names the round square, then there must be something – if only on a conceptual level, a conceptual unity of some kind - which is the carrier of the name and which the expression refers to. Or, to propose that the expression, 'The round square', neither refers to nor names anything. However, by choosing the latter possibility one has in effect given up the view that all words – except for the logical constants – are names, and thereby given up the naming theory of language. However, so it was argued, nobody can deny that the proposition, 'The round square does not exist', is true, nor therefore maintain that 'The round square' is meaningless. For if it did not mean anything it could neither be true nor false. Thus, 'The round square', means something - but it does not name anything.

Russell introduced yet another possibility – and a solution - which at least temporarily saved the naming theory. The expression, 'The round square', he says, is not at all a subject, and consequently does not refer to some mysterious unity. Indeed, the proposition, 'The round square does not exist', does not have a subject. The expression has a grammatical form, however, which (mis)leads one to believe that it does. It is the purpose of a philosophical analysis to reveal its true form, i.e. its *logical* form. The result of this analysis is that what the proposition states is precisely that no subject exists of which it may be predicated that it is both round and square. Thus conceived, the proposition does indeed consist of words which (only) refer to or name existing things, namely the properties round and square, just as it states that both these properties cannot be predicated about a thing. Rephrased in a way in which the grammatical form of the proposition matches its logical form, it goes like this: 'No unity exists which is both round and square'.

Since the logical form of propositions are often misleadingly expressed in normal everyday language, it is of vital importance to scientific and philosophical inquiry alike that a language be constructed in which the logical forms could be made immediately clear. This was the position of Russell – but not of Wittgenstein, his former student. To Wittgenstein there was no point in constructing a new language – only one language existed. What made different spoken languages one and the same language, logically speaking, is that they all have the same underlying logical structure and a set of common logical conditions which they all had to satisfy. It was this logical structure and these conditions that Wittgenstein sought to uncover. Once they had been revealed and understood, the limits of what can be meaningfully and clearly said would become clear.

18.3 Tractatus

The main aim of *Tractatus* is to demarcate the borders of sense, and to argue that these borders are constituted by the limits of *factual* discourse. Thus, everything, which falls within the "logical space" of factual discourse, makes sense, whereas everything beyond its borders is nonsense. Factual discourse itself is constituted by all *propositions* which get their sense - and only have a sense - because their words either represent things existing in the world, or are analysable into something which represents such things.

The world itself is said to be a totality of facts, which themselves consist in the existence of "states of affairs". In their turn, states of affairs are composed of simple or *elementary objects*, which may be named; *names* gets their meaning from being the names of simple objects, or as Wittgenstein says, "A name means an object. The object is its meaning" (ibid., para. 3.203). Propositions composed only of names of simple objects are *elemen-*

tary propositions. It is Wittgenstein's contention that the structure of the objects of the world, or states of affairs, is precisely mirrored in the structure of elementary propositions. Just like names get their sense from the objects they name, the meaning of propositions is constituted by the states of affairs they depict. And just as the object is the meaning of a name, the pictures of elementary propositions are themselves facts, which share a logical and pictorial form with what they represent.

Now, not all factual propositions are elementary; some are compound propositions, which are formed out of elementary propositions, and into which they may be analysed. It is an important assumption of *Tractatus* that all propositions of factual language have a *precise* and *determinate* sense, and that an exact account of what they mean can be given only if they are analysed into their ultimate components, i.e. elementary propositions. In Wittgenstein's theory of factual language and its structure, the notion of elementary propositions serves as a vehicle which guarantees that factual propositions may have precise and determinate senses; they are the rock bottom, so to speak, of factual discourse.

It is the general form of propositions to serve the function of saying something or informing us about something which is the case. And what they inform us about are states of affairs, i.e. ways in which objects are structured. As said above, it is a basic assumption of *Tractatus* that propositions of factual language depict states of affairs in the world, *because they share the same logical structure or form*. For example, the sentence, "this tower is high", depicts the relation between the tower and other buildings. However, a proposition does not only have a *meaning* if it depicts things which exist, but also if it depicts things which could *possibly* exist. In contrast, a proposition which has a meaning, does not necessarily inform us of what is the case, i.e. represents or *depicts* any existing fact or state of affair; it only does so, and thus is *true*, if what it claims to be the case about some state of affairs is satisfied - otherwise it is false.[4]

But how do we get from the "signs" and "structure" of sentences to the "things" represented by them? This, Wittgenstein says, is taken care of by means of applying rules of *projection*, associated with the signs, which translate or transform the structure of the sentences to the structure of the facts. That such projection rules are necessary may be clear, if we consider that in the proposition, "this tower is high", the word "tower" does not look like a tower, nor is the word "high" itself high. It is in virtue of projection rules that a sentence becomes a proposition with a *meaning*, and that it may

[4] For example, the proposition, "Peter has a scar on his cheek", is meaningful because the existence of a scar on Peter's cheek is a possibility. However the proposition is only true if it depicts a state of affair, i.e. the existence of a scar on Peters face; if not the proposition is false.

picture some states of affairs. Knowledge of such projection rules are part of knowing a language, and when applying these projection rules the language user *understands* the proposition, i.e. he knows immediately what the proposition expresses.[5]

To illustrate this idea of depicting and projection, let us imagine that we have a set of drawings of human beings. A drawing of a human being, of course, is not a human being, nor does a human being look like a drawing of a human being; anyhow, a drawing gives us quite a good idea of a real human being - and that means that certain rules of projection apply to the drawing. Now, the drawings may depict human beings which actually exist - and such drawings correspond to *true* propositions. Others may be drawings of fictitious human beings (or human beings whose existence is only probable), and such drawings correspond to false propositions.

However, nothing in the drawings themselves determines whether they picture existing or fictitious human beings; that may only be determined by checking whether a person exists in the world that it depicts. The same goes for propositions. Whether language is about reality cannot be determined only by studying language and the concepts and structure of language. In order to determine whether a proposition pictures existing states of affairs, or only possible ones, we have to check whether facts or states of affairs exist in the world that correspond to the proposition in question. Thus, "acquaintance" with the world, and what exists in the world, is inevitably required in order to determine the truth of factual propositions; and it is required, of course, in order to distinguish factual discourse which makes sense, from metaphysical discourse which is nonsensical.

However, this reference by Wittgenstein to comparison of propositions with what *empirically* exists in the world, which would seem required for determining both the truth and sense of any particular proposition, should not mislead one into believing that Wittgenstein is trying to develop a theory of factual language and its limits based on studies of empirical features of either language or of the world. Far from it. What he is aiming at is to pin down what are the essential, and in his view, *logical features* of language,

[5] Wittgenstein gives the following examples: "A gramophone record, the musical idea, the written notes, and the sound-waves, all stand to one another in the same internal relation of depicting that holds between language and the world. They are all constructed according to a common logical pattern." (ibid. para. 4.014). "There is a general rule by means of which the musician can obtain the symphony from the score, and which makes it possible to derive the symphony from the groove on the gramophone record, and, using the first rule, to derive the score again. That is what constitutes the inner similarity between these things which seem to be constructed in such entirely different ways. And the rule is the law of projection which projects the symphony into the language of musical notation. It is the rule for translating this language into the language of gramophone records." (ibid. para. 4.0141)

indeed of any possible factual language - independently of any empirical features or facts of either language or the world. The most central of these features, says Wittgenstein, is the elementary propositions into which any factual language may be broken down; conversely, elementary propositions serve as points of origin from which it is possible, by repeated use of the logical operator of double negation, neither-nor, to compose the compound propositions of factual discourse. In this fashion, it is possible to work "outwards" from elementary propositions and to calculate the outer limits of any possible factual language.

But if this model of language is to serve as a model whereby factual discourse of the language we actually use may be distinguished from non-factual discourse, and thus from nonsense, it would seem to be required that examples of "names" and "elementary propositions" can be given, and thus that they can be determined. Similarly, it would be required that examples of "simple objects" can be given, as well as examples and accounts of the logical structure or form shared by states of affairs and factual propositions. However, Wittgenstein gives no such examples. In the case of the existence of elementary propositions, Wittgenstein thought it a "matter for the application of logic" to prove that there must be such things, however not to *say* what they were. But nor does he give any indication of how, more precisely, this job may be carried out. Indeed, he says that neither he nor any other philosopher has yet got down to the ultimate components of factual propositions. So, all we have is his specification of elementary propositions as being logically *independent* factual propositions, i.e. propositions of which it applies that the truth or falsity of one elementary proposition never implies the truth or falsity of any other elementary proposition. However, the nature of their elements, the "names" of simple objects, are left just as undecided as the "simple objects" they name.

Pears, in his elucidation of Wittgenstein's position on this point, gives the following reason why he abstained both from giving examples of elementary propositions, and from attempting to demonstrate how an analysis of a particular factual proposition may be carried out:

> "[Wittgenstein] would not have been content with demonstrating that the complete analysis of certain factual propositions happen to contain elementary propositions: he had to prove that the complete analysis of all factual propositions are necessarily composed entirely of elementary propositions. This conclusion could not be established inductively by using logical analysis on a few chosen cases: it had to be deduced from a general theory of meaning." (Pears, 1971, p. 59)

In other words, even if a logical analysis had penetrated to the level of elementary propositions, and thus had allowed Wittgenstein to give examples of such propositions, he would still have needed a general theory of meaning in order to answer the question as to *why* there has to be elementary propositions, i.e. why elementary propositions are central and necessarily in a model of language. Now, according to Wittgenstein's theory of meaning, elementary propositions acquire their sense because they precisely picture or represent the structure of existing things, i.e. simple objects and their combination. "The sense of a proposition", he says, "is its agreement and disagreement with possibilities of existence and non-existence of states of affairs" (para. 4.2). Furthermore, the simplest kind of proposition, an elementary proposition ... "asserts the existence of a state of affair" (para. 4.21).[6] Accordingly, elementary propositions are necessary - in order to depict the existence of simple states of affairs formed by the combination of simple objects in the world.

But why do there have to be simple objects? The answer to this question is given in the paragraphs of *Tractatus* quoted below:

> "Even if the world is infinitely complex, so that every fact consists of infinitely many states of affairs and every state of affairs is composed of infinitely many objects, there would still have to be objects and states of affairs." (para. 4.2211)

> "Objects are simple. Every statement about complexes can be resolved into a statement about their constituents and into the propositions that describe the complexes completely. Objects make up the substance of the world. That is why they cannot be composite. If the world had no substance, then whether a proposition had sense would depend upon whether another proposition was true. In that case we could not sketch out any picture of the world (true or false)." (para. 2.20 to 2.022)

According to this "atomist" assumption about the world and language, it would seem to follow from the existence of elementary propositions that simple objects *must* exist. And so does the entire structure of reality, i.e. it follows from or may be *deduced* from the lattice of elementary propositions, which he believed to be the basic structure of all language. That objects are simple, i.e. "unalterable and subsistent", means that they unambiguously guarantee the existence of what is being referred to, a function which cannot be served by "states of affairs" which, due to variations and circumstances in the world, and thus variations in what may or may not be the case, are both changeable and unstable. In *Philosophical Remarks* (Wittgenstein, 1975, 36), he has the following comment on the necessity of the existence of simple objects:

[6] Thus, "If and elementary proposition is true, the states of affairs exists; if an elementary proposition is false, the state of affairs does not exist" (para. 4.25.)

> "What I once called "objects", simples, were simply what I could refer to without running the risk of their possible non-existence, i.e. that for which there is neither existence nor non-existence and this means: what we can speak about *no matter what may be the case.*"

In other words: objects are necessary in order to secure proper reference. If there were no such simple objects the whole house of cards of elementary propositions *with precise and determinate sense* - and thus the very means of demarcating factual discourse from nonsense - would crumble. But so would logic itself. For although the truth-value of logical propositions (i.e. tautologies or contradictions) is independent of how things are in the world, the existence of logic depends on the possibility of combining factual propositions to form tautologies and contradictions. This requires the possibility of first constructing factual propositions – without which there could be nothing to combine - which in its turn presupposes the necessary truth that reality consists ultimately of simple objects (Pears, ibid. P. 83). This connection between the existence of logic and Wittgenstein's ontology, gives logic an important instrumental role in describing the structure of reality and language: "the propositions of logic describe the scaffolding of the world"; they do this by showing what has to be the case for language to have sense (*Tractatus*, para. 6.124). All that logic presupposes is that names denote objects and propositions have sense (ibid.).

Well, this leaves us with a theory according to which the structure of reality may be *deduced* from the lattice of the structure of elementary propositions, and *also* that the structure of factual language may be *deduced* from the structure of reality - i.e. in virtue of the structure of factual language being a picture or representation of the structure of reality. That is, it leaves us, as noted by Ayer (1985, p.21), with a theory which takes us round in a circle.

Now, Wittgenstein did not actually claim that he could account for how factual propositions and states of affairs shared the same logical form, and thus give an account of how propositions picture or represent states of affairs in the world. On the contrary; just like the fact that "A name means an object. The object is its meaning" could not be *asserted*, it could only be *shown*. For,

> "Propositions cannot represent logical form: it is mirrored in them. What finds its reflection in language, language cannot represent. What expresses *itself* in language, *we* cannot express by means of language. Propositions *show* the logical form of reality. They display it." (*Tractatus*, para. 4.121)

In other words, one cannot properly use language to talk about language, nor use language to *say* that propositions have a certain structure, the elements of which are linked, through the picturing relations, with elements of the world. Well, this is of course precisely what the doctrines of *Tractatus* are largely about. Wittgenstein's way out of this problem is to suggest that the propositions used in *Tractatus* are *philosophical* propositions which serve an *elucidatory* function; they are like steps of a discardable ladder, which one may throw away once one has used them to climb above them. One must transcend these propositions, and then one "will see the world aright" (*Tractatus*, para. 6.54). What one sees from this vantage point is that

> "The totality of true propositions is the whole of natural science (or the whole corpus of the natural sciences) (ibid. para 4.11). ... The correct method in philosophy would really be the following: to say nothing except what can be said, i.e. propositions of natural science - i.e. something that has nothing to do with philosophy - and then, whenever someone else wanted to say something metaphysical, to demonstrate to him that he had failed to give a meaning to certain signs in his propositions. Although it would not be satisfying to the other person - he would not have the feeling that we were teaching him philosophy - *this* method would be the only strictly correct one." (ibid., para. 6.53)

However, what Wittgenstein is saying here, of course, cannot be asserted within *factual* discourse - a fact shared by the basic metaphysical doctrines and claims about language proposed in *Tractatus*. Being well aware of this, Wittgenstein concludes *Tractatus* by saying that although what his theory says is *unassailably true* and *definitive*, it is nevertheless *pure nonsense* (ibid. para. 6.54). This would seem to make it extremely ill equipped, even in his own view, to serve as the definitive basis from which a critique of language can be carried out, which would demarcate sense from nonsense. However, it would also seem to makes it equally ill equipped as a basis from which to determine what may or may not be said about problems and issues within philosophy. It is well known that this was in fact one of the reasons for his dissatisfaction with the theory presented in *Tractatus*, and for his later attempt at suggesting a new approach to language, which would leave him with better opportunities and greater scope for saying something of interest about both language and philosophy.

One obvious problem with the proposition, "we cannot use language to talk about language", and thus of denying the reflexivity of language, is that it has this undeniable ring of the liars paradox. Unless, of course, the language in which this proposition and denial is put forward is a totally different language, having a different structure, and all the rest, than the one being talked about.[7] The inevitability of the paradox in Wittgenstein's attempt to

[7] Cf. Russell's comment in his introduction to *Tractatus*: "What causes hesitation is the fact that, after all, Mr. Wittgenstein manages to say a good deal about what cannot be said, thus suggesting to the sceptical reader that possibly there may be some loophole through a hierarchy of languages, or by some other exit" (Wittgenstein, 1921, p. xxi)

describe how language and the world logically have parallel structures which connect by means of the picturing relationship, and that sense attaches to what we say *only* if what we say is a picture of a fact, is very precisely pointed out by Russell in his introduction to *Tractatus*:

> It is this common structure which makes (a factual proposition) capable of being a picture of the fact, but the structure cannot itself be put into words, since it is a structure *of* words, as well as of the facts to which they refer. Everything, therefore, which is involved in the very idea of the expressiveness of language must remain incapable of being expressed in language, and is, therefore, inexpressible in a perfectly precise sense." (*Tractatus*, p. xx – xxi)

At the outset Wittgenstein – for very good reasons – strongly disagreed with Russell about the need for the development of an ideal language which philosophy could use to express and lay bare the logical structure of propositions of the language we speak, and in which the proper meaning and truth of its propositions may be determined. However, the language developed in *Tractatus* with its simple names and elementary propositions, their logical structure, and all the rest, *is* an ideal language, an *abstraction* of a language, just as ideal and abstract as are the simple object and states of affairs which are claimed to exist in and inhabit the world, and which the words and propositions are supposed to 'name' and depict. And so is the common logical structure of *this* language and *that* world. Between this ideal language of *Tractatus* and the language used to determine the conditions for its truth and meaning, i.e. the language we *speak*, there is no and cannot be any point of "merger". They have nothing in common. For, although propositions in the language of *Tractatus* are themselves facts, i.e. *in virtue* of meaning and being true about things in the world they *depict*, statements about such propositions, although being statements about facts, are not propositions which satisfy the condition for propositions depicting facts of the *world*, i.e. of objects and states of affairs, and thus are not proper propositions by that criterion.

Well, should the result of an analysis with the purpose to reveal the logical structure and conditions for (all) the language(s) we speak be that language cannot be reflexive, i.e. cannot be used by its speakers to talk about language, it will be immediately clear that somewhere the analysis has gone badly wrong. For it is certainly a property of language that it may be used by its speakers to talk about language and its use, for example be used to ask and answer questions of the kind, 'What does that mean?', and, 'What is that called?'. Without this reflexivity it would simply not be a language in which such questions could be asked or answered, nor a language in which the correct implications of its words and the truth of its statements could be determined. In short, *it would not be a language*. What is constructed in

Tractatus, however is not a language, but rather, a 'sign-system' having in common with other 'sign-system' – be they man-made (as e.g. traffic signs) or natural (as e.g. the sign-systems of birds and other animals) – that the notions of 'meaning' of the signs and the 'truth' about what they signify or represent, are notions which figure nowhere within the system themselves. To the extent it is possible to talk at all about the meaning of such signs, this can only be done in a language outside the systems, and then only in terms of some sort of co-variance between the signs and some state of the world, having been observed *and* determined in that language. When it comes to the *truth* of the sign of such system, however, this option of a re-description of signs to terms in the language used for the re-description, does not exit. What may be determined in this language is at most rules and criteria for the existence, once again, of a certain co-variance between signs and states of the world, a notion, however, which is a far cry from the notion of 'truth'.

The obvious and inevitable consequence of Wittgenstein's and other attempts to describe language and its relations with the world in terms of a picture or representations of the world sharing the same structures (logical, causal, or otherwise), is that such a language cannot be reflexive (cf. also arguments for this in Chapter 5). The further consequences thereof is that the *language* being used to describe the conditions for 'language' and propositions in 'language' to correspond to, i.e. to be true and make sense about reality, is a *language* which does not itself fulfil these conditions. The only way out of this problem – as well as the problems of circularity and non-sense to which such attempts to account for how language and reality connects – is to abandon such attempts. As argued in Chapter 5, and illustrated perfectly clearly in *Tractatus*, it is simply not possible to account for the conditions for using language meaningfully and truthfully to say anything about reality and its structure, without at the same time *describing* reality and these structures – however rudimentary - and thus, without presupposing that the language in which we do so may indeed be used to talk sensibly and truthfully about *both* reality and language. To realise this, is to realise that to view the relation between language and reality in terms of a symmetrical structure (logical, causal or otherwise) between independently determinable "parts", is an impossibility. Indeed it is to understand that the relation between language and reality is unanalyzable, and thus cannot even be *shown. It has to be taken for granted.*

18.4 Investigations

The first to go in his review of the doctrines of *Tractatus* were the doctrines about the "atomism" of the world and language, and the "picture" theory of meaning and truth of propositions. Against the view that in the world simple objects exist, which are unique and indivisible, Wittgenstein objects that it does not make sense to talk about objects and their parts, nor to distinguish between what is simple and what is composite, outside the framework of some particular discourse. What in one situation may be "indivisible" is not so in another:

> "But what are the simple constituent parts of which reality is composed? - What are the simple constituent parts of a chair? - The bits of wood of which it is made? Or the molecules, or the atoms? - "Simple" means: not composite. And here the point is: in what sense 'composite'? It makes no sense at all to speak absolutely of the 'simple parts of a chair'." (*Investigations*, para. 47)

But if there are no such things as absolute simple objects, the notion of names getting their meaning from such objects becomes redundant; and with it the doctrine of states of affairs being composed of elementary objects, and of elementary propositions picturing or corresponding logically to such states of affairs. So, the base disappears beneath "the house of cards" of the theory presented in *Tractatus* of truth and meaning of propositions and language.

The departure from his previous conception of language is pronounced in the very first paragraph of *Investigations* in which he brings a lengthy quotation from the *Confessions* of St. Augustine. He comments on the quotation that it gives a picture of language in which "we find the roots of the following idea: Every word has a meaning. The meaning is correlated with the word. It is the object for which the word stands". This, of course, is very much the picture of meaning that he himself had painted in *Tractatus*, but which he now believes only applies to a fraction of sentences. According to *Tractatus* language is the sum total of propositions, i.e. statements of facts. But language is also employed to a host of other uses, for example questioning, commanding, exhorting, promising, warning, etc. In Wittgenstein's new view, what we call language is made up of a multiplicity of various forms.

> "But how many different kinds of sentence are there? Say assertions, question, and command? - There are countless kinds: countless different kinds of use of what we call "symbols", "words", "sentences". And this multiplicity is not something fixed, given once for all; but new types of language, new language-games, as we may say, come into existence, and others become obsolete and get forgotten." (ibid. para., 23)

The aim of his new approach is to show that the diversification of linguistic forms cannot be traced back to the same essential logical form, i.e. *the general form of propositions*. If language has a form it is a minimal one, which does not explain the connections between its various forms. These forms are only connected with one another in a more elusive way, like games, or like the faces of people belonging to the same family; they share what he calls *family resemblance*.

The reason for using the term language-*games* for the various forms of "sentences" and "word", is, first, ... "to bring into prominence the fact that the *speaking* of language is part of an activity, or of a form of life" (ibid. para. 22). And it is to bring into prominence the fact, secondly, that he no longer believed that the meaning of words and structure of sentences were given by their *correspondence* to independently existing structures of the world, but that somehow they had to come from Man himself. Our many different uses of language, each of which has its own structure, says Wittgenstein, are given content and significance by our practical activities in the world; indeed, language is part of the fabric of an inclusive form of life – by which he means our linguistic as well as non-linguistic behaviour, natural expressions, assumptions, beliefs and outlook. This means that there is no "point of support" outside language from which the relation between language and the world may be explained, nor any independent foundation in the world from which the meaning of linguistic discourse may be derived or analysed. The meaning and the necessities involved in linguistic discourse derive solely from within linguistic practices and the rules which govern these practices. Thus, what he now believes is that language, and the forms of life of which language-games are part, determine our view and the meaning we make of reality - because we see things through language and the activities of those forms of life.

According to this view, the meaning of an expression is its *use* in the multiplicity of activities of which language is part (cf. ibid. para. 43). Now, what gives the use of linguistic expressions its stability is rules for their application in the language games to which they belong. These rules, in their turn, are constituted by our *collective* use of them, and established by agreement, custom – and training. This means that the meaning and truth of assertions in the final analysis is a matter of pragmatic and social agreement among language users as to their correct applications.

This position on language of *Investigations* represents a departure from almost all the central assumptions and arguments which made up the model of language of *Tractatus*. Among the differences pointed out by Wittgenstein, there are the following. About the model of language presented in *Tractatus* he now says that it misleads one to think that the aim of a philosophical analysis of language is to uncover

> "something like a final analysis of our forms of language [...] That is, as if our usual forms of expressions were, essentially, unanalysed; as if there were something hidden in them that had to be brought to light. [However] Philosophy simply puts everything before us, and neither explains nor deduces anything. -

> Since everything lies open to view there is nothing hidden in them that had to be brought to light." (ibid., para. 91)

What now "lies open to view", says Wittgenstein, is the variety of *uses* to which expressions can be put in one or other of the many and various language-games, which constitute language. In particular, he vehemently rejects the notion that in "grasping the meaning" of an expression, and thus its link with the thing being expressed, a picture or an image of a thing or state of affairs is "lying before one's *mind*" - let alone, lying hidden somewhere deep down in our minds. Against this view he now proposes that ... "To understand a sentence means to understand a language. To understand a language means to master a technique" (ibid., para. 199). And to master a technique, is *knowing how to do* something - in the case of understanding language it means *knowing how to use* it.

So, to sum up, according to *Investigations* we now have that the meaning of an expression is its *use* in a language-game, i.e. is part of an *activity* or a *practice*; and we have that understanding linguistic expressions is to master a technique, i.e. is an *ability* we exercise or a *practice* we employ. In particular, we have that both the meaning and understanding of an expression can be recognised and measured by *outward* criteria "open to view" in the *public domain* - i.e. in the activity and practices people engage in and how they behave.

And we have that these practices are governed by the *rules* of the language games to which the words and expressions belong. In this sense, Wittgenstein contends, language is like playing a game; and just as playing a game - and knowing how to play it - hinges on how to observe the rules of the game, so the practice of understanding the words and expressions of language hinges on the observance of the rules for their use in different language-games. Furthermore, just as it characterises rules of an activity that it is its function to provide standards for the correct way of performing it, ... "for doing it the right way", rules for the use of language provide standards of correctness in the sense that observance of them constitutes *using language correctly*. The rules for linguistic discourse, just like the rules of any other activity or game, rely on the agreed linguistic practice of a community; indeed, rules provide standards of correctness for those activities and practices, *because* the are based on agreement. No other justification for a given activity or practice, i.e. outside the activity itself, is required or can be given. They are their own justification – in much the same way as in a game, say of chess, there is no answer to the question, "Why does the king only move one square at a time?", other than "If it is *chess* one is playing then that *is* simply the rule".

Now, the idea that language is essentially rule governed does not represent a departure from the views presented in *Tractatus*. But there is a significant difference with regard to his earlier and later view of how the rules of language are involved in the "manufacturing" of meaning and understanding. In *Tractatus* the meaning of words and expressions was thought to be the outcome of a rather automatic application of strictly defined *rules of projection* inherent in the human mind to objects in the world somehow being "fed" into it – and thus relied on processes carried out by some sort of mental *mechanism*. According to the views of *Investigations,* however, it is a mistake to think that something "mental" is involved in the rule-following practices which produce meaning or understanding - neither "mental" in the sense of being consciously on ones mind, nor in the sense of being hidden *deep down* in some "mental machinery". Indeed, it is the views of *Investigations* that following a rule does not involve *knowing* a rule, which one then follows; rather to know a rule *is* to follow the rule - and *shows* or *manifest* itself in the rule-following practice.

And as opposed to the rules of language of *Tractatus*, which were seen as part and parcel of a rigid logical structure depicting the objective structure of reality and *dictating* the correct use of language independently of language users, it is the view of *Investigations* that the rules of language are more like the *signposts* we may find at crossroads which tell us the direction to take. And as it applies to our use of signposts it applies to our use of rules of language that, ... "A person goes by a signpost only in so far as there exists a regular use of signposts, a custom" (ibid. para. 198).[8] Thus, according to *Investigations*, the rules which guide our use of language and afford us with measures of correctness, are constituted by our *collective use of them*; a use, as said earlier, which in its turn relies on *agreement* and *training*. Indeed, so Wittgenstein contends,

> "The word "agreement" and the word "rule" are related to one another, they are cousins. If I teach anyone the use of the one word, he learns the use of the other with it." (ibid. para. 224)

That the rule-following involved in understanding and use of language is a practice, which relies on customs and agreement among language users, means, according to Wittgenstein, that the meaning and understanding of a language is essentially a *social practice*, and thus it means that language can only exists in a community of people. And - to anticipate the main theme of the discussion in the next two chapters - it means, conversely, that there can

[8] That is, rules as well as signposts give directions because a practice exists which establishes the use of signposts or rules. Well, we observe that the guiding function of a sign relies on the agreement that a sign reliably points in the direction of the destination printed on the sign. But, needless to say, that the direction shown by the sign is correct, is of course *not* a matter of agreement; on the contrary, it is a fact on which relies the agreed practice which establishes the use of signposts. I shall return to this point later.

be no such thing as a "private" language. For a person to have or develop a private language, says Wittgenstein, would require that he be able to lay down rules for its use and meaning, but also that he be able to observe these rules in "privacy". However, according to Wittgenstein

> "it is not possible to obey a rule in "privacy" [...] "Obeying a rule is a practice. And to *think* one is obeying a rule is not to obey a rule" (ibid. para. 202).

Although a person may *think* that he is obeying a rule, there is no way that he can know from one occasion to the next whether he is *in fact* obeying the rule - unless he has *means* of checking. And checking that one is following a rule requires, according to Wittgenstein, the availability of *public criteria*.

Because language, as spelled out above, is now considered as part and parcel of *forms of life*, it should be studied empirically in that setting. However, in such studies

> "we may not advance any kind of theory. There must not be anything hypothetical in our considerations. We must do away with all explanations, and description alone must take its place." (ibid. para. 109).

In contrast to *Tractatus*, the purpose in *Investigations* of such studies is not to track down the ultimate limits of language in order to demarcate the *outer* boundaries of what can and what cannot be said, but critically to analyse and describe the very different forms *within* language, and how they may be used in ways which makes sense. Consequently, the interest has shifted from attempts to point out the common essence of the huge varieties of different forms of factual discourse – i.e. differences which were largely ignored in *Tractatus* - to the interest of describing the *network of internal boundaries* between language-games. This description, Wittgenstein says,

> "gets its light, that is to say its purpose, from the philosophical problems. These are, of course, not empirical problems; they are solved, rather,(???) by looking into the workings of our language, and that in such a way as to make us recognize those workings: *in despite of* an urge to misunderstand them (ibid. para. 109). [However ...] It is not our aim to refine or complete the system of rules for the use of our words in unheard-of ways. For the clarity that we are aiming at is indeed *complete* clarity. But this simply means that the philosophical problems should *completely* disappear." (ibid., para. 133)

In this quotation Wittgenstein makes clear that as in *Tractatus*, the aim of the studies of language in *Investigations*, goes hand in hand with the aim of unmasking the metaphysical nonsense produced within philosophy. However, the mistakes from which this nonsense arises, is no longer that philosophers are being led to talk about metaphysical subjects as if they fell within the borders demarcating the logical space of what can be talked factually and thus sensibly about. In *Investigations*, it appears, the very same nonsense is seen to be caused by the fact that philosophers transgress the internal borders of language, i.e. the borders between different forms of language, or language-games - because they talk about subjects with words and sentences which belong within language-games, to which these subjects do not belong.

This point is brought home most forcefully in his discussion and arguments about the differences between *propositions* of language-games concerning publicly observable physical objects, as opposed to the *expressions* of the language-games concerning non-publicly observable internal states - arguments which appear in connection with his refutation of the notion of a "private" language. In the chapter which follows this, I shall concentrate my discussion on these arguments. In particular, I shall discuss his claim that propositions of the form belonging within language-games concerning physical objects, have no counterpart in the language-games of internal states - because the internal states of a person, the pains he feels, or his thoughts, beliefs or hopes, are not the kind of "things" which satisfy the conditions of what may be *described*, or said to be *known*. Of such states, according to Wittgenstein, no linguistic descriptions may be given, only *expressions*; such expressions are learned in the course of the acquisition of linguistic practices, and replace some or other behavioural reactions being "naturally" aroused by these states. When for example we say, 'I feel a pain', we do not describe or put forward a proposition about some pain we feel or know about; we merely give *expression* to the pain. Such an expression no more describes or informs about a pain being felt or known than the original outbursts of the "natural expression" to pain (such as screams, groans, or sights) that the linguistic expression replaces.

Now, despite Wittgenstein's assurance that he will ... "advance no kind of theory or explanation, and that *description* of the workings of our language will take its place", his whole discussion of language and its use is soaked in theory and explanations. His private language argument(s), in particular, is so much soaked in his theory of language and its rules being part of an activity going on in the public domain as to stands in the way of a credible and comprehensive analysis and account of everything lying outside this domain, among them our internal states (e.g. our thoughts, wishes, emotions, images, memory, pains), and of how they may or may not be known to us and talked about.

During my discussion of these claims and theories, I shall argue that the use and understanding of expressions to pains and other internal states necessarily relies on *knowledge* about and correct *description* of such states, just as much as the analysis of the circumstances surrounding the language games in which they may be appropriately used, relies on descriptions and knowledge of such circumstances and games. Indeed, no analysis of these circumstances could be carried out, nor of language games in which the expressions to internal states may be correctly given, which do not rely on and presuppose descriptions and knowledge of those internal states. And, I shall argue further, such descriptions of internal states are necessarily put forward in the very same language and have the same status and form as do descriptions put forward about things and circumstances in the publicly observable physical world.

Later, in Chapter 20, I shall discuss the epistemological consequences of Wittgenstein's view that it is "forms of life", i.e. the practice on which our language games rely and of which they are part, which is the frame of reference for both our correct use of language and our understanding of the meaning of linguistic discourse. In particular, I shall discuss his claim that the fact that our statements and description of the world may be correct or incorrect, does not rely on whether they accord with some objective fact existing independently of our linguistic practices, but rather on whether we observe mutually agreed rules of our linguistic community. These view and claims, I shall argue, imply that the very notion of *truth* becomes relative to what people may come to agree to be true, and the practices in which they engage – a threat, incidentally, of which Wittgenstein was well aware (*Investigations*, para. 241). In short, they imply *relativism* (if not *irrealism*) – a fact which has not been lost on proponents of *Social Constructivism*, who have eagerly referred to Wittgenstein and the theory of language of *Investigations* as an authoritative source of justification for their own untenable epistemological assumptions.

Chapter 19

The external world and the internal

19.1 Introduction

Wittgenstein's arguments for why it nonsense to talk about *knowledge* of internal states, rest on the following two main assumptions. First, it would not make sense to say, for example, "I know that I am in pain", unless it *also* made sense to say, "I am in pain, but I do not know it". However, this condition is not satisfied - for one cannot doubt or not know that one is in pain. Thus, according to Wittgenstein, it does not make sense to say that we have knowledge of things about which we cannot also doubt or be fallible. Secondly, the sentence, "I know that I am in pain", may have the grammatical form and thus *appear* as if it was an empirical, epistemic proposition, but it is not. For the statement, "A knows that p", to be an epistemic proposition requires that an alternative, i.e. "A does not know that p", is excluded. But if the excluded alternative - in the case of pain: "I am in pain, but I do not know it" - is unintelligible, then the sentence "I know that I am in pain" is equally unintelligible; it cannot be a proposition about some knowledge or other of mine. For this reason it would not make sense either to ask "how do you know that you are in pain?", i.e. to ask how you can justify it - as it would, had the utterance, "I know that I am in pain", been an empirical, epistemic proposition (ibid., para. 288 - 290, and p. 221). It would not be a justification to say, "I know it, because I can feel it!". For, according to Wittgenstein, to feel a pain, is to *have* a pain, or to *be* in pain. In order to justify a proposition one has to be able to cite evidence; but that I *have* a pain, cannot be evidence for my *having* a pain; nor can the fact that I *am* in pain be evidence for my *being* in pain.[1]

Put in the terms of *Investigations*, avowals of pains are not propositions which imply knowledge - for, in the *grammar* of the language-games of pain, things like doubts, ignorance or justification simply do not enter the picture. Rather, the grammar of avowal of pain is that of *expressions* of pains - on a par with groans, sighs and sobs. So, if we take, "I know that I am in pain", "I know what I see", or "I know what I think", as epistemic statements on a par with "I know that he is in pain", I know what he sees", or "I know what he thinks", then we are crossing internal borders of language-games and, as a result, produce philosophers' nonsense (ibid. p.221).

The following quotation contains part of the assumptions behind the above argument.

> "In what sense are my sensations *private*? - Well, only I can know whether I am really in pain; another person can only surmise it. - In one way this is false, and in another nonsense. If we are using the word 'to know' as it is normally used (and how else are we to use it?), then other people very often know when I am in pain. - Yes, but all the same not with the certainty with which I know it myself! - It can't be said of me at all (except perhaps as a joke) that I *know* I am in pain. What is it supposed to mean - except perhaps that I *am* in pain?
>
> Other people cannot be said to learn of my sensations *only* from my behaviour, - for *I* cannot be said to learn of them. I *have* them.
>
> The truth is: it makes sense to say about other people that they doubt whether I am in pain: but not to say it about myself." (ibid., para. 246)

But why, then, is it nonsense to say, "I am in pain, but I do not know it", and what it is that makes it nonsensical? Could it, really, be *other* than the *knowledge* we have of pain that makes being in pain something which one cannot doubt or fail to know about? Indeed, the fact that nobody, not even a child, would hesitate to say that the proposition does not make sense, must somehow be built on the knowledge we have of what sorts of things pains are, and thus on knowledge of what *is and what is not the case* about pains. Part of this knowledge is that we cannot be in pain without being *aware* of it, and thus that being in pain *entails* awareness of being in pain, and therefore of *knowing* that one is in pain - if we use the word 'to know' as it is normally used. And part of this knowledge also entails knowing what distinguishes being in pain from being in other "internal" states, such as being depressed, being happy, or just: having a headache as opposed to not having one. If so,

[1] For reasons that I shall discuss later, Wittgenstein deems as nonsense that feeling in pain could be the result of perceiving or observing something taking place on an "inner stage".

it seem to makes perfectly good sense to say that having a pain entails knowing what pain is, and to say that being aware of being in pain - as opposed to being in some other state - entails *knowing* that one is in pain, as opposed to being in some other state.

Well, if it were true, as Wittgenstein suggests, that it is a condition for empirical, epistemic propositions that they are fallible, i.e. they may be either true or false about some matters of fact, it would also have to be admitted that verifying or falsifying the truth of an empirical, epistemic proposition necessarily relies on knowledge of some matters of fact, about which we cannot *at the same* time be fallible, or be in doubt. Without this condition, there would be no way in which we could sensibly undertake verifying or falsifying the truth of a proposition - nor of doubting or deeming it mistaken. In the case of pain, I think we shall have to agree that it is in virtue of knowing what pain is, that I know that I am in pain; and it is in virtue of the same knowledge that I cannot *not* know that I am in pain.

Suppose we compare the two sentences, "I have a penny in my hand" and "I have a pain in my hand". Now, if we follow Wittgenstein's line of argument, I think it could be argued that the *excluded alternative* in the case of having a penny in one's hand, i.e., "I have a penny in my hand, but I do not know it", would be unintelligible. Indeed, it would be just as unintelligible as the statement, "I have a pain in my hand, but I do not know it". If so, again according to Wittgenstein's argument, it would be just as senseless to say "I know that I have a penny in my hand", as it is to say "I know that I have a pain in my hand". However, neither to Wittgenstein nor to anyone else would it be senseless to say "I know that I have a penny in my hand".

It could of course be objected that the two case differ in important respects. One could object, for example, that when it comes to assertions and knowledge about " having an object in my hand" - as opposed to "having a pain in my hand" - I may be mistaken or in doubt in several different ways - and that, consequently, the "excluded alternative" suggested above is by no means the only possible one. Imagine the case where someone without my knowledge slips a penny into my hand - perhaps when I am distracted, or by some other subterfuge. In this case it would be perfectly sensible to say that I have a penny in my hand, but I do not know it - although it could not be said by *me*, but only by a third party. Or, I may be mistaken or in doubt about whether the object I have in my hand is really a *penny*, as opposed to some other small, flat metal object, or a flat, round stone the size of a penny. But, as mentioned above, I cannot be mistaken or in doubt about either without presupposing that I have unmistakable, infallible knowledge of what a penny is, and what it is to have a penny in my hand. But, equally, I cannot have a pain in my hand without knowing what a pain is, and without knowing what it is to have a pain in my hand! So, once again, it certainly makes just as good sense to say, "I know that I have a pain in my hand", as it does to say, "I know that I have a penny in my hand".

The difference in the case of pain, if there is any, is that I cannot have a pain in my hand - *without knowing that I have a pain in my hand.* So, if there is any point in saying that it only makes sense as a joke to say, "I know that I am in pain", it would have to be because to feel a pain entails that one knows it, and that saying so supplies no further information than simply saying, "I am in pain". However, that does not make my saying, "I know I am in pain", *senseless*. On the contrary, it happens to be *true*![2]

Let us for a moment assume with Wittgenstein that sentences such as, "I have a piercing pain in my finger", "I am in a very bad mood", or, "right now I am thinking about ways in which this particular problem may be solved", are not propositions or descriptions of pains, moods or thinking, but rather "linguistic forms" used to give *expressions* to pains, moods and thinking. And let us assume, moreover, that using language to give expressions to such states represents some form of action, which entails making choices among possible correct expressions in a language-game fitting those states. These assumptions granted, I think we shall also have to assume that such actions and correct choices inevitably would have to rely on knowledge of the *conditions* for applying such expressions and, thus, would have to rely on knowledge and correct determinations of the states they are used to express. Without this assumption, I can see no way in which - in our everyday use of language or in linguistic analyses of the kind carried out by Wittgenstein of this usage - we could determine the significance of these different expressions, let alone distinguish and isolate them from other forms, such as propositions about publicly observable objects. Indeed, how could we do so *unless* we could determine, and thus *describe* the states and events to which these expressions are given? And how could we otherwise distinguish between conditions for the correct *use* of such expressions, or *understand* what such expressions express? If, conversely, it is assumed that within the linguistic practice or language-game concerning internal states, no descriptions or references may be made to such states, then we shall have to say that such linguistic analyses and isolation of linguistic forms are carried out *outside* language. However, this would clearly be absurd. Or, we shall have to say that the description and determination of internal states, necessary for our linguistic analysis and isolation of linguistic forms of giving expressions to such states, are carried out in some other language-game than the one within which expressions may be *correctly* given to pain. However, to say so would not be a possibility either, for that would involve transgressing the borders between different language-games, and thus involving us in "philosophers' nonsense".

[2] - as pointed out by Ayer (1986, p. 109).

19.2 Wittgenstein's "private language arguments"

Wittgenstein's contention that what appear to be *descriptions* of sensations and other internal states, are no more than verbal *expressions* given to such states, is reached partly as a result of the arguments already presented, and partly a as conclusion of a further series of arguments against the possibility of the existence of a *private* language of sensations and internal states. Unfortunately, it is not easy to give a coherent account of what has become known as Wittgenstein's "private language argument", partly because it consists of several very different arguments, and partly because there are no obvious links between them. Furthermore, it is not immediately obvious what the roots of the arguments are. If there are any, I think it would be fair to present them as follows:[3]

First, in the traditional empiricist and rationalist philosophical theories, sensations have been taken to be more similar to objects than in fact they are; as a result of taking too far the assimilation of sensations to objects, and as a consequence of taking too far the assimilation of forms of language for sensations to forms of language for objects, the assumption has been fostered that we could indeed talk sensibly about the existence of a *duality* of worlds. This duality distinguishes between, on the one hand, a world populated with physical objects and, on the other, a mental world populated with objects, events and processes, which were conceived to be just like physical objects, events and processes, albeit only non-physical, or mental. Such "inner" objects, it was assumed, are describable just as are "outer" objects; but, as opposed to "outer" objects, which belong to a public world in which they have an independent existence, and thus may be perceived and shared by everybody, "inner" objects are *essentially* owned and, thus, essentially untransferable and unshareable.

Now, it is Wittgenstein's point that if such a mental world existed in us, then for each of us it would be a world of *private* things, and each person's inner world would be a *metaphysically private property*. If so, the meaning of words used to refer to these private things, and thus the language that each of us uses to describe such things, would have to be equally private. Such a language would, Wittgenstein argues, necessarily be a language which is in principle *unintelligible* to others. However, it could be argued that we have a language "*for private use*", i.e. a language in which we can give ourselves orders, blame or punish ourselves, so, why

[3] In my attempt to get to the roots of these arguments - an account of which is not to be found anywhere in Wittgenstein's own works - I have found valuable directions and hints in the aforementioned book by Pears (1985), particularly in Chapter 8, *Sensations*.

> "could we [not] also imagine a language in which a person could write down or give vocal expressions to his inner experiences - his feelings, moods, and the rest - for his private use? - Well can't we do so in our ordinary language? But that is not what I mean. The individual words of this language are to refer to what can only be known to the person speaking; to his immediate private sensations. So another person cannot understand the language." (ibid. para. 243)

Granted that a duality of worlds existed, i.e. that apart from the public world of objects and events, a private world existed of internal and essentially unshareable states and sensations, then it is arguably correct that the language used for describing things in this world would have to be just as *private* as the things being described. And granted, furthermore, that the view of the duality of worlds implied the assumptions of these worlds being totally *detached* from one another, then a language for describing what exists in the one would have to be correspondingly detached from the language for describing what exists in the other; indeed, there would be no shared concepts or notions in the two languages, nor any links between descriptions of *what* is described in the two languages. The language used for describing what exists in the privacy of our minds would be detached from the language for describing publicly observable things and events, and no links whatsoever would exist between the two languages. If so, the language for internal states and sensations would not only be necessarily unshareable but also necessarily *unteachable* - which obviously it is not.

This is, and here I think we shall have to agree with Wittgenstein, the unavoidable and untenable consequence of the "dual world" position adopted by *empiricists* and *rationalists*. However, to Wittgenstein these consequences of the empiricist and rationalist position arise primarily as a result of taking too far the assimilation of sensations to material objects, i.e. of assuming that the private states and sensations of a person may be conceived of as object-like things, and thus as things which exist and may be described in the same way as material objects. The consequences of taking too far the assimilation of sensations to objects is, therefore, similarly taking too far the assimilation of *forms* of language for sensations to *forms* of language for objects. Now, one way of avoiding the untenable consequences of the "dual world" position of rationalism and empiricism, would be, first, to demonstrate how internal states and sensations *differ* from material things, and to argue how these differences affect the way language may be used to give expression to such states. Another way, being part of the same strategy, would be to set up a list of links between the language for sensations and the language for material things, by means of which a language for sensations could be "recovered" and made into some form or other of a language, which is both shareable and teachable. This, it seems to me, is in fact the strategy adopted by Wittgenstein.[4]

In this pursuit, Wittgenstein starts by pointing out that a language for sensations - and indeed for the whole range of other internal states - cannot be learned without the existence of *public criteria*. As he puts it: "An inner process stands in need of outward criteria" (ibid., para. 580). Nor may the language for the "inner" be taught to someone with whom one cannot communicate about publicly observable things. In this sense a language for sensations and internal states cannot be a self-contained language – because it depends on an auxiliary language, i.e. the language of public things. Now, it so happens that sensations and internal states have many links with the world of physical things. In the case of sensations, say of pain, the public criteria or links that Wittgenstein has in mind (indeed, the only ones he seriously considers), are "natural expressions" (groans, sighs, moans, cries, etc.), as well as other behavioural responses to pains, such as grimacing and clenching one's fists. With these links, a language may be learned for sensations and internal states, in which we may communicate to others the states we are in, i.e. by giving verbal *expressions* to such states.

But how do we learn to do so? Well, a person, say a child, who is in pain will display various sorts of pain-behaviour, which by adults will be recognised as natural expressions of pain. The links between pain and its natural, behavioural expressions can then be used by a teacher to teach a child in pain to use the word 'pain'. In Wittgenstein's own words:

> "Here is one possibility: words are connected with the primitive, the natural, expressions of the sensations and used in their place. A child hurts himself and he cries; and then adults talk to him and teach him exclamations and, later, sentences. They teach the child new pain-behaviour. "So, you are saying that the word 'pain' really means crying?" - On the contrary; the verbal expression of pain replaces crying and does not describe it." (ibid., para. 244)

There is a problem, however. For, although a child in pain may display various sorts of pain-behaviour, which may serve as the necessary outward criteria for teaching the child to use the term 'pain' to express feelings of pain, they are by no means sufficient for the teacher to recognize the child's behaviour as being his or her natural expressions to pain, nor for recognizing that the child is in pain. They cannot be if it is also true, as Wittgenstein

[4] An alternative strategy would of course have been to give up the basic assumption of the two world legend itself, i.e. the assumption of the possibility of a private world of internal states, which is completely distinguishable and detached from the public world of physical things and events. The advantage of this alternative strategy is that the problems of the rationalist and empiricist "dual world" position, which Wittgenstein sets out to solve, are completely avoided - they simply do not occur. Arguments for this strategy will be presented in later sections of this chapter.

himself argues in *Philosophical Remarks* (1929/1975) that we can only feel our own pains, but not the pains of others, and therefore we cannot be sure that people, when exhibiting the behaviour we do when we are in pain, are in fact feeling pain. He gives the following example.

> "Suppose I had stabbing pains in my right knee and my right knee jerked with every pang. At the same time I see someone else whose leg is jerking like mine and he complains of stabbing pains; and while this is going on my left leg begins jerking like the right though I can't feel any pain in my left knee. Now I say: the other fellow obviously has the same pain in his knee as I've got in my right knee. But what about my left knee, isn't it precisely the same case here that of the other's knee. [Consequently] The two hypotheses, that others have pain, and that they don't and merely behave as I do when I have, must have identical senses if every possible experience confirming the one confirms the other as well. In other words, if a decision between them on the basis of experience is inconceivable." (ibid. p. 93 and p. 94)

Well, if the two hypotheses have *identical* senses, and therefore it makes no difference to assume that others are in pain or that they are not, when they exhibit the kind of behaviour we do, when we are in pain, then neither hypothesis makes sense. Consequently, to say that we can recognize from the behaviour of others that they are in pain when they exhibit the same behaviour as we do when we are in pain, is equally nonsensical.

However, in the text just quoted Wittgenstein does not mean to say that we can *never* know whether others are really in pain; what he does mean to say is that to ascribe pains to others is really nothing over and above that of ascribing pain-behaviour – for all we have in the case of others is their behaviour, so all we can ascribe to them is pain-behaviour. We never actually ascribe e.g. toothache to other people, says Wittgenstein, but only behaviour being akin to one's own when one has toothache oneself. We cannot ascribe or attribute to others what we *have*, for example, our toothaches "in thought", so to speak.[5] But if I cannot ascribe to others the pain I have, how then can I ascribe to others my *pain-behaviour,* i.e. ascribe to others my *reaction* to pain – and thus my reaction to something that I may only ascribe to myself, but not to others? And if, in the example above, all a teacher can do is to ascribe to the child some sort of behaviour, but not the feelings of pain he himself has when he exhibits behaviour akin to the child's, how then can the child's behaviour be sufficient criteria for the teacher to recognize it as *pain*-behaviour, i.e. as a natural reaction to pain - as opposed, say, to some other feelings, like happiness, anger or embarrassment?

[5] Or, as Wittgenstein remarks: "Philosophers who believe you can, in a manner of speaking, extend experience by thinking, ought to remember that you can transmit speech over the phone, but not measles" (ibid. p. 95).

Now, let us turn to the second part of the strategy of showing why it is mistaken to take too far the assimilation of private internal states, like pains, to public things like material objects. In other words, let us turn to that part of it which involves demonstrating that internal "things" residing in the private domain, are not at all object-like things, about which propositions may be put forward or descriptions may be made. What kind of things, then, are the internal states to which both natural responses and linguistic expressions may be given - how may they be characterized? On this issue Wittgenstein has very little to say - indeed, he cannot really say a great deal about it without putting himself in a paradoxical situation. For, if he did maintain, as both he and the empiricists and rationalists do, that internal states and mental phenomena are fundamentally private phenomena, then they are the sorts of things about which propositions and descriptions could only be put forward in a language, which is equally private. So, although Wittgenstein does not deny the existence of internal states or sensations, and thus does not deny the existence of *that* to which natural reactions and verbal expressions are given, he is not in a position to say *what* they are. The nearest he comes to saying what a sensation is, is that "it is not a *something*, but it is not a *nothing* either", a conclusion, he contends, which makes sense because "a nothing would serve just as well as a something about which nothing could be said" (ibid. para. 304).[6]

But now I am anticipating what is actually the *conclusion* of a series of arguments which Wittgenstein puts forward against the possibility of a propositional language of internal states and mental phenomena. Such a language, as we have seen, would to Wittgenstein have to be a private language - so the strategy he adopts is to argue that a propositional and descriptive language for internal states and sensations cannot be developed *in privacy*, that is, concepts and descriptions cannot be developed by a person of things which may only be observable to himself. A language thus based would be a language in which notions and concepts could not have a meaning - not even for the "creator" of the language. It would be a language, so the argument goes, which *nobody* could speak - for one cannot speak or understand a language in privacy, i.e. teach oneself the *rules* of a language, or *follow* the rules of a language in privacy. Nor can one teach oneself the meaning of words for sensations in the *ostensive privacy* of the mind.

Now, before continuing my discussion of Wittgenstein's arguments against the possibility of a propositional language of internal states and sensations - which to Wittgenstein must indistinguishably be arguments against the existence of a *private* language - I should like to stress that I entirely agree with Wittgenstein that there can be no such thing as a private language, nor of private uses of language. There are overwhelmingly good

[6] However, this is, to say the least, to beg the question.

arguments for this, which I shall present in Chapter 20 - in which it should hopefully *also* become clear why I find Wittgenstein's strategy and arguments against the notion of a private language both wanting and misleading. For the present, though, I shall discuss and criticise Wittgenstein's "private language argument" solely on the premises he himself presents.

Central points of this argument and these premises appears in paragraph 258:

> "Let us imagine the following case. I want to keep a diary about the recurrence of a certain sensation. To this end I associate it with the sign "S" and write this sign in a calendar for every day on which I have the sensation. - I will remark first of all that a definition of the sign cannot be formulated. - But still I can give myself a kind of ostensive definition. - How? Can I point to the sensation? Not in the ordinary sense. But I speak, or write the sign down, and at the same time I concentrate my attention on the sensation - and so, as it were, point to it inwardly. - But what is this ceremony for? for that is all it seems to be! A definition surely serves to establish the meaning of a sign. - Well, that is done precisely by the concentration of my attention; for in this way I impress on myself the connexion between the sign and the sensation. - But "I impress it on myself" can only mean: this process brings it about that I remember the connexion right in the future. *But in the present case I have no criterion of correctness* (italics added,) One could like to say: whatever is going to seem right to me is right. And that only means that here we can't talk about 'right'."

What Wittgenstein seems to be saying here is, first, that we cannot give ostensive definitions of sensations; this, I think, is something we shall have to agree on - if what he means by sensations is a "something/nothing" about which nothing can be known, and thus about which nothing may be the case or correct. Secondly, even if we could, we could not get the connexion *started* between a sensation and a term for it - because, according to Wittgenstein, we would have no *criterion of correctness*; and if we have not, everything would count as correct - and thus nothing would.

But, seriously, what *criterion* for correctness is available the first time a connexion is established between a term and a thing in material reality? It seems to me that, normally, we do not really have any means of establishing a connection between a term and a thing, and of doing it correctly, *other* than to "concentrate our attention" on the thing, and to try to "impress on ourselves" *what is the case, or correct, about the thing*, to which we intend to apply the term. What "criterion" would be needed, then, other than our knowledge of *how to use the notion of "correct" correctly* when determining what is the case, or correct, about the thing in question?

Now, it could be that when Wittgenstein coined the doctrine: "the meaning of language is its use", he meant its use as established by some *already* existing practices and customs of the speech community, and thus that the existence of signs, the meaning of which has been established *by virtue* of such practices and customs, serves as indispensable criteria against which the correctness of use of signs may be checked. But how did such practices get started in the first place, i.e. before uses had been agreed among speakers which could serve as criteria for correctness?

More likely, though, he means that because in "this case", the thing in question does not belong within the domain of publicly observable things, a "definition of a sign for it cannot be formulated", which may serve as a definition or *rule* for its future correct use. For, according to Wittgenstein, no definition of a sign may exist in anybody's mind that cannot be checked against public criteria, nor may a rule for the correct use of the sign be followed that is not part of the rule-following practice of a speech community. Now, if this is what Wittgenstein means, he cannot avoid getting into serious problems. For, granted that the word 'pain' is and only can be an *expression* of pain - on a par with natural behavioural reactions caused by pain - how then could it be used by anyone and have a meaning in the absence of that to which it is an expression, and by which it is caused? How, for example, could it be used by someone to ascribe *pain* to someone else, as in the ascription, "He is in pain", or, "He has a pain"? Indeed, how could it be used to ascribe pain to someone else as e.g. in the example of a teacher intending to teach a child the correct use of the word 'pain', *unless* it is assumed that it is a word that the teacher may use descriptively to *refer to* and *mean* precisely that to which the child's crying is a natural reaction, i.e. the child's feelings of pain? Now, it will not do to object that in this case of teaching, public criteria in terms of behavioural reactions exist, which may function as a basis for the speech community to form rules for the correct use of the word 'pain'. This objection does not address the fact that *that* for which these behavioural criteria may so function, is itself something which is not publicly observable, namely the feelings of pain of the person displaying this behaviour – and thus is something for which neither a child nor anybody else "have any criterion of correctness" (op.cit.). And if so, it is very difficult to see how it would be possible for anyone to acquire from others, or to obey with others, rules for giving expressions to one's pain, let alone to develop with others a language game – in any serious sense of this term - of expressing and talking about pains.

This criticism would seem to be a serious threat to Wittgenstein's model and views of the nature of our discourse about internal states, but also to his insistence on the necessity of the existence of public criteria of correctness for the use of terms to express such states. And, by implication, it a threat to his insistence that it is the practice and agreement of the speech community which, on the basis of such criteria, bestow the meaning and correct use of our utterances and terms. For, as pointed out by Ayer,

> "what is the community except a collection of persons? And if each of those persons is supposed to take his orders about meaning solely from the others, it follows that none of them takes any orders. The whole semantic house of cards is based upon our taking in each other's washing, or would be if there were any laundry to wash. On this interpretation, Wittgenstein's argument, so far from proving that private languages are impossible, proves that they are indispensable." (Ayer, 1986, p. 74)

Later, in para. 288 of *Investigations* Wittgenstein extends the above argument by saying "I need a criterion of *identity* for the sensation". So, it may be that all Wittgenstein means is that in recalling the connection between one's sensation and a sign correctly, one cannot check the identity of a sensation with a *memory* of a sensation, or check a memory of a sensation with another; that is, one cannot *rely* on one's memory to furnish a criterion of correctness. This, it seems, is what he is actually saying in the following paragraph:

> "Let us imagine a table (something like a dictionary) that exists only in our imagination. A dictionary can be used to justify the translation of a word X by a word Y. But are we also to call it a justification if such a table is to be looked up only in the imagination? - "Well, yes; then it is a subjective justification." - *But justification consists in appealing to something independent*, [italics added.]. - "But surely I can appeal from one memory to another. For example, I don't know if I have remembered the time of departure of a train right and to check it I call to mind how a page of the time-table looked. Isn't it the same here?" - No; for this process has got to produce a memory which is actually *correct*. If the mental image of the time-table could not itself be tested for *correctness*, how could it confirm the correctness of the first memory? (As if someone were to buy several copies of the morning paper to assure himself that what it said was true.)
>
> Looking up a table in the imagination is no more looking up a table than the image of the result of an imagined experiment is the result of an experiment." (ibid., para. 265)

Apart from insinuating that one cannot rely on one's memory to furnish a criterion of correctness, Wittgenstein in this paragraph makes the additional claim that justifying the correctness of recognition (i.e. of re-identification) consists in appealing to 'something independently'. I shall discuss the latter claim first.

If access to *'something independent'* were required for justifying recognition, not only of sensations, but also of things in material reality, i.e. if this requirement for the correctness of recognition applied *in general*, justification of recognition would require the existence of something which might function as "exemplars" or "standards", against which checks may be made. But, is it really this kind of standards that Wittgenstein has in mind, when he talks here about "something independent"? Hacker, who in his book, "Wittgenstein, Meanings and Mind" (1993), has made the most extensive analysis to date of Wittgenstein's "Private language arguments", believes that this is exactly what Wittgenstein has in mind, and that support for his belief is to be found in notes made by Wittgenstein prior to writing *Investigations*. Thus, Hacker says:

> "In order to give a name to a sensation or sense-impression in a putative private language, the private language theorist must show how the name is to be defined or explained. For a sign is a name only in so far as it is given a rule-governed use, and the role of a definition is to determine for future occasions how the expression is to be used (LPE 291[7]). In explanations of word-meaning by public ostensive definition, a sample typically plays a crucial normative role in this respect. Having a sensation, say a toothache, seems to provide one with a mental sample which will function as a standard for subsequent applications of the word 'toothache'. [...] This, Wittgenstein argues, is an illusion. 'The private experience is to serve as a paradigm, and at the same time admittedly it can't be a paradigm' (LPE 314). Why not? A sample must function as a standard of comparison for subsequent applications, must be preservable or reproducible; and it must be possible to lay a sample alongside reality for match or mismatch. Hence it must be possible for there to be a technique of projection manifest in the use of the sample as a standard of correct use. These requirements are not satisfied in the case of sensations or impressions - although they seem to be." (ibid. p. 47)

Now, it is obviously untenable to assume, as does Hacker, that to experience or recognise things, and to use words correctly for what has been experienced and recognised, requires the existence of samples that may serve as independent standards of such recognition and correct use. In fact, Wittgenstein himself comes very close to arguing why - in some other notes. Thus, in *The Blue Book* (Wittgenstein, 1958) he warns us against the mistake of believing that in order to recognise an object or a property, we need to compare it with a sample; in most cases no samples are available. However, he also warns us against believing that because no such samples are available, the sample has to be a "mental image". For one thing, he says, we constantly recognise familiar things or properties without the aid of any images. Secondly, to explain recognition by resorting to images only leaves us with

[7] This reference is to: Wittgenstein's Notes for Lectures on "Private Experience" and "Sense Data", ed. R. Rhees, *Philosophical Review*, 77 (1968), pp. 275 - 320.

the problem of explaining how *images* are recognised. Wittgenstein brings home this point by giving the example of someone being asked to imagine a yellow patch. "Would you still", he asks, "be inclined to assume that he first imagines a yellow patch, just *understanding* my order, and then imagines a yellow patch to match the first?" (ibid., p. 12).

I can agree with Wittgenstein that this way of talking about the recognition of an object, i.e. as a process whereby one compares one's recognition or experience of the object with a sample - be such samples *actual* objects or *imagined* - is quite untenable. In the case, for example, in which an object is recognised as exactly the same object as one having been seen before, it is obvious that this recognition cannot consist in comparing the object with a sample - for the "sample" would have to be (numerically) identical with the object being perceived and recognised. To justify the recognition by matching it with a sample could in this case only consist in looking *once more* at the very same object - in which case one would be no better off that the man who "buys several copies of the morning paper to assure himself that what it said was true". Nor in the case of recognising an object as some particular kind of object, say a cup, can this recognition consist in comparing the recognition of the cup with a sample-cup, as it would only leave us with having to explain how this sample-cup is recognised. That is, we are back to the very same problem which was pointed out by Wittgenstein in the case of a sample being an *imagined* object residing in our memory. Unless, of course, one believes that there is a difference when the standards or samples concerned are real, material objects. That is, unless one is prepared to assert that, unlike imagined sample objects, sample objects in material reality may be recognised independently of being recognised, and so on. There are no reasons to think that Wittgenstein would be willing so to believe or assert.

To the best of my knowledge Wittgenstein does not offer any alternative account of the processes involved in correctly recognising an object. However, it will obviously have to be agreed that recognition of an object must involve *remembering* what makes an object some particular kind of object. This takes us back to the other claim made in the quotation of paragraph 265, i.e. the claim that - in the case of recognising or experiencing sensations - one cannot rely on one's memory to furnish a criterion of correctness. But if it is true, as Wittgenstein contends in *The Blue Book*, that in most cases we do not have samples which may serve as standards for the correctness of recognition of objects, and that, anyway, it does not make sense to talk about standards against which matches of recognition may be made, then we shall simply *have* to rely on our memory to furnish the "criterion of correctness" for recognition. There is, quite simply, no other option. So, why cannot our memory serve the same function when it comes to recognition or experience of sensations and other internal states?

The reason given by Wittgenstein is, as we have seen, that because we do not have independent *and* public criteria in the case of sensation, there is no check on whether we are in fact re-identifying the same sensation, and also that we might forget how we used the word on previous occasions. However, what he believes to apply to our ability (or rather lack of ability) to re-identify things and events when it comes to internal states, and to our ability to remember terms previously used when referring to such states, could apply equally well in the case of publicly observable states and events and our ability to remember terms previously used to refer to them. Luckily, however, both individual language users and communities of language users are perfectly capable in general of re-identifying states and events they have previously encountered, and of remembering the terms they have used to refer to them. Otherwise, there could be no knowledge about things, nor any use of language to communicate about such things, and thus no language.

If, therefore, our ability to re-identify publicly observable states and events, and to remember the terms we use for such states and events, is a necessary condition for language and use of language about such things to exist, what reason would we then have to assume that such abilities were lacking when language users talk about and describe non-publicly observable states and events? None at all. If, on the other hand, definitive grounds for such doubt could be given, both concerning our ability to re-identify sensations and other internal states, and for remembering expressions and terms used for such sensations and states from one occasion to another, then we should have to agree that no basis existed for linguistically describing such sensations and states. However, such grounds granted, there would be no basis either for the development of language-games in which *expressions* would be given to such states. For, how could e.g. *giving expressions to pain* ever be a well defined language-game, unless it were assumed that feelings of pains belong to the sort of things that language users were perfectly capable of recognising and re-identifying, and assumed, furthermore, that language users are indeed capable of remembering the rules for how and on what occasions expressions of pain are correctly given, and thus of remembering the *conditions* concerning the correct application of such expressions?

Now, what could be the conditions for giving expression to various sensations other than our awareness of those various sensations? What, for example, could be the condition for using a pain-expression other than feelings of pain? Since this possibility has been ruled out by the assertion of Wittgenstein's that a sensation is a something/nothing of which noting can be said or determined, and which, anyway, cannot be recognised or re-identified, it seems that the range of what could possibly serve as conditions or criteria for correctly giving (verbal) expressions to sensations has been pretty much exhausted - and each in turn dismissed. Now, there is of course the possibility that when Wittgenstein says that verbal expressions given to sensations *replace* the original, natural behavioural reactions, he means that the language user chooses the appropriate verbal expression on the basis of the behavioural expressions which on *previous* occasions would have been displayed as expressions of that particular sensation. So, could it be that such previous behavioural expressions, although now absent or being suppressed, may serve as criteria for the correct choice of verbal expression? Well, this would not make sense either, for if "no-one can learn of one's sensations *only* from one's behaviour" (cf. para. 246, op.cit), then one cannot use one's behaviour *only* (present or previous) as a criterion for the correct use of expressions of one's sensations. In notes written prior to the publication of *Investigations*, Wittgenstein rejects this possibility on the following grounds: "One wishes to say: In order to be able to say that I have toothache I don't observe my behaviour, say in a mirror. *And this is correct*," but to this he adds, "but it doesn't follow that you describe an observation of any other kind. Moaning is not the description of an observation. That is, you can't be said to *derive* your expression from what you observe" (Wittgenstein, 1936/1968, p. 319).

It would seem that what was intended by Wittgenstein to be arguments against the notion of a *private* language for describing "things" populating our "inner" worlds, has somehow turned into arguments against the existence of sensations and other internal states as something which we may talk sensibly about *at all*; arguments which, by the same token, inevitably undermines any notion of the existence of a language-game of *giving expressions* to pains as well. How did that happen, and where in Wittgenstein's thinking about these issues did things go wrong? I shall attempt an answer in the next section. However, to find the roots of the mistakes Wittgenstein makes, does not make what he says more sensible, let alone true; but it may make more understandable what made him say it.

19.3 Sensation of the internal as opposed to observation of the external

> The atmosphere surrounding this problem is terrible. Dense mists of language are situated about the crucial point. It is almost impossible to get through to it. (Wittgenstein, 1936/1968, p. 306)

Time and again in *Notes for Lectures on "Private Experiences" and "Sense Data"*, from which the quotation above is taken, Wittgenstein's ever present opponent is objecting that there must be more to 'seeing green' than just giving a name to a "sample", and more to 'being in pain', than just giving expressions to (the sensation of) pain.

> (O:) "But aren't you saying that all that happens is that he moans, and that there is nothing behind it". (ibid. p. 302). "I don't just *say* 'I've got toothache,' but *toothache makes me say this.*" (ibid. p. 315)

And just as invariably Wittgenstein answers:

> (W:) "I am saying that there is nothing behind moaning. (ibid. p. 302). "moaning is replaced by 'I have a pain'; but moaning stands for nothing at all; it is not a statement (ibid. p. 301). Some things can be said about the particular experience and beside this there seems to be something, the most essential part of it, which cannot be described (ibid. p. 275). As it were: There is something further about it, only you *can't say* it; you can only make the general statement. It is this idea which play hell with us". (ibid. p. 276)

However, his opponent persists,

> (O:) "Surely if I call a colour 'green' I don't just say that word, but the word comes in a particular way, or if I say 'I have a toothache' I don't just use this phrase but it must come in a particular way!" [...] "surely seeing and saying something *can't be all*!" (ibid. p. 319)

To which Wittgenstein comments:

> (W:) "Here we make the confusion that there is still an object we haven't mentioned. [However,] Moaning is not the description of an observation. That is, you can't be said to *derive* your expression from what you observe. Just as you can't be said to derive the word 'green' from your *visual impression* but only from a sample." (ibid. p. 319)

We noticed that in the case of 'having a toothache', the expression given replaces a natural reaction to "something"; however, to the observation of this "something" no *sample* corresponds, so the *object* being left out, or not yet mentioned, is - "nothing". And where there is "nothing" behind a sensation, no description is made of anything. So, what is left undescribed is the sensation, for "you can't be said to *derive* your expression from what you observe". In the case of 'seeing green', the word '*green*' is derived from a sample-*object*, which makes it a description of that object. The "object" being left out as undescribed is, again, the *sensation*; for in the case of 'seeing green', "you cannot derive the word '*green*' from your *visual impression*.

Later in his notes it seems to become clearer what shrouds language concerning this problem in a dense mist and creates the terrible atmosphere surrounding it. It is none other than the traditional division of the world into two separable and detachable "realms": a "realm" of *private* sense data, experiences and sensations, and a "realm" of *public* material objects and "samples". In the conclusion of his notes he remarks,

> "Privacy of sense data. I must bore you with a repetition of what I said last time. We said that one reason for introducing the idea of the sense datum was that people, as we say, sometimes see different things, colours, e.g., looking at the same object. Cases in which we say "he sees dark red whereas I see light red." We are inclined to talk about an object other than the physical object which the person sees who is said to see the physical object. It is further clear that we only gather from the other person's behaviour (e.g., what he tells us) what the object looks like, and so it lies near to say that he has this object before his mind's eye and that we don't see it. Though we can also say that we might have it before our mind's eye as well, without however knowing that he has it before his mind's eye. The 'sense datum' here - the way the physical object appears to him. In other cases no physical object enters." (ibid. p. 316)

And in an earlier passage he writes,

> "But aren't you neglecting something - the experience or whatever you might call it -? Almost *the world* behind the mere words?" But here solipsism teaches us a lesson: it is that thought which is *on the way* to destroy this error. For if the *world* is an idea it isn't any person's idea. (Solipsism stops short of saying this and says that it is my idea.) But then how could I say what the world is if the realm of ideas has no neighbour?"

So, where are we? From the first quotation it appears that the notion of a "realm" of private sensations and sense data is somehow indispensable, and from the second that the notion of "ideas" (about the world) in the privacy of our minds, needs to be supplemented with a neighbouring "realm" of the world. For, obviously, to talk about the world is not to talk about ideas. And we begin to understand more clearly (or do we?) why the "object" drops out in the case of sensations of internal states and events, such as pains and emotions; for the "objects" of the "internal" *are* internal states, and thus do not "point" to anything else, i.e. to something outside or in any neighbouring "realm". So, clearly, they have to be "something" observed *about* which nothing is observed or may be said.

However, if we turn to our sense data, be it of "the 'green' of an object", or of any other property of things, in the world of material objects and samples the story is no different. For, such sense data, residing in the privacy of our minds, are just as private, and thus just as *unshareable* and *incomparable* with the sense data and "visual impressions" of other persons, as are sensations of internal states residing in the same "realm". One begins to understand Wittgenstein's exasperation with the problem he has at hand:

> "The 'private experience' is a degenerate construction of our grammar (comparable in a sense to tautology and contradiction). And this grammatical

monster now fools us; when we wish to do away with it, it seems that as though we deny the existence of an experience, say toothache." (ibid. p. 314)

For, if doing away with the grammatical monster of 'private experience' means denying the existence of *experience*, this denial not only affects the existence of the experience of toothache but, equally, the experience or sense data of objects in material reality and their properties. By the same token, it means denying the existence of material objects and their properties *as something which we can talk sensibly and correctly about*; for, how can we talk sensibly and correctly about objects without or independently of persons experiencing objects? Indeed, the consequences are irredeemable. However, the consequences of *not* slaying this grammatical monster and the division of "realms" it represents, are just as irredeemable. For, if this division implies the assumption that no necessary relation exists between anyone's *private* world of experiences or ideas, and the world of material objects and samples, it makes no sense at all to suggest that an independent criterion of correctness for those individual experiences or ideas could be found in a *publicly shared* world of material objects and samples. And if, furthermore, no independent conclusive criterion of correctness and sameness of private experiences and sense data resides with any *individual observer*, the suggestion that such a criterion could reside in the *general agreement of most observers*, would be equally nonsensical. This division and the grammatical monster of private experiences it entails, means the end of any sensible notion about agreement on correctness among communities of persons, but also of any sensible notion of a publicly shared world of independent samples and objects as well.

We have to agree with Wittgenstein that "it is impossible to get through to the problem" on the basis of the "degenerate construction of private experience". What sets the agenda for this construction and the problem being posed, however, is the division between the internal, *private* world of sensations and that of the external *public* world of observed facts. It is *within* this division that Wittgenstein struggles with the problem and mistakenly believes that a solution may be found. However, he is stuck in the division, a fact of which he is well aware, and stuck with the notion of "private experiences", none of which he sees any possibility of getting rid of. And rightly so. For, as I shall argue, the only way we may "get through to the crucial point", or even better, get rid of the problem altogether, is by realising that it is the division of the world into a public and private "realm" which *is* the problem.

So, let me start the clearing away of the mists which surrounds the language in which the problems are posed, by recapitulating what has been said in previous chapters about this division. First, let it be clear that we can indeed distinguish between, and thus "separate" the activities of perceiving, experiencing, observing and describing objects from the objects that these activities concern, and which exist independently of these activities. However, we cannot talk about *either* without referring to the other. Thus, we cannot talk about *objects* without referring to the observation and experience of persons of objects, nor independently of descriptions of objects being put forward by language users. Conversely, we cannot talk about our *observations, perceptions* and *descriptions* of objects without referring to actual objects existing independently of these observations, perceptions and descriptions. To talk about "sense data" or "experiences" of objects independently of talking about the objects being perceived or experienced, would be just as nonsensical as to talk about descriptions of objects without referring to the objects that the descriptions concern.

This interdependency or *necessary* relation between our perception and experience of objects and the objects being perceived and experienced, means that although we may distinguish between perceiving and experiencing objects and the objects being perceived and experienced, the perception or experience of objects is not something which exists and may be determined and described *apart* from determining and describing the objects being perceived and observed. Even when we are talking about our different perceptions or experiences of an object, for example, talking about how an object and its properties appears differently to us from our different vantage points, we are not talking about our perception or experience of the object *as opposed* to talking about the object, but we are inevitably at the same time talking about and describing the objects which we see.

Now, it may be objected that it makes perfectly good sense to say that, when different persons observe or perceive a thing - for example the cup on the table - they may have different experiences of it, but also to say that such different experiences of different persons are necessarily *private*. For, nobody except I myself have access to my experience of the cup. Furthermore, the fact that depending on vantage points relative to the cup, the same person may have different experiences of its shape, colour, and other properties and features of the cup, would seem to imply not only that we may distinguish between such different experiences of the cup and the actual cup on the table, but also that such experiences of objects are perfectly *detachable* from the object existing independently of being perceived and experienced. But it does not imply that. In none of the cases in which we talk about the different parts, features or properties of the cup, observable to us from different vantage points, are we talking about our *experiences* of parts, features, or properties of the cup *as opposed* to the actual parts, features or properties of the cup "out there". As argued in Chapter 4, nobody could ever make such a distinction, nor describe what one experiences or perceives without describing and referring to it as *that* which one experiences or perceives. And any of the descriptions of my experiences of the cup will inevitably be, and coincide with, my descriptions of the cup. In none of the cases am I describing some - private - *experiences* (or sensations or sense data) of the cup - as opposed to describing a - publicly observable - *material* cup on the table. If it were true that no one else has access to what I experience on the table, it would have to be equally true of myself; for not even I myself have access to my *experiences* of the cup - as distinct from the actual cup. In this sense, I do not have access to something distinguishable from what is not perfectly accessible to others as well.

But nor is what I experience and observe "private" in the sense of not shareable with or observable to others. If it were, it would surely not make sense to say *both* that the differences between the experiences of an object by different persons or by the same person in different circumstances may be pointed out, determined and described, *and* that such differences of experiences are *private* – in the sense of unshareable and unintelligible to others. The fact that the experiences of objects by different persons may be compared - and deemed different – presupposes that everyone involved may together determine and put forward correct descriptions in a *perfectly public language* not only of the objects that their experiences concern, but also of their different experiences of it. Without this presupposition, it would not make sense to talk about differences in the experiences of objects of different persons, nor could there be any comparison of such differences.

In conclusion, we may certainly distinguish between the act of experiencing and perceiving a thing on the table, and the actual thing on the table, and even distinguish between different experiences of the same thing, e.g. relative to different vantage points. But we cannot talk about either without presupposing that a necessary relation exists between experience of things and the actual things - a relation, however, which at the same time *logically* denies us the possibility of talking about either without talking about the other. Because we are not free to talk about our perception and experience of objects independently of talking about and referring to the objects that our perception and experience concerns, we cannot talk about our perceptions and experiences of objects as things which exist in a private "realm" of things being detached from the "realm" of objects.

So, to return to where we parted company with Wittgenstein, the reason why in the case of 'sense data' and experiences of objects, no object drops out and no sensations are left undescribed, let alone left *private*, is that a *logical* relation exists between descriptions of objects and descriptions of 'sense data' or experiences of objects, and that a *necessary* relation exists between such descriptions, 'sense data' or experiences and the objects they concern.

19.4 Internal states and sensations of the "internal"

But what is the situation when our experiences concern our internal states and sensations, e.g. our thoughts, images, memories, emotions, or our so called bodily sensations, such as pains, hunger, or itching? What are the conditions for talking sensibly and correctly about such states, and for talking about our experiences, observations and descriptions of them? In this section, I shall argue that fundamentally, epistemologically, the conditions are not different from the conditions which apply in the case of observing and describing material objects, and for describing our experiences of such objects.

I shall begin my arguments by first recapitulating some of the points of Wittgenstein's refutation in *Investigations* of the possibility of a private language of sensations or internal states. This refutation involves the argument, firstly, that the meaning of sensation words cannot be learned ostensively in the privacy of the mind, nor may such words be used consistently or correctly - for no public and independent criteria exist for the meaning and use of such words. And, by implication it involves the argument, secondly, that no language for mental phenomena could exist in isolation, i.e. detached from a shared language with public criteria. In the case of sensations and internal states the public criteria are *behavioural expressions*. As an example Wittgenstein observes that when a person is in pain, he will behave in various characteristic ways, and that the word 'pain' becomes linked with the description of this behaviour - and thus linked with the shared language for material objects.

And, the argument continues, because meaning and correctness of use of words relies on criteria, which are publicly observable and depends on rules agreed by the speech community, then the language used *for* sensations and internal states cannot be *about* sensations and internal states, i.e. it cannot be descriptive. To the extent that we do have a shared language concerning sensations and internal states, this language can only be a language of *expressions* which replaces what is publicly observable and describable, i.e. the behavioural expressions or reactions to sensations and internal states. Without such links to natural, manifest behavioural reactions to internal states, there could be no (shared) *language-games* of giving expressions to internal states and sensations.

Well, to this I shall suggest that the crucial point of these matters may be stated much more simply: We cannot have or learn a language of pains or other internal states without first having a language about objects and things in material reality, i.e. of something which is publicly observable. Indeed, is it not quite obvious that we cannot talk about pains or other so called "bodily sensations" without having a language in which we may talk about material things? After all, pains happens to be the sorts of things which are being felt in *bodies*, and thus in material objects, so how could we talk about pains without at the same time being able to talk about material objects, including bodies? Likewise, how could we talk about internal states, such as emotions, memories or images, without talking about persons who have them, i.e. persons who know that they are persons with bodies existing in material reality? To talk about pains without bodies would be unintelligible; there could quite simply be no language for pains *before*, or as *separate* from, a language of bodies and objects. Presumably, Wittgenstein would not have overlooked this rather obvious fact, and treated what are clearly *descriptions* of sensations and internal states as *expressions* of such states, if he had not concentrated so exclusively on the links between words for sensations, like e.g. 'pain', and natural reactions to sensations of pain, like e.g. moaning, crying, or wailing. What he overlooked or ignored was the numerous ways in which the language for sensations and internal states is *logically* "linked" with the language of objects and bodies.

When we feel pain, or "have a pain", as Wittgenstein puts it, the pain is "had" or felt somewhere localisable in our bodies; we do not just "have a pain", but we have toothache, pain in our finger, or stomach-ache, headache etc. And there are other ways in which pains are linked with the world of material objects. Pains in our bodies often occur (although not always) under certain conditions of contact between our bodies and other objects in material reality, - e.g. when being pricked by pointed objects like needles, or when being cut by sharp instruments like knifes; or when our bodies, or parts of our bodies, are injured in other ways, e.g. when being hit by heavy falling objects. Correspondingly, we characterise or describe our pains as e.g. "pins and needles", "sharp", "stabbing", "shooting" or "dull".

Having realised how internal states and sensations are related to bodies, it becomes obvious why it does not make sense to say that e.g. pain is "something" existing in a vacuous, detachable mind, and obvious why it does not make sense to say, "I have a pain, however not in my body, but in my mind", or say, "I am in pain, but my body is not". Putting it this way makes no more sense than saying that the cup I see on the table does not exist on the table, but in the privacy of my mind. Conversely, it becomes much more obvious why it makes perfectly good sense to say that the pain in my finger is just as much part of, i.e. a property of my finger, which I may refer to, as is the nail at the end of the finger, or the shape of the finger. This is to be understood in the sense that it is part of my being a person with a body that I may occasionally feel pain in my body and its various parts. Indeed, it is just as much part of or a property of my body, and of this body being *my* body, that I may feel pain in it, as it is a property of my body, that it, just like other material objects, has a particular extension, shape, and colour. And it is part of my being a person with a body, and part of knowing that my hands, feet, and legs are *mine* that the pain in my body can *only* be felt by me. And it is part of my being a person that it is *I* who may be aware of, and who may observe and know the pain in my body, and not my body which knows or is aware of the pain, nor my legs, feet, or fingers. Furthermore, when I bang my big toe on the leg of a table, the resulting pain in my toe exists for me as something, which is just as real as the leg of the table, and as something which I may localise just as effortlessly as I may localise the leg of the table, or my big toe.

Thus, talking about feelings of pain, or of "having pains", independently of talking about the bodies of persons in which they are felt or "had", would be just as unintelligible and nonsensical as talking about the observations and experiences of tea-cups on the table, independently of talking about the actual tea-cups on the table.

Well, it could be objected that despite the existence of a necessary relation between the notions of e.g. 'pains', 'bodies', and 'bodies of persons', there are significant differences in the conditions for describing sensations and internal states belonging to persons with bodies, and of describing the material properties of our bodies or of other objects. One of these differences is that unlike objects, and even unlike the bodies of persons, the thoughts, images, memories, emotions of persons or the "bodily sensations" of persons, such as pains, hunger, or itching, do not exist without being observed. Although a person does not exist without his body, and cannot refer to him or herself without referring to his or her body, there is a sense in which it is correct for a person to say that *my body*, just like other material objects, but unlike pains in my body, exists independently of whether or not I am *aware* of its existence. Thus my body, arms, legs, fingers, and finger-nails, etc., exist when I do not observe or attend to them, or if I cannot see or feel my body, e.g. because I am anaesthetised, unconscious, or asleep. This does not apply to the so called "bodily sensations" such as pains and hunger. I cannot say: "I have a pain in my finger, but I cannot feel it", or say, "the pain in my finger caused by a prick of a needle, is still there, but I am not aware of it". Nor does it make sense to say, "right now I am thinking very hard on how to reach a solution to a particular problem, but I am not aware that I am thinking about this problem, and how it may be solved".

Thus, it seems that the *asymmetry* of the relation between observing something, and the actual thing being observed, which applies in the case of material objects, does not apply in the case of internal states and sensations. Indeed, the dependence of the existence of internal states and sensation on being observed would seem to imply an *identity* of observing and experiencing such states and sensations, and those states and sensations themselves. If so, it could well be argued that it does not really make sense to say that observations and experiences of such states have an object independently of the observations and experiences of them.

And yet, it would not be correct to say that the existence of pain in my finger, e.g. subsequent to having been pricked by a needle, *relies* on, or is *dependent* on, my being aware of it - in the sense, for example, that it would cease to exist just in case I choose not to be aware of it (unfortunately, it does not!), or if I decide to direct my attention to something else. Nor does the existence of the pain rely on or depend on my observing or being aware of it in the sense that I would feel a pain in my finger just in case I so decided. In this respect the existence of pains does not exist only *in virtue* of being observed or experienced. However, my feeling and thus *knowledge* of the pain in my finger depends on my having observed the pain in the finger - but so does my perception and knowledge of the needle giving rise to pain. That is, neither my knowledge about the needle, nor my knowledge about the pain in the finger, exists independently of my having observed the needle and the pain respectively.

So, in spite of the fact that "internal states" and "bodily sensations" do not exist *unobserved*, it makes good sense, nevertheless, to say that such states and sensations exist as something which we may *observe, have knowledge about, refer to* and *describe* - in much the same way as we may refer to material objects, which do exist independently of being observed. When I say, "I have a pain in my finger", I am not talking about an *experience* or *sensation* of pain residing somewhere in my mind, but I am talking about and referring to the pain in my finger, in much the same way as when I say, "I have a nail on my finger", I am not talking about my *experience* of the nail on my finger, as opposed to talking about the nail on the finger.

And although we may talk about sensations and internal states, e.g. about pains, thoughts, emotions etc. of persons, as things which are different from material objects and properties of objects and bodies, the language we use to talk about such sensations and internal states is not a different or separate language (let alone, a private language), but, necessarily, the same language which we use for describing objects and bodies and their material properties. The words 'pain', 'finger' and 'nail' used in the expressions, "I have a pain at the end of my finger", and, "I have a nail at the end of my finger", logically belong to the same language, i.e. the language in which the meanings of the terms "finger" and "end of my finger" are the same in both expressions, and in which the difference of meaning of 'pain' and 'nail' is determined. But nor are the "grammar" or "forms of language" used for giving expressions to internal states and sensations different from "forms of language" being used for describing material objects and bodies. When saying, "I have a pain at the end of my finger", I am talking about and referring just as much to something being observed to be the case about my finger as I am doing when I say "I have a nail at the end of my finger". The reference of the word 'finger' or 'end of my finger' is the same in both expressions; it is *my* finger, e.g. the index finger on my right hand, about which different things are being predicated. If this was not the case, the strings of words "I have a pain at the end of my finger" could not be any more informative than a cry; on the contrary, it would be far less informative.

The result of the discussion so far seems to be that the *interdependence* or necessary relation between observation or description of "things" and the things being observed, known and described, apply equally whether the things are objects in material reality or internal states and bodily sensations. In this respect the condition for talking about observing, knowing and describing things, and for talking about the existence of those things as things which may be *observed*, *known* and *referred* to, must be the same whether we are talking about material objects or internal states and bodily sensations. This result is by no means surprising; it is far more surprising that it was ever questioned that the conditions for describing and expressing ourselves in language could be different or vary depending on *what* we use language to describe and express ourselves about. If, in contrast, mental or internal states were things to which we could only give linguistic expressions on a par with natural reactions to such states, but of which we could have no knowledge, nor any language in which we could talk about and describe them correctly, then we would not *have* the notions of mental or internal states. Nor would we have a language or any of the concepts we use to characterize and distinguish such states from external, material things and states, and none to describe how the properties of mental states of person differ from those of material objects. If so, we would of course have been spared of all the problems to which the traditional division between Mind and Matter gives rise. However, the way in which Wittgenstein attempts to get rid of these problems, i.e. by proposing that the language we speak divides into different language games for talking about and describing material things and for expressing mental states, being sharply separated by internal borders, only re-introduces this division and its problems in a new guise.

19.5 The "internal" and "external" of a person

A person is someone with a body who exists in a world of material objects and other persons. But a person is also someone who may distinguish himself from the rest of material reality and other persons. So, it makes good sense to say both that a person is someone who is "detachable" from the rest of the material world, and that he is just as much part of the world as are other things. Not only does the body of a person, i.e. his arms, legs, hands and feet, belong to the world "out there", but even so do things "inside" the skin of his body (his bones, muscles, blood-vessels, brain, his elementary tract, and so on). And what in this respect belongs to the "internal" of a person, is just as much part of what is "external" and publicly observable as is the rest of material reality.

We notice that the distinction mentioned here that a person may make between himself and the rest of the world, is an internal/external distinction which does *not* coincide with the traditional Cartesian distinction between, on the one hand, what belongs to material reality, and thus what is publicly observable, and on the other, what belongs to the mind, and thus what is not publicly observable. However, could such a distinction not be made - i.e. between that part of a person which is publicly observable and what is not - and thus what in this respect may be said to be internal to the person as opposed to what is external? All that such a distinction would require is that what is observable only to the person himself is defined as the internal of the person, whilst that of the person which is observable to everyone else as well, is defined as the external of the person. According to such a division, what would count as the external would be the body and actions of the person, his or her speech and other forms of communication with other persons; and what would count as the internal would be the emotions, pains, thoughts and images of a person and, presumably, that of the person - the "agent" or the "I" - who "generates" the actions and speech of the person.

Now, this is a distinction which has certainly been attempted (among others by radical behaviourists within psychology, although the "internal" of a person is only poorly defined, i.e. as all that of a person which cannot be the object of scientific inquiry because it is not publicly observable). However, it is a distinction which is not only highly unpractical, but also impossible. For one thing, a person being someone who knows how to use the pronoun "I" to refer to himself, is someone who also knows that one cannot refer to oneself without referring to one's body. I cannot use the term "I" to refer to "myself", nor to something being "me" independently of referring to parts of me which are just as observable to everyone else as they are to myself. Nor can I say ,"*This* is "I", or my "ego" as distinct from *that* which is my body", or say "I and what is "me" is something being detachable from the rest of my person". Indeed, I cannot thus detach "I" from "my body" any more than I can detach from my body the pain I feel in my body.

But neither can I say, "this is my action and speech - but the agent "inside" me, who produces this action and speech, is distinguishable from the part of my person, who carries out the action and who does the talking". Indeed, none of the things I can think of which belong to what is only directly observable to myself, such as my thoughts, emotions, images, and the like, are things which reside in parts of me, which are "detachable" from the rest of my person. Nor are any of them things which are distinguishable from the person I am, i.e. a person who at the same time exists in some concrete situation in reality. On the contrary, a person is precisely someone who knows that one finds oneself in particular situations in a reality which is publicly observable, while at the same time being able to record what is only directly observable to oneself, e.g. moods, pains, thoughts, beliefs and whatever else goes on in one's mind. And just as we cannot have a language and terms for pain which is not part of the language for bodies, and thus for objects in material reality, we cannot have a language for mental states of persons, such as their thoughts, beliefs or knowledge, which is not part of or independent of a language for other parts of the person. Nor, conversely, can we have a language for these other parts of persons, their bodies and the world of which their bodies are part, without referring to those mental states of persons, their beliefs, thoughts and knowledge, and thus without referring to persons having or entertaining mental states.

Thus, there is no such thing as transcendental "mentation", nor of a transcendental "I" or "Ego" apart from or detachable from the rest of the person, i.e. an "I" or "Ego" existing apart from the person with a body who acts in the world, and who communicates with other persons. And there is no transcendental "I" or "Ego" who monitors the emotions, pains and thoughts inside "me", or who initiates my thoughts, actions or speech. But there is a person of "flesh and blood" who has knowledge of the world around himself and of himself, and who may think and reflect on the knowledge he has about the world and himself. A person, however, who knows that there are things about himself which only he may directly observe and know about - but who also knows that the things which only he may observe and know about himself, are indistinguishably related to that of himself which is perfectly observable to anyone else. Without *this* knowledge of persons about themselves, there would be no persons, nor any shared knowledge of reality of persons, and thus no language.

19.6 The status of descriptions of internal states

In previous sections I argued that, fundamentally, the conditions for using language to describe internal states and sensations are not different from the conditions for using language to describe objects in material reality. However, it has to be admitted that the procedures for determining the truth and correctness of statements about internal states, i.e. things which are not publicly observable, are different from the procedures for determining the truth and correctness of applications about material objects, i.e. things which are publicly observable. If, for example, I am discussing with somebody whether the statement, "this line is longer than that line", is a true statement about two particular lines on a piece of paper, we will be able to settle the discussion by measuring the lines with a ruler. That is, we will both be able to take part in a procedure for determining whether the consequences of the statement hold true for the lines in question, and to observe the result of a test as to the truth of the statement. This is not the case when somebody says e.g. that the knock he received on his head by a falling object was painful. We may discuss what the *implications* of such a description are, at whatever length necessary, and in the course of the discussion arrive at better and better agreement about what they may be. But in the case of the unfortunate sufferer of a blow to his head by a falling object, the procedures for determining whether a "pain-description" is a true description of what he feels in his head, can only be carried out by the sufferer himself; for only he knows and may observe what he feels in his head. It cannot be determined or tested by or together with e.g. bystanders, who may have witnessed the event, and who on the basis of their own experience may rightly surmise that the person hit by the heavy object is in pain – and who, therefore, have no reasons to doubt that he is using the "pain-description" correctly.

But if internal states such as pains are the sorts of things which are not publicly observable, how then can we be sure that when different people talk about pains, they are talking about the same kind of things. How, more precisely,

> (1) "Can I be sure that when I use the term "pain", I use this term to refer to the same kind of "thing" as others do when they use the same term?" Or, conversely, "when other people use the term "pain" do they then use this term to refer to the same kind of "thing" as I do when I use the same term?

This would seem a perfectly sensible question. But at the same time it appears to be a question which, for obvious reasons, is impossible to answer. However, it is also a question which it is impossible to *ask*. For this question, and the various ways it may be put, is a question in which the very condition for putting it forward is itself questioned. When asking, "do I use the term 'pain' to refer to the same thing as others do when they use the term", I am obviously asking this question in a language of which the term 'pain' is part – and thus a term which is supposed to have a meaning and correct use, which I and other people with whom I share the language know *in virtue* of being speakers of that language. In other words, it is a term that I and other speakers of the language know how to use correctly, and who therefore also know and may determine what kind of "thing" the term may be used to refer to. If I did not know this - and did not *presuppose* that all other speakers of the language knew this – neither I nor they would have any idea what I was asking about, and no further discussion of the question would seem possible.

Admittedly, we may have all kinds of difficulties in giving *adequate* verbal descriptions of the pain we may feel, and we often feel uncertain about the choice of appropriate terms. Is this pain, for example, a sharp, piercing, dull, shooting, tender, searing pain, or is it a nagging or stabbing pain? Indeed, we may have great difficulties in conveying precisely to others the suffering we endure when we are in pain. However, these problems and difficulties of adequately describing pains are not relevant to the question being discussed, nor do they invalidate the point just made. Indeed, these problems and difficulties of adequate descriptions of pains, and discussions about such problems, could not take place unless the people discussing them had a concept of and a term for pain which they shared, and knew of what it may be used correctly.

Thus, the question is *obtuse* in the sense that putting it forward presupposes that we know the meaning and correct use of the terms of the question - but then we are asked to put this knowledge within a parenthesis and "forget" that we do, or pretend that this knowledge is immaterial for a discussion of what the terms may be correctly used to refer to - or whether they may be used correctly to refer to anything at all. But naturally, it is a question which can only be asked granted we have already learned a language which we may use to talk about pain, and thereby granted pain to be the kind of "thing" which we may talk correctly about and refer to. Hence, if we do ask this question, the answer is logically implied: an affirmation would be redundant, while a denial would be contradictory.

This, I think, would probably have been obvious if the same question had concerned the use of the term 'cups' instead of 'pains'; indeed it would probably have been so obvious that we would hesitate to ask it. For if we did ask this question about the use of the term 'cups', we might as well ask the same question about all other terms in our language - and we would be well on the road to asking whether we can be sure that we may use terms in our language to refer to and talk correctly about *any* objects in material reality, and thus in effect whether we may use language to talk correctly about anything. However, although in *particular* cases we may be in doubt as to whether a particular term may be correctly used to refer to some particular thing, i.e. a thing which has been identified in a shared public world, we cannot doubt that as language users taking part in this discussion, we do know (other) correct terms for the thing, nor doubt that we know how to use *these* terms correctly. One cannot doubt the necessity of these conditions for settling the question under discussion - conditions, of course, which apply equally when we use language to investigate language, or when used to talk about any other matter - unless, of course, one has been seriously contaminated with scepticism, and mistakenly assumes that one may use language to question the very condition for using language.

Now, it might be objected that in the case of the use of the term 'pain', we do have a special case. For one thing, pains are exactly the sorts of things that cannot be identified in a shared, public world. Only the person in pain may identify his pain. Only he can experience his pain, and there is no way that he may "show" his experience of the pain to others. And if, instead of the original question (1), we had asked

> (2) "Can we be sure that different people *experience* pain the same way? Or, can I be sure that when I experience pain, my experience of pain is the same as the experience of pain of other people?",

then it would have to be admitted that this is both a sensible and a serious question. For, surely, the very possibility of answering the question of whether different people in fact use the term "pain" to refer to the same sort of "thing" (as in question (1)), relies on the possibility of an answer to the question of whether people actually experience pain the same way (as in question (2)).

Well, if question (2) is the sort of question to which an answer *could* be given, it could not consist in a test of which the *experiences of pain* of different people were compared. The question could *only* be settled by verbal means, e.g. by comparing the *descriptions* that different people may put forward about their experiences of pain. Consequently, the question would have to be reformulated for example as follows:

> (2a) "Would other people use the term "pain" to refer to what I now feel in my finger, if they could experience the pain that I feel in my finger"?

However, this brings us back to where we began - i.e. to a question which cannot be asked, let alone be discussed, without presupposing that we have already learned a language with terms not only for pains, but also for experiences of pains, and that, *in virtue* of being users of this language, we know what these terms may be used to refer to. Thus, it is presupposed that about the experience of pain something is the case, or true, and that something else is not the case, or false, and that any one of us is able to determine when it is the case, and when it is not the case, i.e. determine when we experience pains. But it *also* presupposes that as users of this language, and thus able to ask this very question, we do have terms for such experiences, the correct implication of which we know and may correctly apply.

There are good reasons, then, why it is impossible both to answer and to ask the question, "can we be sure that different people use the term 'pain' to refer to the same kind of 'thing'?" Whatever may have urged philosophers to ask this question, it obviously implies and presupposes assumptions which at the same time are called into question. To assume the tenability of this kind of question, makes no more sense than assuming that we share a language in which we may communicate about objects in the world, and at the same time assuming that objects in reality do not exist as the sorts of things which may be talked correctly about.

I am not saying that the fact that we have a language with terms for both objects in reality and pains in bodies *proves* the existence of objects in reality and pains in bodies. What I am saying - and have expressed in the *Correctness Principle* - is that we cannot begin to discuss or investigate language and the use of language to talk about objects or pains, let alone discuss the possibility of there being any objects or pains we may observe, feel and talk correctly about - without presupposing that it is indeed the case. In short, what I am saying is that one cannot take part in this kinds of philosophers' discussion without committing oneself epistemologically.

That it is possible and, for some purposes, even makes sense to compare the use and status of sentences containing terms for internal states, say, 'pain', with sentences containing terms for publicly observable objects, say, 'penny', relies on the presupposition[8] that we know *both* what sort of thing a penny *and* a pain is, and that we are able to determine and describe the circumstances under which the terms 'pain' and 'penny' may be correctly used. Such investigations of language and use of language logically rely on epistemological commitments which cannot be ignored, nor explained, nor done away with by such investigations.

19.7 Conclusion

In this chapter I have tried to argue, first, that it does indeed make sense to say that we have knowledge of and may use language to describe and talk correctly about our internal or mental states. This, as I have tried to show, is simply a precondition for any discussion of the status and nature of our non-publicly observable internal states, and of the way in which we use language correctly to express ourselves about such states. Indeed, it must be just as much a precondition for discussing and talking about such states, as it is for determining and describing the publicly observable circumstances in which they occur, and the criteria for using language correctly to express ourselves about them. So, to the extent it is assumed to be at all possible to engage in discussions about internal states and ways of using language for them, we shall have to assume that conditions exist which *logically* must be the same as those which apply for describing and talking about publicly observable states and things in material reality. That is, in the cases of either of such states it must be a precondition that we have knowledge of and a language in which we may talk correctly about them.

[8] - as does any interesting distinction between the two sentences.

Secondly, I have argued that we cannot talk sensibly about, nor determine the differences which exist between, non-publicly observable internal states and publicly observable material states, let alone talk sensibly about the differences in the knowledge and description between different persons of either states, without presupposing that the language we use for such discussion and determination is indistinguishably the language being used to talk about our publicly shared world. And I have tried to show that any *general* questions or doubts as to whether different persons use language and its terms in the same way when talking about their internal or mental states, are questions and doubts which in themselves question the very conditions for putting forward such doubts or questions – just as would any *general* questions or doubts as to whether different persons use language and its terms in the same way when talking about things in material reality. So, to the extent it makes sense to discuss any *concrete* differences in the experience, knowledge and use of language of different persons concerning internal or mental states, we shall have to presuppose that such discussions are carried out in a language which is both public and shared among language users. To suggest, as Wittgenstein does, that language divides into two different and sharply separate language games or forms, namely, a language game for *descriptions* of publicly observable things existing in material reality, and another for *giving expressions* to – as opposed to describing – internal or mental states, only makes his own discussion about internal states and those language games completely redundant. It may have been the aim of this suggestion to solve and get rid of the intractable problems arising from the traditional division between Mind and Matter; however it only cements this division and creates problems of an even more intractable and nonsensical kind.

Thirdly, I have argued why it is not possible to treat the internal or mental states of a person *in isolation*, i.e. as states being "detachable" from the rest of the person, i.e. his or her body and the verbal and non-verbal acts carried out by the person. And I have shown that it only makes sense to distinguish between the "inner" and the "outer" parts of a person on the assumption of the existence of a *necessary* relation between the two, i.e. none of these parts of a person could exist without the other, nor could any of the concepts we use to characterize either part be well defined without reference to the concepts used to characterize the other. Hence, when it comes to persons, what is usually conceived of as the "inner" of the person is very much part of the "outer" of the person – and vice versa - just as the terms used to talk about the one are logically related to the terms being used to talk about the other.

But if internal or mental states, such as beliefs, knowledge, thoughts, feelings and emotions are not publicly observable, then how can we talk about them in a language we share? How do we ever come to learn and share a language to talk correctly about something which is obviously non-public? These are some of the issues I shall address in the next chapter in which I shall attempt to pin down the basis and conditions for this inter-subjectivity of knowledge and language of persons; i.e. conditions which apply equally for knowledge and description of publicly as for non-publicly observable things or states.

Chapter 20

The inter-subjectivity of knowledge and language

20.1 Introduction

In this chapter an attempt will be made to pin down the conditions for the inter-subjectivity of language and cognition of persons. By this I mean the conditions which provide persons with the possibility to talk about and communicate their knowledge and experience to others - not only of things existing in publicly observable material reality, but also of their non-publicly observable thoughts, emotions, feelings and other internal or mental states; conditions, furthermore which provide persons with the possibility to talk about and determine their individual differences concerning their knowledge and experience of both that which is publicly observable and that of the persons which is only directly observable to themselves. In other word, I shall attempt to pin down the conditions which provide a person with the possibility for being a person, i.e. someone who may be different from other persons.

The result of my analysis of this inter-subjectivity between persons is that it relies on a notion of 'truth' which is such that what is true or false, correct or incorrect, is also true or false, correct or incorrect for other persons - and thus that the notion of 'truth' presupposes a notion of 'other persons', or just of 'others'. I have formulated this dependency of the notion of truth on the notion of 'others' in a principle of inter-subjectivity, the implication of which may be stated thus: Granted that other people could have the knowledge and experience I have, and could describe what I know and experience, they would describe what I know and experience the way I do. Or, the way I describe what I know and experience would also be correct for others, could they know and experience what I do.

This principle of the inter-subjectivity of the notion of truth and its logical dependence on a notion of 'others' would seem self-evident. However, had the implications of this principle been fully understood, many problems within philosophy and psychology concerning persons and their knowledge and experience - both that part which is personal and that which persons have in common with others - would probably have been posed differently. In the last section of the chapter I shall discuss, in view of this principle, the relativism which runs through Wittgenstein's later theory of language, as well as the irrealism of the so called Social Constructivist Movement, whose followers claim support for the epistemological position underlying their theories in views defended by Wittgenstein in *Investigations*.

20.2 Personal versus public knowledge and experiences

It has to be admitted that it is somehow puzzling that pains and other internal states, which are only directly observable to the person who have them, and thus cannot be shared by others, are nevertheless things which we may talk and communicate about in a language we do share with other persons. And puzzling how a crucial part of the knowledge that a person has of himself, of his mind and body - and, not the least, of his mind and body being his - rests on observations and experiences to which only the person himself has direct access. And yet, neither this knowledge, nor the observations and experiences on which it rests, are private. For, in spite of not being directly accessible to others - and in this sense not being shareable by others - we both can and do communicate with others about such knowledge, observations and experiences, and thus, undeniably, we do have a language in which we may talk both meaningfully and correctly about them. So, shareability in the sense of being publicly observable and known cannot be a condition for the possibility of inter-subjective communication among persons about things which only they may directly observe, experience and know about.

In the case of observing, experiencing and communicating about objects in material reality, our situation is arguably significantly different. Take for example two people sitting on either side of a table with cups and plates, a teapot, a bowl of sugar and bottle of milk. All these things exist in a shared public world and are perfectly observable to both persons; they may together determine what is on the table and also whether the descriptions they put forward about them are correct. And yet, it could be argued that how these things are observed and appear to them from their different vantage points on either side of the table is different, i.e. due to the fact that the parts and features of the things which are directly observable to the one, are not the same as those which are so observable to the other. To the person sitting on one side of the table the sugar bowl and milk bottle will be to the left of the teapot, while to the other they are to the right of it, and so forth. However, this does not present any serious difficulties since, first (as argued in Chapter 16), it is part of our knowledge of things having been identified as particular things, that they will appear differently when looked at from different vantage points - and that, generally, things do appear differently when observed with regard to different possibilities of observations and action. Thus, having identified the things on the table as cups, plates etc., the persons at the table know that if they move to different positions, the things will look differently because other properties and parts of the things will now be observable to them. And these differences of perception and experience do not represent any serious problem since, secondly, none of the particular ways of perceiving and experiencing the things on the table, and none of the descriptions by either person of their experiences of the cups, plates, etc., are unique to him or her. Indeed, it is assumed that they are not, just as in general any person and language user will assume that if other people could look at the things from his or her vantage point(s), they would observe what he or she does, and report that they perceive the same features and properties of the things and describe what they perceive as he or she does. So, although two people sitting opposite each other at a table do not share exactly the same perceptions or experiences of the things on the table, they both know and take for granted that if the other person could be in their position, the person would perceive and observe what they do, and describe what they perceive and observe as they do. If we could not count on this assumption, communication and action between persons about things in the world would be impossible.

But there are numerous other ways in which the knowledge and description of persons concerning things in publicly observable reality may differ. Just think of the differences due to differences of our background, education, previous history of experiences, and the opportunities to act with, observe and describe such things which are or have been available to us. Examples are *legion* – I only have to think of the knowledge I have of aeroplanes, their construction and how to fly them compared to that of a pilot. Or, conversely, think of the knowledge I have as a clinical psychologist about the transference phenomena occurring between client and therapist during psychotherapy compared to that of an aeroplane pilot, who has not encountered such phenomena, and who does not have the language and terms to describe them that I have. And yet, despite the fact that our knowledge of these and numerous other matters are not exactly the same, and probably never will be, we are indeed able to communicate the knowledge that each of us has about aeroplanes and transference problems and those other matters, and thus to share our knowledge of what in this respect is personal to each of us. But if the condition for the inter-subjectivity and shareability of knowledge and description in the actual case as well as between persons in general, is not and cannot be that persons have exactly the same knowledge and experience of things, nor the same possibilities of describing things, since this condition is only rarely if ever met due to their different background, education, history of experiences, and so forth, on *what* then relies this inter-subjectivity of knowledge and description of persons?

It relies no doubt on the fact referred to in Part II of this book that, apart from differences in our knowledge, experience and background, we do share a substantial amount of knowledge and description of the world in which live and act, of the things with which we may act, of ourselves and of the persons with whom we may co-act. However, to say so does not of course add anything to our notions of 'shared knowledge and description', nor does it suffice to account for the inter-subjectivity of knowledge and description that we do not share with others, and which is *personal*. It does not do so unless it is assumed that vantage points, background, and situations we may be in are *in principle* shareable, and thus that other person could be or could have been in the same situations. But nor does it suffice independently of assuming that, granted other people had had the same background, or had been in the same situations that we ourselves have been or are in, then they would have the knowledge of the situation and the things that we have, and describe them the way we do. However, this suffices, indeed it will *have* to suffice, to say that it must be fundamental to the cognition and experience of persons that although other people may not be in our situations, and may not have, or may not have had, exactly the same experiences and knowledge that we have or have had, they would - *could* they be, or *had* they been, in our situations. Likewise, it suffices, and will *have* to suffice to say that to be language users and to share a language with other persons *logically* implies and presupposes that other language users, granted that they could be in our situation and have the experience, knowledge, background, points of view etc. that we have, would use language to describe what we experience, know of, etc., in those situations as we do. Or, that they would consent that the descriptions we put forward about these things and this knowledge are correct and correctly applied.

Now, if we can agree that these presuppositions are fundamental to the cognition, use of language and communication of persons, and indispensable for any meaningful discussion among persons about what they know and how they describe what they know, I think we shall also have to agree that this inter-subjectivity of cognition and language relies on a notion of 'truth' which logically implies that what is true or false, correct or incorrect, is also true or false, correct or incorrect for other persons, or just for 'others'. Indeed, I think that the inter-subjectivity of knowledge and language of persons may be boiled down to such a notion of 'truth' – and therefore that our notion of truth both implies and presupposes a notion of 'others'. The assumption of this notion must be the rock bottom, the point of departure from which any discussion about our knowledge and description must be based and proceed - whether such discussions concerns our knowledge and descriptions of things in publicly observable material reality, or our personal so called internal states, such as our emotions, thoughts - or feelings of pain. No analysis of our cognition and use of language in general, nor of the differences which exist between different persons concerning their cognition and description of that which they cognize and describe can go beyond these assumptions and presupposition; nor, as will hopefully be clear in the discussion which follows, can we prove or deny them without conceding them. During this discussion it should also be clear why the notion of "private languages" describing "private experiences" is incompatible with the conditions which necessarily apply for the cognition and use of language of persons.

20.3 The principle of the interdependency of the notions of 'truth' and 'others'

Although we may sometimes find it difficult to give adequate and satisfactory descriptions of our internal or mental states, we do seem to know, and assume that we know, what it implies to describe, for example our different emotions as happiness, anger, or sadness, just as we seem to know, and assume that we know, what it implies to describe our thinking as attempts e.g. to solve some particular problem and finding various solutions to it. Nor do these occasional difficulties seriously shake our certainty about the existence of such states - any more than similar difficulties sometimes encountered in giving adequate and satisfactory descriptions of things and events in material reality would shake our certainty as to the existence of those things and events.

More importantly though, the lack of possibility of public examination, test or direct comparison of such states for determining, on the one hand, whether our internal states and our experience of them are in fact the same, and, on the other, whether we use the same terms to describe them, prevents us from assuming that we may indeed talk about such states – or even discuss our individual differences concerning such states. In other word, the lack of public criteria or standards does not prevent us from assuming that we do share a language in which it makes sense to talk about such states, and also that terms exist in this language with which they may be correctly described. Let me give an example. Let us suppose I am discussing the taste of a particular apple with someone, and let us suppose that the issue of the discussion is whether the apple tastes sour or sweet. I, for example, maintain that to me the apple tastes sweet, while my opponent maintains that to him it tastes sour. Well, we are obviously disagreeing about the taste of the apple (and may not come to an agreement), indeed, we may even realize that we are not agreeing. But what does it take to realize that we disagree about the taste of the apple?[1] Apparently not that we have procedures or independent criteria for discernment in the situation, because we do not. What it takes and presupposes to discuss our taste of the apple and to realize that we disagree about it, is that we both know what it implies to say of something that it tastes sweet, and what it implies to say that it tastes sour. That is, it presupposes that we know the implications and use of these terms, and thus we know and presuppose that 'sour' and 'sweet' refer to experiences of the tastes of things, which we both know of. Lastly, but not least, it presupposes that when describing our experience of the taste of the apple, we describe it as others would, could they have our experience. Otherwise, we could not make experiences of taste - or for that matter any other of our non-publicly observable states and experiences - the subject of everyday conversation, still less the subject of an inquiry into individual differences or similarities. There would simply be no way that we could talk about or determine individual differences, disagreements or similarities between different persons about such experiences and states.

Now, it may be objected that there are significant differences between what may be rigorously determined, referred to and talked about and what cannot, and that it does not make sense to maintain that all knowledge and description is equally correct or true, or that assertions of what we know may be put forward with the same kind of certainty – if only in all instances of knowledge and description the same notions of 'true' and 'correct' are employed. Just think of the considerable variations which exist in how things are described by different persons, and by the same person in different situations, not only when the things concerned are "internal" states and events, but even things in material reality. Not only our intentions, purposes and possibilities of observation and action, but even our moods and temperaments may at times determine what we experience and how we experience ourselves and the rest of reality, and thus what we know of, and how we describe ourselves and the rest of reality. So, would it not then be reasonable to reserve the notion 'correct use of language', and 'true knowledge and assertions' for cases in which no such individual differences and variations exist, and to reserve the terms 'exist' and 'determinable' for things and

[1]- as opposed to assuming that one of us must be using the term "sweet" or "sour" wrongly, or does not know the implications of such terms.

events about which no uncertainty prevails - because they belong to what is materially and publicly observable?

This solution has been attempted, notably by the logical positivist and by the radical and logical behaviourists in psychology and philosophy respectively, who aimed to establish a firm epistemological basis for scientific research. In this pursuit they argued that only *that* existed objectively, and hence could be the object of scientific research, which could be determined by rigorous public criteria and standards, and that only knowledge and description of what had been thus determined and observed, could be said to be meaningful and true. In effect, any determinate notions of the truth and meaning of statements and knowledge would have to derive from observations fulfilling such criteria and determinations. Consequently, what cannot be subject to rigorous public scrutiny and observation fulfilling such standards and criteria does not exist, nor can description of things which cannot be so observed be true; hence, descriptions and the existence of knowledge of such things may be discarded as nonsensical.

However, against such arguments we only have to consider that the very process by which we determine a situation, and what in this situation is materially and publicly observable, presupposes that something is the case or true about the situation and things being determined and observed - which is also the case or true to other people – in casu the people involved in the observations. Thus, it is not because situations exist or may be arranged, in which things are publicly observable, and which we may come to agree to describe in particular ways, that certainty "emerges" of what - for everyone involved - is true or correct about things and situations. It is the other way round; for, no such determinations of correctness of knowledge and description of things and situations could be agreed upon, let alone be established and function as criteria or standards for correctness among language users, unless it was presupposed that when determining these standards, we already have a concept of 'truth' which we know how to use correctly; a concept of 'truth', furthermore, which is such that what is true or false, correct or incorrect, is also true or false, correct or incorrect for others.

This is a presupposition and condition on which, logically, rests the possibility of distinguishing that which is publicly observable from what is not, and on which rests our determination and characterization of individual differences and variations in that part of our knowledge and experiences which is personal. Thus, the point, so easily overlooked, is that even in a situation in which the things and events being described are publicly observable, it is logically implied and presupposed that the descriptions put forward and being understood by others that these others, being in the same situation in which we are, will have the knowledge we have about these things and events, and will describe them as we do. In general, it is presupposed that our notion of 'correct' or 'true' is such that what is true or correct, is also true or correct for others. It is because of this logical implication of the notions of 'true' and 'correct', fundamental to use of language and communication, that we may talk correctly and rigorously - not only about our knowledge of what is publicly observable, but also about that which is not - and thus *that we may distinguish between what is and what is not publicly observable*. So, from an epistemological point of view, there is no need to restrict what may be talked about, and talked *correctly* about, to things which are materially and publicly observable. In particular, there is no need, nor any grounds for assuming that the notions of 'truth' and 'meaning' applying to knowledge and description concerning things that are not publicly observable, are different from those applying to knowledge and descriptions of things which are publicly observable.

So, rather than attempting to avoid or do away with the problem of individual differences and the fact that part of the knowledge of persons of themselves and of reality is *personal*; and rather than to characterize this knowledge as uncertain and what it is about as being non-existent, a proper epistemology should be able to account satisfactorily for such differences of knowledge and personal experience. For such an epistemology it would be obvious to ask how it is possible to talk about *standards* for what count as "correct" descriptions and "publicly" existing things, unless it is presupposed that different persons - in spite of the differences and variations in their knowledge, background, opportunities for observation and action - are able to distinguish between conditions under which such standards apply, and in which they do not. And ask how we could ever determine what are the conditions for individual differences and variations to exist in knowledge and descriptions of the same things, unless such differences could be *rigorously* determined, *correctly* talked about and referred to - and thus unless we presuppose that we do have correct descriptions and are using language correctly when discussing and communicating about these differences. That is, without presupposing that both the knowledge shared by different persons, and the knowledge of persons which is personal, are perfectly sensible issues of inter-subjective discourse.

Let me illustrate this point by giving the following example. One of my friends tells me: "I am terribly depressed; everything looks so grey and colourless - even the trees and flowers looks grey and colourless". Now, could it not reasonably be argued that at least in this case we are not talking about a "public" issue, but rather of something "private", and also that it is a situation in which it would make no sense to maintain that my friend is still using language correctly? Is it not a situation in which any well-defined notions of correct or true assertions have been suspended? Not at all. For one thing, I do understand what my friend is saying. I am perfectly able to communicate with him about his - in this case - curious experience of the colours of trees and flowers. But a condition for maintaining that I understand what he says, and for communicating with him about his curious experiences of the colours of trees and flowers is, naturally, that he still uses language correctly when talking about his experience of these things. That is, it is a condition that he knows the correct implications of the various colour categories, and that he knows how to apply them correctly. And it is a condition that what he is talking about is something he may refer to, and about which true and false assertions may be made, i.e. his curious experience of the colours of trees and flowers. Thus, it is a condition that both of us are still using the same language - indeed the very same language that he and I use under normal circumstances to talk about other ordinary everyday matters; and it is a condition that we are together able to determine *what* he is talking about. If we are able to do so, and thus able to talk about his experiences, however curious, it has to be maintained that he is using language correctly when describing his experiences.

However, it is quite clear that his description of the colours of the trees and flowers is not of *general validity*, and I do not take his description as an attempt on his part to produce descriptions of general validity - i.e. descriptions which would be correct under normal everyday conditions of observation. But an important part of the message he is trying to convey to me - and which I understand - is precisely that his situation is not normal, i.e. that his depression affects his perception and descriptions of things in ways which differ from how he normally perceives and describes them. A fact of which he himself is perfectly - and probably painfully - aware.

So, although my friend may feel eminently on his own with his unusual and personal experience during his depression, his experiences are not *private*, nor are his descriptions of his experiences. They are perfectly understandable to others because it is presupposed - by us and by him - that he is using language when describing what he experiences as others would - could they be in his situation and experience what he does. Indeed, our communication about what he experiences relies on the presupposition that what to him is the case or true about his experiences would also be the case or true for others, had they his experience.

Let me conclude my discussion of this example by saying that it shows that individual differences in the cognition and description of reality of different persons may exist and be determinable, i.e. cognition and description of reality of persons may exist and be determinable which is *personal*. However, it also shows that a condition for these differences between persons and their cognition and description to exist, is that persons and language users, despite such differences, share a vast amount of knowledge and correct descriptions of reality. And it shows, furthermore, that the possibility of determining and of talking correctly about such differences relies on the presupposition that, when we describe what we know of or experience, we use language to describe it as others would, if they had our knowledge and experience; however, this in its turn both presupposes and implies that the notions of 'correct' and 'true' of persons are such that what is correct or true, is also correct or true to others - i.e. the notion of 'true' and 'correct' is not well-defined independently of a notion of 'others'. This I shall call the principle of the interdependency between the notion of 'truth' and the 'notion' of others. If the presuppositions and implications of this principle did not apply, no personal differences could exist, nor any possibility to determine or talk sensibly about such differences. Conversely, granted the presuppositions and implications of this principle, a person is someone who may be different from other persons. Indeed, a person is someone who *is* different from other persons.

Now, suppose my friend finds himself in the even more serious situation in which it can no longer be maintained that about the things he experiences something is the case or true, and thus that his cognition of these things does not imply knowing that something is the case or true about them; and suppose that in the language he uses to describe what he cognizes and experiences, assertions and descriptions have no correct implication nor use. Well, if he did find himself in this situation, I think we shall have to agree that it cannot be maintained that he is still a person having knowledge about things, nor be maintained that his mental "goings on" amount to cognition. But neither could it be maintained that he had a language, or that he could be a language user. It would be a situation in which the fundamental conditions for cognition, language and use of language would no longer exist, and communication and co-action with him would be impossible. But it could not be maintained either that he had a *private* language in which he might "communicate" his cognition and experiences with or for himself. This applies for his cognition and use of language as a whole, had his cognition as a whole been damage in this way, and for parts of it, if only parts of his cognition and use of language were thus affected.

Suppose my friend, having in some fashion recovered, tells us: "I have knowledge about some particular "things", but what is the case, true or correct about them, is not the case, true or correct for others, - or I cannot be certain that it is". Now, what could he possibly mean by that? Does he mean that if others could know of the "things" that he knows of, and thus could know what is the case or true about them, then it would not be true or correct for them? Well, if that is what he means, he is obviously contradicting himself.[2] Or, could it be that he means that these "things" of which he knows, are in principle inconceivable to others, because what is true or correct about them, is true or correct in a sense which is different from the sense in which the statement he puts forward about his (exclusive) knowledge may be? That is, the notion of true and correct in the case of his unique knowledge of these "things" is different from the notion of true and correct which others have, and which he himself has in other cases, for example when communicating to others his unique knowledge. In other words, does he mean that this concept of true or correct, which applies to his knowledge about these particular "things", is a concept which is special to him, in the sense: *private*, and consequently, that his knowledge about these "thing" is equally private?

Now, for such a claim to have any bearing - even for himself - would seem to require that he be able to account for how his "private" concept of true and correct differs from the one he shares with others, i.e. account for it in the language in which the claim is put forward. But if he could do that, his "private" concept of true and correct would *not* be private, nor inconceivable to others. Likewise, in order to maintain that what he knows to be the case or true about these particular "things" is not the case or true to others, would seem to require that he be able to determine what would be the case or true to others about these "things", and how it differs from what is the case or true to him - again in the language in which the claim is made. However, if he could do that, then what to him is true or false about the "things" would be perfectly conceivable to others, and what he knows about them would be perfectly expressible in terms of the language which he shares with others.

So, we may conclude that my friend is either contradicting himself or talking nonsense - or both. This would have been immediately obvious if instead he had said "I have discovered that the statement, " ... . ", is true or correct, but it is not true or correct to others, or I cannot be certain that it would be true or correct to others". What is obvious is that he mistakenly thinks that one could share a language with other people, and also that in this language the notions of 'true' and 'correct' could be different for different people.[3]

I think we shall have to agree that for the same reason that nobody may claim to possess private knowledge or a private language, and thus a private notion of truth, no such private language or knowledge may be ascribed to others - neither "in toto", nor in part. For, how could we justifiably ascribe a private language or knowledge to others without being able to account for

[2]- for, to say that others may know of the things that my friend knows of means that the things being known are things about which the same is the case or true; otherwise they would not be the same things.

[3] There is of course the possibility that what he means is merely that he is not sure of the correct implications and application of the statement – but that is a quite different matter.

both such a language and knowledge in a language which is *not* private, and into which this knowledge and language must somehow be translatable? This, together with results of the discussion of other examples, should suffice to shows that *I cannot know for myself what others might not know as well*; and to show, once again, that to be a user of a language I share with others, means that I cannot know what is true or correct to say about a "thing", which may not be true or correct to others, could they know what I know about the "thing".

At this point it would be relevant, I think, to bring in a much quoted paragraph from Wittgenstein's *Investigations*, and compare the points he argues with the ones presented above. In the paragraph in question Wittgenstein says,

> "If I say of myself that it is only from my own case that I know what the word "pain" means - must I not say the same of other people too? And how can I generalize the one case so irresponsibly?
>
> Now someone tells me that he knows what pain is only from his own case! - Suppose everyone had a box with something in it; we call it a "beetle". No one can look into anyone else's box, an everyone says he knows what a beetle is only by looking at his beetle. - Here it would be quite possible for everyone to have something different in his box. One might even imagine such a thing constantly changing. - But suppose the word "beetle" had a use in these people's language? If so it would not be used as the name of a thing. The thing in the box has no place in the language-game at all; not even as something: for the box might even be empty. - No one can 'divide through' by the thing in the box; it cancels out, whatever it is.
>
> That is to say: if we construe the grammar of the expression of sensation on the model of 'object and designation' the object drops our of consideration as irrelevant." (Wittgenstein (1945/1953) 293, p. 100)

Well, for a start, Wittgenstein does not only imagine that there is a word for pain in "people's language" - he knows it for a fact; indeed when - in the example above as well as in general - he is discussing the issue of pain and the meaning of the term 'pain' he is deeply involved in a "language-game" of which the term 'pain' - to everyone involved in that game - is an expression given to a sensation, i.e. pain somewhere in one's body. So, he knows and presupposes that he and everyone else taking part in this game know the meaning of the term 'pain', and know to what it may be applied, *in casu* pains in ones body. And he presupposes, therefore and necessarily, that sensations or feelings of pain are the sorts of things which may exist and to which one may refer. Without presupposing this, and thus presupposing that others would use the term 'pain' to refer to the feeling of pains that he may have in his body - could they feel what he feels - he would not have a clue of what he himself is talking about. The whole discussion of the "language-game" of giving expression to pain relies on this presupposition and, thus, that he may indeed generalise from himself to others, when they use the term 'pain' - just as they may generalise from their use of the term to his and the use of the term by others.

Now, having made explicit the presupposition on which the discussion and issues raised in Wittgenstein's example must obviously be based - as must indeed any other discussion of the meaning of the term 'pain' - we observe that he begins the example by suggesting - for the sake of argument - that he only knows from his own case what pain is, and what the term 'pain' means. And he suggests that the same might apply to others. Thus, the example suggests that he and they possess knowledge about their own feelings of pains, what their feelings of pain are like, and what the term 'pain' means for them. Indeed, for the sake of argument he asks us to presume that it makes sense to say so, and that within the "language-game" of pains such statements are perfectly understandable to everyone. But, obviously, it does not make sense to say anything of the kind - in any "language-game" - unless the implications of the term 'pain' is shared by everyone involved, and unless this term is used to refer to the same sorts of things. Without these presuppositions the example disintegrates into nonsense.

But instead of saying exactly that - and contrary to the presupposition on which his whole discussion is based - Wittgenstein goes on to suggest that because my feelings of pain, and those of others, are not publicly observable, *my* feelings of pain may be completely different from the feelings of pain of others and, thus, I and others may use the term 'pain' to refer to completely different things. What Wittgenstein is saying, then, is that it makes sense to say that differences could exist in how different people feel pain, and even that this could be said in a "public language"; yet this needs not be a language in which we may talk *correctly* about pain, even less a language in which it is possible to determine the differences which exist between experiences of pain of different persons. However, to say so comes uncomfortably close, if not all the way, to saying the sort of thing that my friend was saying in the earlier example: "I know that I feel pain in my body and also that to me it would be correct to use the term 'pain' to refer to these feelings, but it would not be correct for others, or I cannot be certain that it would be". But if we say that, we have yet to understand what it means to take part in and to share with others the "language-game" of pain, and to understand that such statements violate the very presupposition on which this and any other "language-games" necessarily relies.

However, instead of concluding thus, Wittgenstein contends that there can be no such things as pains, nor of knowledge of pains - neither from our own cases nor in general. So, Wittgenstein's analysis of the language game of giving expression to pain leaves us in a situation in which nothing exists but "words" which are arbitrarily used, and games which are played according to arbitrary rules. For, in our own cases all we have - and others in theirs - are words or expressions, yet not of anything. For, as he says, "if we construe the grammar of the expression of sensation on the model of 'object and designation' the object drops out of consideration as irrelevant." Exit my pains as well as those of yours; henceforth, we may all live happily without them!

What is so deeply puzzling - indeed so much so that it takes the genius of a philosopher to think it - is how this Wittgensteinian construction of a "language-game" of pains and other internal states and sensations could ever have been take seriously and thought to have anything to do with language and use of language. Indeed, it would seem to be one of the biggest philosophical mysteries of the twentieth century.

20.4 Social Constructionism and the relativism of Wittgenstein's later works

So far, I have tried to argue that language and use of language by persons would be inconceivable independently of a notion of 'truth' which is such that what is true and false, correct and incorrect, is also true and false, correct and incorrect for other persons. This notion of truth, I have argued, must be the condition on which relies the inter-subjectivity of the cognition and use of language of persons, and thus the possibility of persons to communicate what they know of themselves and the world in which they act and co-act. And I have argued, furthermore, that this condition logically implies that the notion of 'truth' is not well defined independently of a notion of 'other persons', or just 'others', and thus is a notion which in a fundamental sense is social.

Well, it would seem to be almost self-evident that to be able to communicate about what one knows, implies knowing that one's knowledge, categories, conceptual systems, and descriptions are indeed "inter-personal", i.e. that one shares such conceptual systems with others. Although I may know of things which others do not (yet) know of, or know different things about them than others do, to know and say so presupposes that if others had my knowledge, my cognition, my experiences etc., and could describe what I describe, they would have the knowledge of these things which I have, and describe them as I do. Or just: what is the case, correct or true, is also correct or true for others. Conversely, it would not make sense to say that what I know to be the case, true or correct, would not be the case, true or correct to others, nor understandable to others or translatable into the language I share with others. This would not make sense, since I would be unable to substantiate such a claim - i.e. specify the nature of the difference of my knowledge as opposed to that of others - save in a language I share with others. In other words, we cannot know for ourselves what others may not know just as well.

Hence, that knowledge and language is fundamentally social, and that persons have a notion of 'truth' which is such that what is true is also true to others, does not rely on the fact that persons may together come to agree on and make conventions, criteria or rules for what may be considered correct or true about the things or situations in the world in which they find themselves. On the contrary, no social conventions, criteria or agreement about the truth of knowledge and correctness of descriptions of these things and situations could be established among persons, unless prior to establishing such conventions, agreement and criteria, they had together determined and identified the thing or situations to which these conventions and criteria apply and, therefore, unless they already had a notion of 'truth' of which it is presupposed that what is true or correct is also true and correct for others. Therefore, to say that our knowledge, language and our notion of 'truth' is fundamentally social does not mean that the notion of 'truth' is a social phenomenon, i.e. a product of socially agreed practice. On the contrary, social phenomena and practice, including the development of conventions, criteria, rules or agreement on how to use language and its terms, depend on notions of 'correct' or 'true' and 'incorrect' or 'false' which are shared.

However fundamental - and almost embarrassingly banal - this epistemological condition for language and cognition of persons may seem, it has been widely overlooked by proponents of the Relativism, Pragmatism and Anti-realism underlying so many present day theories about the social and cultural "construction" of cognition, language and communication. The epistemological position on language and cognition defended by Wittgenstein in his later work (*Investigations* and *On Certainty*), to which I shall turn in a moment, is a case in point. Now, there would be little point in addressing Wittgenstein's later views and arguments, had they not lately been given a renaissance by proponents of Social Constructionism within the social sciences (Psychology, Sociology and Anthropology), and used to promote and underpin the cognitive and linguistic relativism of this position. So, it seems, there are good reasons to address the points and arguments of Wittgenstein's later work and critically to assess the epistemological consequences of the position he attempts to defend.

But first, let me give a brief sketch of the main points of Social Constructionism and its epistemological implications, as presented by one of its prime promoters within psychology, Kenneth J. Gergen (1985). In his own words, the "central contours" of what he sees as an "emerging body of thought of the Social Constructivist movement", may be summarized thus:

> "Social constructionist inquiry is principally concerned with explicating the processes by which people come to describe, explain, or otherwise account for the world (including themselves) in which they live." (ibid. p. 266)

As rightly pointed out by Gergen, the way people describe, explain or otherwise account for themselves and the world they live in, may be different for people living in different societies and cultures; indeed, vast differences in the knowledge and description of the world may be observed within the same society or culture at different times in history. Cultural and historical relativity in the sense of differences in the social practices, co-operation, and customs of different cultures and societies, is a well observed fact - of which Gergen gives many example from psychology and anthropology; and so are the differences in the socially agreed standards, rules, conventions and institutions which guide these practices, verbal as well as non verbal, of different cultures and societies. It is arguably correct, furthermore, that such social practices, institutions, agreed standards, rules, etc., may bring about constraints in the opportunities available for people belonging to different cultures and societies to observe and describe the world they live in, and thus that differences may exist in what people in different societies and cultures may come to know of and how they may be able to account for this world and themselves.

However, to Gergen and other proponents of Social Constructionism, this observed historical and cultural relativity has significant consequences for epistemology - in that the very notions of 'knowledge', 'true', 'observation' and 'empirical validity' themselves must be the products of socially agreed practices, customs, co-operation and all the rest. According to Social Constructionism, these notions, and thus all other notions of our cognition and language, are dependent on - indeed, *exclusively* dependent on - our socially and culturally agreed practices, rules and conventions. Thus,

> "Social Constructionism views discourse about the world not as a reflection or map of the world but as an artefact of communal interchange. [...] The terms in which the world is understood are social artefacts, products of historically situated interchanges among people. From the constructionist position the process of understanding is not automatically driven by the forces of nature, but is the result of an active, cooperative enterprise of persons in relationship." (ibid., p. 266)

According to Social Constructivism, then, historical and cultural relativity implies cognitive and linguistic *relativism* and, thus, a position which "asks one to suspend belief that commonly accepted categories or understandings receive their warrant through observation. Thus, it invites one to challenge the objective basis of conventional knowledge." (ibid., p. 267)

However, what Social Constructionism *does* invite one to believe is, conversely, that

"The degree to which a given form of understanding prevails or is sustained across time is not fundamentally dependent on the empirical validity of the perspective in question, but on the vicissitudes of social processes (e.g., communication, negotiation, conflict, rhetoric). [...] Observation of persons, then, is questionable as a corrective or guide to descriptions of persons. Rather, the rules for "what counts as what" are inherently ambiguous, continuously evolving, and free to vary with the predilections of those who use them. On these grounds, one is even led to query the concept of truth." (ibid., p. 268, italics added)

When the epistemological implications of this position become "fully elaborated", says Gergen,

"it becomes apparent that the study of social process could become generic for understanding the nature of knowledge itself. [And consequently] In similar fashion, epistemological inquiry along with the philosophy of science could both give way, or become subsumed by, social inquiry." (ibid., p. 266)

Now, Gergen is well aware that this "constructionist orientation", may appear as "rampant relativism". However, he avoids suggesting what a language or linguistic practice would be like, which allowed the implications of Constructionism to be accounted for in ways which are not "relative" to historical and cultural practices and their transformation. "Constructivism", he contends, "offers no foundational rules of warrant and in this sense is relativistic". However,

"this does not mean that "anything goes". Because of the inherent dependency of knowledge systems on communities of shared intelligibility, scientific activity will always be governed in large measures by normative rules. However, constructionism does invite the practitioners to view the rules as historically and culturally situated - thus subject to critique and transformation." (ibid., p. 273)

Gergen, and many others who subscribe to his Social Constructivist Movement,[4] seems to think that one can defend a relativistic position, which does not necessarily imply that "anything goes" - if only the scientific community at any given time adheres to norms that warrant "what goes", ... norms, of course, which are relative to and may change according to the

[4] For a recent, very elaborate and extremely clear account of the epistemological and ontological assumptions of this so called movement, see Burr (1995).

historical and cultural circumstances at those times, and so on. Thus, these emerging views of Social Constructivism, and their far reaching consequences for epistemology, says Gergen, have been followed by a "steadily intensifying concern with the constraints over understanding engendered by linguistic conventions". And with other Constructionists, he proposes that "Wittgenstein's Philosophical Investigations must be viewed as seminal in this regard" (ibid., p. 267). So, let us turn forthwith to what Wittgenstein actually said and argued about these matters.

I shall start my presentation of the relativism to be found in Wittgenstein's later works that Gergen seems to be hinting at by first recapitulating some of Wittgenstein's views on language and its use. Whilst it was the view of *Tractatus*, that language has a unique essence, a single underlying logic, which is explainable by means of an account of the structure of language and its relation to the structure of the world depicted by language, Wittgenstein's attention in *Investigations* was drawn to the fact that an explanation of the relation between the structures of language and the world cannot be given, save in language. This means that there is no "point" outside language in which such explanations may be given. Furthermore, he realized that the claim in *Tractatus* of a denoting relation between words and objects in the sense that "names mean objects", amounted to an account of the relation between language and reality in which language is accounted for in terms of reality. But perhaps even more important, Wittgenstein came to realize that language is not a complete and autonomous system, the structure of which may be investigated independently of human activities and practices, our dealings with each other and the world we live in; on the contrary, language is deeply woven into such practices. Our many different uses of language, each of which has its own structure, says Wittgenstein, are given content and significance by our practical activities. Indeed, language is part of the fabric of an *inclusive* form of life - by which he means our linguistic as well as non-linguistic behaviour, natural expressions, assumptions, beliefs and outlook.

According to this later view of Wittgenstein, meaning does not come about in a denoting relation between words and things, nor in a picturing relation between propositions and facts; rather, the meaning of an expression is its use in the multiplicity of activities of which language is a part. What governs the use of an expression are rules for its application in the language-games to which it belongs. These rules are constituted by our social practices and collective use of language, and established by agreement, custom - and training (*Investigations*, especially para. 198 - 206); indeed, they are rules which constitute standards of correctness for such use and practices, because the are based on agreement. No other justification for a given activity or practice, i.e. outside the activity or the practice itself, is required or can be given; they are their own justification - in much the same way as in a game, say of chess, there is no answer to the question, "Why does the king only move one square at a time?", other than, "if it is chess one is playing then that is simply the rule".

In the case of use of language this means, according to Wittgenstein, that nothing *external* to our collective activities, or "form of life", is involved in following the rules of language - just like nothing external to the game of chess determines the correct moves of chess. Indeed, it is not possible to point to something independent, to objective facts existing outside language and our linguistic practices - if all we may point to are practices performed and cases of such practices; there are only the constraints which lie in the fact that one is not following a given rule, and thus not talking correctly, if one's activity fails to conform to the community's agreed practice in that case. Thus, that our statements and descriptions of the world (and of our social practices and actions in the world) may be correct or incorrect, does not rely on whether they accord with some objective facts existing independently of our linguistic practices, but rather on whether we observe the mutually agreed rules of our linguistic community.[5] It is the "form of life" - i.e. the practices on which our language-games rely, and of which they are part - which is the frame of reference for both our correct use of language and for our understanding of the meaning of linguistic discourse. In accordance with

[5] This view of Wittgenstein - in effect the view that it makes sense to talk about the correctness of our conceptions and descriptions of facts in reality without presupposing the independent existence of facts in reality, which these conceptions and descriptions may be correct about - is brought out relatively clearly, so it seems, in remarks like the following: "If the formation of concepts can be explained by facts of nature, should we not be interested, not in grammar, but rather in that in nature which is the basis of grammar? - Our interest certainly includes the correspondence between concepts and very general facts of nature. (Such facts as mostly do not strike us because of their generality.) But our interest does not fall back upon these possible causes of the formation of concepts; we are not doing natural science; nor yet natural history - since we can also invent fictitious natural history for our purpose. I am not saying: if such-and-such facts of nature were different people would have different concepts (in the sense of a hypothesis). But: if anyone believes that certain concepts are absolutely the correct ones, and that having different ones would mean not realizing something that we realize - then let him imagine certain very general facts of nature to be different from what we are used to, and the formation of concepts different from the usual ones will become intelligible to him", Ibid., p. 230). Well, an immediate and obvious comment to these remarks would seem to be that it only makes sense to talk about facts of nature as *facts of nature* - be they the currently realized or different ones which may be imagined - on the condition that facts of nature are the sorts of things which exist independently of our conceptions and descriptions, but equally important that facts of nature exist as something about which we may have correct conceptions and descriptions. Otherwise it would not make sense to talk about what we realize as being facts of nature; but nor would it make sense to talk about realizations or conceptions as being realizations or conceptions of such facts.

this view Wittgenstein talks about the "given" or the "foundation" as being our "form of life".

Now, if the rules for use of language, which both *guide* us in what we say and *ensure* the correctness or truth of what we say, are the product of our agreed social practices and beliefs, and that no other constraints exist on the meaning and correctness of use of language, then it follows that the notions of 'correct' and 'true' themselves must be the products of such agreement and beliefs. Wittgenstein was aware of this problem - and has the following response:

> "'So you are saying that human agreement decides what is true and what is false?' - It is what human beings say that is true and false; and they agree in the language they use. That is not an agreement in opinion but in form of life." (*Investigations*, para. 241)

According to Wittgenstein, then, the only constraints on use of language are internal ones, founded on the agreement of beliefs and linguistic practices of the speech community; justification of our uses of language, i.e. what makes what we say either true or false, therefore, can only be a matter of those practices and beliefs that are part of "form of life". "Form of life" is the frame of reference, the "bed-rock" of language and its use - and no explanation, nor any justification is needed or can go beyond mere pointing to "forms of life":

> "If I have exhausted the justification I have reached bedrock, and my spade is turned. Then I am inclined to say: "This is simply what I do" (ibid. para. 217). [...] What has to be accepted, the given is - so one could say - forms of life." (ibid., p. 226)

However, it seems to me that this "pointing" to forms of life is a rather empty gesture. For if justification of the truth or falsity of what we say about "the world" is a purely "internal" matter, founded on agreed beliefs and opinions about "the world" which make up our forms of life, how could what is right or wrong, true or false, fail to be *other* than agreement of opinion? And how, then, could the very notions of 'true' and 'false' themselves fail to be other than the products of such agreement of opinion?

In *On Certainty* (Wittgenstein, 1969) where these views are further developed, Wittgenstein argues that in every language-game there are propositions and rules for descriptions which are "foundational" or "certain" (ibid., para. 79 - 83, 90 - 92, 105, 162, 211, 411); they have to be taken for granted - because the meaning of propositions of the language-game to which they belong would cease if the rules were violated or the propositions doubted. These propositions are foundational in that they constitute the framework of our language and practices and make up the system within which all testing of our beliefs takes place: "the questions that we raise and our doubts depend on the fact that some of our propositions are exempt from doubt, are as it were like hinges on which those turn" (ibid., pare. 341). The propositions "I have two hands", and "physical objects exist" are examples of such propositions in the language-games of our current "form of life" (ibid., para. 358, 403). Such propositions are "grammatical" - by which Wittgenstein means that they are logical propositions which have the special role in our discourse of forming part of the very conditions for its meaningfulness. Doubt, says Wittgenstein, is only possible within the framework of a language-game. In order to doubt whether or not I have two hands, or whether physical objects exist, I must understand what is meant by "hands" and my "having" them, and "physical objects" and "exist". However, this immediately makes the doubt meaningless:

> "The fact that I use the word "hand" and all the other words in my sentence without a second thought, indeed that I should stand before the abyss if I wanted so much as to try to doubt their meaning - shews that absence of doubt belongs to the essence of the language-game, that the question "How do I know ..." drags out the language-game, or else does away with it." (ibid., para. 370)

This point is illustrated by the example of a pupil learning history; the pupil has to accept the language-game before he can question whether something is true or something exists. (ibid. p. 310-15). If the pupil continuously doubted whether the world had existed for longer than a few hours or years, the business of learning history would be impossible. Such doubts are "hollow", says Wittgenstein (ibid. para. 312), for they try to make the entire language-game impossible - and therefore makes the doubt itself impossible. This makes him assert that,

> "If you are not certain of any fact, you cannot be certain of the meaning of you words either." (ibid. p. 114)

This would seem a perfectly reasonable claim - until we remind ourselves that, according to Wittgenstein, the foundational role of "grammatical" propositions about facts consists in their indubitable certainty *in practice*, i.e. in action, and thus that "it belongs to the logic of our [...] investigations that certain things are in deed not doubted", ibid. para. 342). In other words, this is a reasonable claim until we remind ourselves that, according to Wittgenstein, what constitutes the *certainty of facts*, is the very same as that which constitutes the *certainty of the meaning of words*. For example, the meaning of the proposition "I have two hands", is given by its use in the language-game in which we talk about hands. And the certainty of its meaning is guaranteed in virtue of our observance of the agreed rules of this language-game, part of which is our adherence to the "foundational" beliefs that such things as hands exist in the world (ibid. para. 79 - 82). However, nothing external to the agreed observance of these rules and adherence to these beliefs determines the meaning of words, nor their correct use. If we doubted the correct use of the proposition, and thus its validity, we would violate the rules and do away with the beliefs of that language-game; it couldn't be played. However, the *fact* that my two hands happen to exist - independently of what I and others may agree to say or do - is immaterial to the certainty or *truth* of the proposition, "I have two hands"; for, according to Wittgenstein, it does not make sense to talk about what exists as facts of reality apart from or independently of our language-games, beliefs and other agreed practices.

Thus, in Wittgenstein's scheme of things, the *certainty of the fact* of my having two hands, both *relies* on, is a *product* of, and is *guaranteed* by the same language-game and beliefs which guarantee the certainty of the *meaning of the proposition*: "I have two hands". If so, the claim, "If you are not certain of any fact, you cannot be certain of the meaning of your words either", is quite simply redundant. For, within the positions held by Wittgenstein of an inclusive, self-justifying "form of life" as the basis both for our use of language and of which language is a part, there is no way that the certainty of the fact of two hands could be different from, or could come about as something other than, the certainty of the meaning of the words "two hands".

In summary, language and use of language, as conceived of in Wittgenstein's later work, are like games in which we cannot go wrong; we can only go (cf. also Grayling, 1988, p. 103). Just as it applies to any other game, say chess, which we either play - or fail to play - so it applies to taking part in a language-game that we either play it - or we do not play it at all. It is a position according to which doubt is only possible within a language-game in which things are taken for granted as facts by agreement, and according to which, therefore, *any* doubt necessarily becomes meaningless. Indeed, it is a position in which the notions of 'doubt' and 'false' are *indiscriminable* from the notions of 'void' and 'meaningless'.[6] And it is a position from which it follows that such claims as "physical objects exist", and "my two hands exist" have the same status as "God exists", because it applies equally to such claims that their correctness does not depend on anything

[6] Cf. the following remarks: "To say of man, in Moore's sense, that he knows something; that what he says is therefore unconditionally the truth, seems wrong to me. - It is the truth only inasmuch as it is an unmoving foundation of his language-game. (ibid. pare 403)"; "I want to say: it's not that on some points men know the truth with perfect certainty. No: perfect certainty is only a matter of their attitude," (ibid. para 404); "The truth of my statements is the test of my understanding of these statements" (ibid. para. 80); "That is to say: if I make certain false statements, it becomes uncertain whether I understand them", ibid. para.81.); "What counts as an adequate test of a statement belongs to logic. It belongs to the description of the language-game" (ibid. para. 83).

existing independently of the language-game in which they are put forward. Both the correct use and the validity of those claims is a matter purely internal to the language-games to which they belong. According to this position, conversely, the *existence* of such things or "facts" as physical objects, hands, and God are dependent only on the agreed use and meaning of the notions of 'physical objects', 'hands', and 'God'. That is, all it takes to talk truthfully and correctly about matters of fact and what exists, is our commitment to the rules of language which are based on our agreed beliefs and assumptions as to what *count* as matters of fact and what exists.

It is a position, furthermore, according to which the notions of 'truth', 'exist', 'reality' and 'facts' are relative to our forms of life and, therefore, just as prone to the fate of change, as are those forms of life which, at any given moment, we may decide to agree on - or cease to agree on. For although forms of life are the "givens" and "foundational" for our linguistic practices, what is foundational and taken for granted is so taken - subject to agreement and customs; and agreement and customs may change over time. Forms of life, says Wittgenstein, are like the beds and banks of a river which guide and determine the course of the flow of water; just like the beds and banks of a river erode, and cause the direction of the flow of water to change over time, so do forms of life change over time, and thus our beliefs and notions of 'true', 'correct', 'reality' and 'facts'. (ibid., para. 65, 96 - 99, 256) And just like the shape of the bed and banks of a river at times t_1 and t_2 may be "solid" and "fast" at those times, yet different, so may the beliefs and notions of language which are the "givens" and "foundational" at t_1 be different from those which are "givens" and "foundational" of the form of life at t_2 (ibid. para. 96 - 97). What is foundational and given, then, is only *relatively* foundational and given; indeed, it may develop and change in quite arbitrary directions.

Now, no one, I think, would deny that our knowledge and descriptions of things in material and social reality may develop and change over time. According to the arguments presented in this book it makes good sense to say, furthermore, that how we describe things in the world, which exist independently of our descriptions and knowledge of them, may differ according to the situations in which we find ourselves, i.e. depend on our possibilities of observation, action, and purposes in those situations. So, in this sense, it makes perfectly good sense to say that how we use language to describe the world and this knowledge, is deeply "woven into the fabric of our actions and behaviour", and into the ways in which we relate to things in the world. However, the problem with Wittgenstein's position is that far from suggesting how such development, differences and changes of knowledge and description occur and may be accounted for, his whole idea of form of life *prevents* the detection of differences and changes. For, that form of life, according to Wittgenstein, is self-justifying and inclusive means not only that form of life cannot be justified or put into question from *within* itself, but also that it cannot be accounted for from *without* itself. This means that if now the fundamental facts and beliefs taken for granted according to the agreed practice of the form of life at, say, time t_1 happened to be different from those of a later form of life at, say, time t_2, the agreed meaning and correct uses of words in the language-games to describe the facts and beliefs at those different times would similarly be different - and so would be the notions of 'true' and 'correct'. Hence, what count as "foundational" and "facts" and, therefore, as correct descriptions of facts of the form of life at t_1, could not be the same as those of the form of life of t_2; but nor could the differences between what is correct or foundational at t_1 and t_2 be determined from within the form of life at t_2 – for from within the forms of life of t_2 they would quite simply be incomprehensible and meaningless!

In order that any differences or changes be detectable between the forms of life in question would require and presuppose a point of vantage outside either from which a comparison could be carried out. Indeed, it would require and presuppose the existence of cognitive schemes and a language external to forms of life with concepts of 'true', 'correct', and 'fact' which are not relative to any particular form of life. Thus, it will also and clearly have to be a language with descriptions and concepts of 'true' and 'false', which have a different status from that which applies to descriptions and concepts of 'true' and 'false' of languages which are part of forms of life. However, this requirement of a special language *outside* form of life to account for the occurrence of changes and development of linguistic practices and cognition occurring *inside* form of life, would of course be unacceptable, and I think, in itself a sufficient reason for abandoning Wittgenstein's attempt at a general explanation of language and use of language in terms of and as part of self-justifying and inclusive forms of life.

The same argument could also be stated this way. Granted it were possible to make claims to the effect that the conceptual schemes and language-games of the form of life at t_1 is different from those at t_2, the notions of 'correct' or 'true' of the language in which the account is given could not be different from the notions of 'correct' or 'true' of *either* language game, nor therefore could the notions of 'correct' or 'true' of the language game of t_1 be different from those of t_2. If we said otherwise, we would be saying both that it is possible to know of and correctly describe that which is the case, true or correct at t_1 and t_2 and also that what is the case, correct or true, at those times would not be correct or true outside these times and their forms of life - and therefore could not be correctly described outside those forms of life. This, of course, is self-contradictory and just shows once more why it cannot be maintained that differences exist between conceptual schemes, language and use of language at t_1 and t_2 which are fundamental in the sense suggested by Wittgenstein. What he is suggesting, in effect, is comparable to what my friend in the example mentioned earlier was maintaining, i.e. that he knew of and was able correctly to describe "things", but what is true or correct about them, might not be true or correct to others. However, for the same reason that any determinable differences between what is the case and true for my friend as opposed to others would require that his notions of 'true' or 'correct' are not different from, but have to be the same as the notions of 'true' or 'correct' of the language in which this determination of difference is made, it can now be argued that any meaningful claims put forward about the difference between what is the case or true in forms of life at t_1 and t_2, require that the notions of "true" or "correct" of forms of life at those times are the same as the notions of 'true' or 'correct' of the language in which the claim is put forward. This being the case, notions of what is true or correct at t_1 and t_2 cannot be relative to forms of life at those times, nor can they be different from one another. This rules out Wittgenstein's relativistic assumptions of form of life and, once more, that form of life thus conceived could be foundational for language and use of language.

But it also immediately rules out the assumption that historical and cultural relativity of cognition and description implies the assumption of cognitive and linguistic *relativism*. Indeed, as far as epistemology is concerned, there is nothing particularly problematic in the fact of historical or cultural relativity, nor in the fact that differences exist in social practices and values, and ways in which people live and form communities in different cultures or at different times in history. Those are facts which are quite open to observation and description. Nor is there anything epistemologically problematic in assuming that the differences we may determine in the knowledge and descriptions of reality and things of different cultures and societies derive from differences in social practices, customs and values of those cultures and societies. This is to be understood in the sense that social practices, customs and values may bring about constraints in the possibilities available of observing and exploring things in material and social reality, and thus bring about differences in what is known about reality by people living in different cultures and societies, and differences in how this knowledge may be correctly described.

However, what *is* epistemologically problematical, is to assume that the variety and differences in knowledge and description of reality in different cultures and societies can be *fundamental*. And what *is* epistemologically problematical is to draw the conclusion from this variety that then there are as many version of "truth", "reality", "facts" and "values" as there are different conceptual schemes and beliefs, or "forms of life". Just how problematical and why becomes clear when we consider the assumption of cognitive relativism that the only way we may make sense of and come to understand an alien culture or society is by translating the aliens' beliefs, concepts, and practices into our own terms, practices and beliefs. However, what in this way we might have managed to make sense of and understand, may have very little - if anything - to do with the alien form of life as conceived of by the aliens. Or, even worse, the cognitive schemes and practices of an alien culture or society may be so different from our own that we cannot even recognize them - and if we could, we would not have the slightest inkling of what they were like from "within". This cognitive relativism of "forms of life" is consistent with everything Wittgenstein says about this notion, and is captured in remarks spread around in his later work. For example, "We don't understand Chinese gestures any more than Chinese sentences" (Zettel, para. 219, see also para. 350) or "If a lion could talk, we could not understand him" (*Investigations*, part II, p. 223).

The assumption of Wittgenstein and other proponents of cognitive and linguistic relativism of fundamental differences in the cognition and language of different cultures and societies does not make sense, nor is it tenable. If this assumption of cognitive relativism were true (whatever "true" may mean), it would have to be admitted that it is an assumption which can only be *relatively true*, since, according to cognitive and linguistic relativism, the notion of truth, like any other terms of language and cognition, is dependent on and changeable with opinion and beliefs, and thus cannot be anything fixed. Indeed, everything Wittgenstein says about forms of life, beliefs, social practices, customs, rule-following, etc., and what of such forms is "foundational", "the given", can itself only be relatively true. Unless, of course, he supposes that the language-game in which he talks about the existence of different forms of life, their language-games, rules and all the rest, is not itself a language game in which the meaning of the notions being used are "relative" and the descriptions put forward about them are only "relatively" true. And unless he supposes, furthermore, that in this language game, terms exist which may be used to talk in non-relative ways about the objective facts and things of material reality which surrounds these different forms and games and in which they are situated. Unless he suppose this and, furthermore, that these facts, things, and the rest of material reality surrounding these different forms of life, *exist* outside this language, and hence exist *independently* of his descriptions and cognition of them and as things to which he may refer and talk correctly about, he could not talk sensibly about the existence of any of these different forms of life, their language games, practices, or rules. However, he cannot suppose so without at the same time denouncing the assumption of the general relativism of language and cognition. This is why the position of cognitive and linguistic relativism is so difficult for its proponents to defend; when they try they simply cannot help involving themselves in contradictions. But nor can they help involving themselves in talking nonsense when defending the anti-realism which this position implies. Inevitably so, for by holding such a position they commit themselves to notions of 'reality', 'facts, 'propositions about facts', 'true', and 'correct" which are arbitrary, and thus totally void.

Let me summarize in this way the reasons why equating cultural or historical *relativity* with cognitive and linguistic *relativism* does not make sense. To the extent that we may detect "forms of life" in other cultures and societies *as* forms of life, that is, may detect social practices, co-operation, linguistic communication, as well as rules and institutions which guide these social practices, we logically presuppose that people belonging to those societies and cultures do indeed have knowledge about the situations in reality in which they find themselves and of the things that their social practices in those situations concern. Thus, we presuppose, logically, that they know what is the case, true or correct about these things, situations, social practices, and all the rest. Conversely, we presuppose that it applies to these situations, things etc. that they exist as things about which something is the case, true or correct, and of which they may have knowledge. And we presuppose, logically, that in the language used by the people to communicate and describe their knowledge, descriptions exist of these things and practices which have correct implications and uses. That is, we presuppose that the *Principle of Correctness* applies to their knowledge and language - just as much as we have to presuppose that it applies to ours. If we - and they - did not so presuppose, they could not even begin to co-operate, let alone begin to establish agreement about rules and institutions to guide their societal practices and co-operation. And granted that they do have this required knowledge and a language in which they may communicate about and correctly describe what they know, we - and they - inevitably, logically, presuppose that the notions of 'truth' of their cognition and language is such that what is true or correct is also true and correct to others. Last, but by no means least, to the extent that we may detect and, thus, correctly describe any of the social practices, customs, knowledge, use of language and descriptions within those cultures and societies, we inevitably, logically, presuppose that *we* may know of and correctly describe what *they* know of and describe - i.e. that we may indeed share their knowledge, description and terms, as well as their notions of "true" or "correct".

These presuppositions must be the logical conditions on which rest any claim as to the existence of forms of life *as* forms of life in alien cultures and societies, as well as for detecting and correctly describing any *difference* between our own "forms of life" and those of alien cultures and societies. In other words, the existence of determinable cultural or historical differences of "forms of life" rely on *mutual accessibility* between "forms of life". So, far from supporting cognitive and linguistic relativism, the existence of determinable cultural or historical differences of cognition and language rules out cognitive and linguistic relativism. Indeed, it is a fundamental mistake to think that cultural or historical relativity of cognition and language invalidates any of the general epistemological conditions for cognition and language – as Gergen and his followers want us to think. On the contrary, the existence of determinable cultural and historical relativity of cognition and language presupposes that *general* epistemological conditions do exist which are shared across cultures and history, and hence are not "relative", nor "cultural" nor "historical".

Chapter 21

The conditions for people to be and function as persons: Summary and consequences

Since the renaissance the notion has prevailed of the existence in reality of a duality of worlds, a public material world and a subjective mental world, each the preserve of fundamentally different objects and phenomena. With this division inevitably arose the insoluble question as to how the two worlds were related, and, just as inevitably, questions as to the status of our knowledge of the objects and phenomena "within" them. To Descartes, who was the first to address these questions in a systematic way, it seemed obvious that the knowledge we have about our thoughts, to which we have immediate access, is the only knowledge of which we may be absolutely certain. Any attempts to establishing certainty as to the knowledge of the physical world had to be based on knowledge and experience belonging within our mental world, and somehow, the certainty of the existence of the physical world had to be built on that basis. However it appeared to Descartes that although the certainty as to the existence of our thoughts could not be denied, nor the existence of someone who did the thinking, it was still possible to doubt the existence of everything else. Descartes writes,

> "I noticed that whilst I thus wished to think all things false, it was absolutely essential that the 'I' who thought this should be somewhat, and remarking that this truth '*I think, therefore I am*' was so certain and so assured that all the most extravagant suppositions brought forward by the sceptics were incapable of shaking it, I came to the conclusion that I could receive it without scruple as the first principle of the Philosophy for which I was seeking.

> And then, examining attentively that which I was, I saw that I could conceive that I had no body, and that there was no world nor place where I might be; but yet that I could not for all that conceive that I was not. On the contrary, I saw from the very fact that I thought of doubting the truth of other things, it very evidently and certainly followed that I was; on the other hand if I had only ceased from thinking, even if all the rest of what I had ever imagined had really existed, I should have no reason for thinking that I had existed. From that I knew that I was a substance the whole essence or nature of which is to think, and that for its existence there is no need of any place, nor does it depend on any material things; so that this 'me' that is to say, the soul by which I am what I am, is entirely distinct from body, and is even more easy to know than is the latter; and even if body were not, the soul would not cease to be what it is." (Descartes, 1628/1911, p. 101)

What may now seem so deeply puzzling, is how Descartes could conceive of the notion 'the existence of an I who can think', and even of an 'I' who exists as a substance distinct from his body, and at the same time maintain that neither the notion 'exist', nor the notions of 'I' or 'think', 'substance' or 'body' had any implications whatsoever. To Descartes it seemed to make perfectly good sense to say that he knew that he existed, but also that he could use the term 'exist' about something of which nothing *else* could be said to be the case or be predicated. And it seemed to him to make good sense to maintain that he could think and talk about something, *in casu* about himself, his body and about other material things in the world, and at the same time that he could think that the things he could think and talk of did not exist. Thus, to Descartes it would be equally true to say both, "I can think of the cups and glasses out there on the table", and, "I can think that the cups, glasses and table out there exist nowhere at all". Indeed, it appeared obvious to Descartes that he could use the words 'I', 'my body', 'substance', 'cups', 'table', 'glasses', 'out there' and 'nowhere' correctly, but that these words had no implications, nor did anything exist to which those words might be correctly applied. But to say so, is to say both that such terms may be correctly used - and also that such terms are completely void. No wonder Descartes found that from this position, no amount of further thinking on his part could annihilate the contradiction; and no wonder that he came to realise that an additional principle had to be introduced about the existence and knowledge of material reality, which could restore for him the relation so miserably lost during his thinking, between his thinking and the words he used to express his thoughts, and that to which they referred - and thus restore the very content and meaning of the thoughts and words, he seemed to be able to produce and use with such proficiency. What saved the day and restored the world was a second principle of how an infinitely perfect God in his goodness had implanted in us some true knowledge of the physical material world that He had Himself created.

But however clearly Descartes saw the necessity of such a principle for the possibility of a division of reality into a physical and mental realm, he never realized that we cannot determine and talk consistently about the content of either realm independently of referring to and describing the content of the other. That is, he failed to realise that we cannot talk about objects in the realm of physical things independently of or without referring to our cognition and perception of them, and thus without presupposing that such objects and their realm exist as things we may perceive and know of, and use language to refer to and talk correctly about. Neither, conversely, can we talk consistently about our perception and experiences of objects in the material physical realm, independently of or without referring to such objects and their realm, i.e. objects and a realm which exist independently of our cognition and description of them. Had he realised this, and thus realised that none of the notions and terms we use to describe objects and phenomena in either realm can be well defined independently of referring to notions and terms we use to describe objects and phenomena in the other, the history of philosophy over the last four hundred years would have looked very different indeed.

Now, it is well-known that Descartes' philosophical principle about the existence of material objects was based on the *clarity* of his notion of an infinitely perfect being, and on the *clarity* of the thought that the falsity of the existence of material objects was incompatible with the goodness of God. From there on he was allowed, so he contended, to take as certain and true the existence of all other things he could think of with equal clarity. And he started to make a list of all those things, from which, gradually, re-emerged a world of material objects - and with it his own body and the perceptions, emotions and feelings residing within it. But, as pointed out by his critics, the problem with the criterion on which his philosophical principles were based was that it was too easy for others to claim that they did not see what he saw with the same clarity as he did. However, both the division of reality into two independently determinable parts, and Descartes' approach to bridging the gap between the two, remained unchallenged by those who followed him, just as did the assumption - widely held among philosophers even to-day - that individuals begin with knowledge and experience of their own self and the content of their minds, i.e. their sense data, perceptions, thoughts and emotions. From this supposed personal or subjective experience it is believed to be possible to work toward knowledge of the external physical world as well as toward knowledge of other individuals, their selves and minds. In short, it is assumed that one could begin with the "subjective content" of individuals, living mentally in isolation from one another, and that from this content, of which we have absolute certainty, we may work toward true knowledge of the nature of what causes this content and the rest of the objective order of reality. George Herbert Mead was among the first to deliver substantial arguments against this assumption of how persons, from "splendid" solipsistic isolation, come to acquire knowledge about themselves, others and reality. In *Mind, Self, and Society* (Mead, 1934) he rightly pointed out that, "if *per impossible*", there was only one individual in the universe, that individual could not have a language, nor would it be aware of its existence, since to be aware of oneself is to "look" at oneself from the standpoint of another.

Mead was undoubtedly right in claiming that it would be impossible for an individual growing up in isolation to develop cognition of the kind we have, may account for and understand, and impossible to develop a language in which he might describe his cognition. And he was also right in claiming that there are both psychological reasons as well as reasons of principle for this. With regard to the latter we are now in a position further to strengthen Mead's original claim. Let us suppose with Mead that to be aware of oneself is to "look" at oneself from the standpoint of another. Well, such "looking" both means and presupposes that if one could be in the position of other persons and look at oneself as they do, then one would see and come to know oneself the way others do. Without the presupposition that what, from the position of others is known by them to be the case or true about me, would also be the case or true for myself, could I be in their situation and have their knowledge, there would be no sense in talking about seeing oneself from the point of view of others. Hence, to be aware of oneself as a person presupposes that one shares with others a concept of 'truth' which is such that what is true for others is also true for oneself – and vice versa – i.e. presupposes that the notion of 'truth' of persons is inter-subjective. Indeed, it requires a notion of 'truth' which is *inherently* inter-subjective in that it logically implies a notion of 'others'. This notion of truth, I argue, must necessary precede the possibility of anyone "looking" at himself from the standpoint of another.

Now, let us turn to the case of an individual, being the only one existing in the universe, and the possibility of such an individual developing a language in which he could describe for himself his cognition of the world. The logical problem in asserting that such an individual could have developed knowledge of and a language to describe the world, i.e. knowledge and a language which were *comparable* to our own, and thus a cognition and language which would be understandable to us and translatable into our language, is that this assertion cannot be made unless we are *also* prepared to assert that the notion of 'truth' of his language and cognition is the same as ours, and thus that it is a notion of 'truth' which logically implies a notion of 'others'. Indeed, no such assertions could be consistently made unless we are also prepared to assert that he has a notion of truth which is such that what is correct or true, would also be correct or true for others - *should others exist.*

Apart from the reasons *of principle* for assuming that it is not possible for an individual having lived in isolation from birth to develop cognition and language of the kind we know of, there seems to be empirical support for assuming that for a child to develop into someone being aware of himself as a person, it is necessary that he be received and understood by others as a person. Thus, empirical research of early mother-infant communication seems to show that for a child to develop knowledge about himself, reality and other people around him, and later on to acquire a language to communicate this knowledge and that of others, depend on the mother's (or other care-person's) indefatigable efforts and willingness to understand and see the child's behaviour as being *intentional.* And it seems to rely on the mother's efforts to understand, not only what goes on "inside" the child, but to interpret the child's reactions to her, and his actions with things, as expressing his knowledge about things, and his attempts intentionally to act upon them.

Empirical investigations also seem to show that the child up to a certain age - presumably due to an over-generalization of what he sees as his mother's apparently unlimited knowledge about his experiences, intentions, needs and actions - believes that other people experience things exactly as he does. In effect, he believes that others are in the same situations as he himself and share his point of view, and the knowledge he has about things in those situations - and even that they may "have" his thoughts and feelings. Thus, the child seems to over-generalise the fact that "what he knows may also be known by others" to mean that others do indeed find themselves in exactly the same situation as he does, and having the same knowledge he has, and having access to his thoughts and feelings. Only later does the child learn that other people may perceive the situation differently from how he does, and learn that other people may perceive the same situation from points of view, which are different from his. And only later does the child realise that parts of himself, his feelings and thoughts, are only directly observable to himself, and also that this part of himself is what makes him uniquely *him,* and thus a person being both physically and mentally distinct from other persons.

If we go back to the story being told by most philosophers even today, namely that a child starts with a "subjective", though *certain,* knowledge "from his own cases" about himself and the world, it becomes an insoluble problem as to how this knowledge comes to accord with the knowledge of reality of other persons, i.e. the sort of knowledge which is shared by the community in which the child grows up. Fortunately, to look in this way at the problem of how the child comes to acquire the knowledge of his community, is to turn the issue on its head. For, if it is right what empirical investigations of the child's initial development of knowledge seems to indicate, and what from a logical point of view must necessarily be the case, the child does not start out with private knowledge "from his own case", but with knowledge of which it is assumed by the child that it is indeed shared by others - i.e. by his mother. The problem, it would seem, is rather to account for how the child later in his development comes to appreciate that, although what he knows may indeed be known and shared by others, others may not be in exactly the same situation as he himself. Thus, the problem seems to be to account for how the child begins to learn to appreciate the notion of 'different points of view', and how he begins to learn and appreciate the difference between what is and what is not observable to him as opposed to others, i.e. that the knowledge one may have of a situation and about oneself may be *personal*. In short, to account for how the child begins to learn that to be a person is to be someone who may be different from other persons.

However, what cannot be accounted for or explained - but which has to be presupposed and taken for granted - is that for a child to be able to learn this from other people in the community in which he grows up, and be able to take part in their "forms of life", the child must have a notion of 'truth' which is such that what is correct or true is also correct or true for others. This must be a presupposition which is fundamental for people to act and together identify the situation they are in, and for identifying or determining the things and acts performed in these situations. With this presupposition in mind, I can agree completely with the following considerations of Hamlyn's,

> "The acquisition of knowledge, however, is in effect the initiation into a body of knowledge that others either share or might in principle share. This is because the standards of what counts as knowledge [...] are interpersonal. The concept of knowledge, truth and objectivity are social in the sense that they imply a framework of agreement on what counts as known, true and objective. It is not that what is true, etc. is what society determines as such. To suppose *that* is to commit a fallacy which runs through a good deal of the so-called sociology of knowledge; there may well be a viable sociology of beliefs, but not, properly speaking, a sociology of knowledge. Nevertheless, if something is claimed as knowledge it must be possible in principle to show that it is so by following recognised procedures, or something of the kind, whether or not the person making the claim can himself go through these moves. All this implies a context in which others come into the picture in such a way that agreement with them is not only a general possibility but presupposed at certain points. It is for this reason that the epistemological approach which starts from the position of the individual alone set against the rest of the world is so wrong." (Hamlyn, 1978, pp. 58-59)

What is so wrong with those epistemological approaches and what they are lacking - be they generic constructivist approaches, or naturalist biological or functional computational approaches - is not just a social context of others, which enables the individual to confront and compare his knowledge with the knowledge of others with the purpose of determining e.g. whether his knowledge is in accordance with theirs, and thus may "count as" objective or true, or whether it relies on one's subjective dreams, illusions, or imagination. What is lacking is precisely a concept of 'truth', which logically implies and presupposes a concept of 'others', and which makes it possible for persons together to develop procedures for determining the objectivity and truth of their knowledge, and for everyone to compare his knowledge with the knowledge of others. In the final analysis, what is lacking is an inter-subjective concept of 'truth' which makes it possible for a person to *be* a person, i.e. someone who may share an incredible amount of knowledge with other people, but who also has knowledge about himself and the world which is personal, and thus is someone who also differs from other people; someone with whom we may agree - and disagree.

The existence of such a notion of 'truth', which is inherently inter-subjective or social, has to be presupposed and taken for granted as a *principle* for human cognition and language – just as must the general principle of the correctness of language and cognition (cf. Part II). It is not a notion which we learn or acquire from others, nor from the "forms of life" we engage in with others, but, on the contrary, is a precondition for people together to develop and engage in "forms of life", and to learn from each other. And, as applies to the correctness principle for cognition and language, the principle of the inter-subjectivity of the notion of 'truth' cannot be justified or proved, nor denied or doubted - without being conceded. As the examples in Chapter 20 showed, any attempt to develop tests to prove it, and thus prove that our knowledge of ourselves and the world is in fact the same, would have to presuppose that we already have a notion of truth which is such that what is known to be true or false for oneself, would also in principle be true or false for others. Conversely, any denial or doubt as to whether people, with whom we may share knowledge and a language in which we may talk about what we know, do in fact share the notion of 'truth', would be nonsensical and contradictory. For, in order to substantiate such denial or doubts, it would have to be presupposed that the language being used to express and determine any difference in our and their notions of 'truth', is a language in which the notion of 'truth' necessarily is shared by us and them.

This is just another way of saying that solipsism and "privacy" of the cognition and use of language of persons is ruled out for reasons of principle. And it is to say that any epistemology and any psychological theory about human cognition and language in which this principle of the intersubjectivity of the notion of 'truth' as well as the correctness principle for cognition and language are not fundamental parts, can be rejected out of hand. Both principles are required in order to talk consistently about cognition and use of language by persons – and of that which this cognition and language concerns. And they both apply and have to be taken for granted whether our knowledge and description concern things in publicly observable material reality, or concern our internal or mental states, such as our thoughts, images, emotions or feelings of pain. Of such mental or internal states it has to be presupposed that they, just as publicly observable physical things, exist as things about which something is the case or true, and as things about which it may be known by us that something is the case or true - just as it has to be presupposed that in the language we speak and give expressions to such states, propositions may be put forward about them which have correct implications and uses. And no matter whether our knowledge and perception concern publicly observable material things or our mental states and personal experiences, that which in a concrete situation we know to be true and the case about the things we perceive and experience, would also be known or true for others, could they be in our situation and perceive and experience what we do. This means that the notion of 'others' must be a fundamental epistemological notion, just as fundamental as the notion of 'truth' to which it is logically related.

For too long the Cartesian dual world legend has played havoc in the minds of philosophers and psychologists, and stricken even the best with a blindness so impenetrable as to prevent them from seeing that an alternative indeed exits, which allows us to talk both correctly and consistently about the 'mental' and the 'material' world, the 'inner' and 'external' realm, and how they divide and differ. It is part and parcel of the alternative presented in this book that we can indeed distinguish between mental phenomena and material objects, i.e. between having knowledge about, perceiving, experiencing and describing things, and the material things existing independently of these mental activities, just as we can distinguish between knowledge and experiences of publicly observable things, and knowledge and experience of that which is only directly observable to ourselves. But it is also part and parcel of this alternative that just as it is impossible to talk consistently about our knowledge, experience and description of material things without referring to and talking about these things, so we cannot talk about that which is 'internal' and non-publicly observable without talking about and referring to that which is 'external' and publicly observable. Just consider the impossibility of a person talking about what belongs to his 'internal parts', his thoughts or feelings, independently of or without referring to the 'external parts' of himself. Or, the impossibility of talking about his 'internal states', and of those being *his*, without referring to himself as a person with a body who, at the same time finds himself somewhere in physical, material reality. Or consider the impossibility of a person talking about his 'internal' experiences of things in material reality without referring to the things being experienced as things which exist 'outside' himself in material reality. Or, consider the impossibility for a person talking about his feelings of pain, without referring to his body and parts of his body in which pain may be felt, or conversely, talk about his body independently of referring to his body as being the sort of thing in which only he may feel pain. In particular, it is part and parcel of the alternative offered in this book, that we cannot talk consistently about our knowledge, experience and thoughts of material things independently of referring to and presupposing that we may talk just as correctly about such knowledge, experience and thoughts as we may talk about the things that these "mental activities" concern; and it is part and parcel of this alternative that we cannot talk consistently about that which is 'internal' and non-publicly observable independently of referring to what is publicly observable, nor without presupposing that we may have knowledge about and talk just as correctly about that which is 'internal' as about that which is 'external'. Indeed, none of the distinction and opposition between the 'mental' and the 'material', or between the 'internal' and 'external' can be consistently made independently of presupposing *both* such dependencies and reference *and* correctness of knowledge and description. That is, independently of presupposing that a necessary relation exists between them. For this reason, it makes just as little sense to ask how or whether there is any relation between the 'internal' and 'external' of a person as it does to ask how or whether there is a relation between "Mind" and "Matter".

21.1 The necessary relation between the personal and the public knowledge of persons

A person is someone who may be different in various ways from other persons, indeed a person is someone who may realize that he is *in fact* different in these ways from other persons, and thus someone who may determine *how* he is different from others. For example, persons may differ from other persons because they do not have exactly the same vantage point in, nor therefore exactly the same knowledge of, the particular situations in which they find themselves, and differ because they do not bring into these situations the same history of experiences, educational background or social training. It is therefore not only correct to say, but we *have* to say that the cognition of persons, and the very outlook they may have upon themselves, others and the world they live in, is uniquely *personal*. What has been so tremendously difficult to understand is that although the knowledge and descriptions of persons may be personal, what is personal may be so only because it is *in principle* accessible to other persons. And difficult to understand that we can only be persons who may *realize* that we are different from one another, and *determine* how we differ, because we share with other persons extensive knowledge and correct descriptions about ourselves, others, and about many other things in concrete, everyday situations. Conversely, what may be difficult to understand is that what is common ground to us, and what make us so similar to one another, is at the same time that which ensures that we may be different from one another, i.e. that we are *persons*. So, although we have to say that the knowledge of a person is uniquely personal, it has to be added that no knowledge may be personal, which may not be made accessible to other persons, and thus be part of the knowledge of other persons.

Now, for a person to recognize that he is different from other persons with whom he finds himself in some particular situation, requires not only that he may see himself and others from his own vantage point, but also that he may look at himself and others from their vantage points. Indeed, to recognize that and how one differs from other persons both requires and implies that, cognitively and linguistically, a *symmetry* exists in the relation between oneself and other persons from whom one differs. Well, this may be just another way of saying that there cannot be any *general* problem about the minds of others - no matter how different their cognition and language may be. Indeed, the more differences we may determine between our own cognition and that of other persons, be those differences personal or cultural, the more similarities we necessarily presuppose exist between our own cognition and that of others.

As an aside, do these arguments not also apply when we talk about differences between the cognition of persons and, say, of animals - and am I not saying that no differences may be determined between our own cognition and that of animals, unless we are willing to assume that our cognition and that of animals are in fact very similar? Or even worse, am I not saying that we cannot claim that differences exist between our own cognition and that of an animal, unless we can look at our cognition and the cognition of the animal both from our own vantage point and from that of the animal? This is indeed what I am saying, however, there is nothing wrong in so saying - provided we make clear the sort of "differences" we are talking about. If we hold, as Nagel rightly does (1974), that we cannot put ourselves in the position of animals and look at the world from their vantage points, and hold in addition that *profound* and *fundamental* differences exist between the cognition and communication of persons and that of animals, who do not have a language and with whom we cannot otherwise communicate about their cognition, and with whom, furthermore, we only in very limited and rudimentary senses may be said to be able to co-operate, then we may of course say, i.e. as a figure of speech, that there is a world of difference between our own cognition and the cognition of animals, whilst at the same time contending that virtually no similarities exist. However, the point being made here is that *if* fundamental differences are assumed in the cognition of persons and that of other living beings such as animals, we have at the same time, logically, precluded ourselves from determining any *further* differences - and at least only relatively few, if any, *concrete* differences - due to the fact that similarities in the cognition and action of persons and animals do not exist, which would allow us to claim access to the cognition of animals, or allow us their assent as to the correctness of our understanding and description of their cognition. Such assent would require not only that we and the animals could together identify the things we cognize and act on, but also that these things are the same as those, which in other situations or circumstances may be conceived differently (cf. the Principle of Identity). And it would require, furthermore, that we assumed that animals share with us a notion of 'truth' which is such that what is correct or true is also correct or true to others. After a century of intense research into the cognition of animals, nothing has been found which supports the assumption that animals may have notions of 'identity' and 'truth' in this sense; nor is there any evidence that by training they may develop the notions of 'identity' and 'truth', so essential to the cognition, co-operation, and use of language of persons.

Now, as said above, in order for someone to recognize that and how one differs from other persons both requires and implies that, cognitively and linguistically, a symmetry exists in the relation between oneself and other persons from which one differs. However, there is an *asymmetry* as well. Although we may communicate and make accessible to others what we ourselves know, our knowledge of things and of our feelings, thoughts and experiences, are personal in yet another sense: no one except I can know what I know, the way I know it; and no one may be the agent of my thoughts, or have the images I have; and no one can feel the pains I have in my body, or feel my emotions of happiness or grief - or just feel what it is to be the person I am. It is perhaps not difficult to understand therefore why it has been so obvious to assume that each of us lives in our own private worlds. And obvious to forget that although, psychologically, a person may feel eminently on his own in being thus different and separate from others, he cannot be a subject on his own independently of other persons from whom he may differ. Nor can he be different from others, and make accessible to himself and others how he and the situation in which he finds himself differs from that of others independently of presupposing that what he knows to be true or false about himself, reality and all the rest, would also be true and false to others, could they be in his situation and experience things the way he does. That is, independently of the presupposition that he shares with others a notion of 'truth' which is inter-personal, a notion which imply a notion of 'others'. And because what of oneself is "internal" and only *directly* known or observable to oneself (such as one's thoughts, feelings, emotions, images, etc.) is necessarily related to and part of what of oneself is "external" and observable to anyone (i.e. one's body, action and expressions), then what in this sense may be personal is just as much available for descriptions in the very same language we use to describe what is publicly observable.

The inter-subjective notion of 'truth' of persons makes possible that their cognition and use of language may be "perspectual", indeed that it inherently is so. That is, makes it possible - despite the fact that we do not have exactly the same vantage points as others, and do not bring into a situation the same knowledge, but also knowledge which is personal, we may still count on the fact that we can make ourselves understood to others, and make available to others the knowledge we have. Any epistemology and any psychological theory of cognition and language of persons, which cannot account for and which does not imply this inherent "perspectual" property of our cognition and use of language, can be rejected out of hand. It quite simply has nothing to do with cognition and use of language by persons.

21.2 *Equality* as a necessary condition for communication and co-operation between persons

So, whether what we know is known only to ourselves, and we describe what we know only to ourselves, we always do so in the logical presence of "others", who may know what we know, and for whom what is the case or true of what we know, is also the case or true for them. This, and the general condition of correctness of knowledge and language of which it is part, we may call the *epistemological matrix* of being persons and language users - meaning that which logically guarantees that in knowing what we do and talking about what we know, we are in principle and from the outset understood by others. These are the necessary conditions on which rest the possibility for persons to communicate and co-operate with each other, i.e. for persons to have knowledge about and together identify the situation in which their communication and co-operation takes place, as well as for knowing and identifying the things that the communication and co-operation concern.

That the knowledge and descriptions of a person is knowledge and descriptions which *in principle* may also be the knowledge and description of others, means that being and functioning as a person, is to be a person in a community with other persons. And it means that - in spite of the personal differences existing between persons in how they cognize and describe any particular situation or matter - to be and to function as a person with other persons, with whom one may determine and talk about these differences, logically implies that as persons we may *in fact* make ourselves understood to other persons, and conversely, that others may make themselves understood to us. In this respect a person is a person *on equal terms* with other persons, and no person is more or less of a person than any other person. That is, he is a person who has knowledge of and who in the language he shares with others may correctly describe himself and the situations in which he finds himself in material reality with other persons - and who, like others, may ask questions and reflect on this knowledge, and who may come to know and understand much more about himself, others and the situations in which he and they find themselves.

And just as much as determining the differences between persons rely on the presupposed, general logical condition of being inherently understood by others and on making available to others how we differ from others, so does the determination of whether in any particular situation we are *in fact* being understood by others, or whether we are not. Conversely, without the general logical condition of persons being inherently understood, there could be no distinction between being understood or not being understood, and no determining of when or that we are in fact being understood and when or that we are not.

Now, however much to be a person logically implies being in principle and from the outset understood by others and understanding others, whether we are in fact understood by others, and others are in fact understood by us in any of the concrete situations in which we find ourselves, necessarily requires a *commitment* and an *intention* to understand others, and to make ourselves understood to others in those situations. This may be immediately obvious when we consider that a person is someone who is different from other persons since, in *addition* to the knowledge and descriptions he shares with others about the concrete situations in which he and they find themselves, a person also has knowledge of those situations which is personal. So, in order for co-operation and communication to succeed in spite of such differences, it is necessary for the persons involved to be aware that such differences may exist, and to take those differences into account.[1] For this reason it makes good sense to say that to be and function as a person with other persons with whom one may communicate and act, *logically* requires the intention and commitment of a person to make himself understood, and of others to understand him. Also in this respect the conditions for being and functioning as a person are the same for all persons, and every person is on

[1] Depending on the nature or amount of the differences, for example by making explicit what they know about the situation, and what are the implications of their descriptions of it.

equal terms with others. And in this respect a person is someone who in his encounters and communication with other persons is therefore entitled to be *received* and *treated* as a person.

We may contend then that part of the condition for being and functioning as persons among other persons is a commitment to recognize the equality of persons with respect to these expectations and intentions. However, this means, as far as I can see, that *ethical requirements* must be part of the necessary conditions for people to be and function as persons, indeed for language, use of language and communication to occur and develop between people. Thus, if we could not count on and presuppose that in virtue of being persons we are inherently and in principle being understood by others, - and vice versa - and, furthermore, count on and presuppose the commitment and intentions of others in fact to understand us - and vice versa - there could be no communication and co-operation between persons, nor any language and use of language. Conversely, if people did not have an implicit commitment to understanding others, and in particular to realise and understand the differences between themselves and others - and vice versa - there could be no persons.

The conditions for being and functioning as persons are logically given to any person in virtue of being a person, and they are honoured - whenever persons in their co-operation and communication in concrete situations *intend* to acknowledge them, i.e. whenever they intend to honour their commitment as persons to receive and treat other persons as persons. But am I not contradicting myself now? Am I not saying both that the conditions for someone to be a person are logical, and also that for someone *to be* and *function* as a person in the final analysis depends on the willingness of others and himself to acknowledge what is logical? Well, this is what I am saying; however, I do not think that I am contradicting myself - for the following reasons.

As should be clear from the discussion concerning our cognition of objects and things in physical material reality, conditions of a logical nature exist for talking consistently and correctly about such objects and cognition, and obvious, as numerous examples from traditional and current philosophy have shown, that such conditions may not be honoured - however with no serious consequences for the existence of such objects and things. Indeed, neither the objects and things themselves, nor the conditions for talking consistently about them cease to exist because philosophical theories fail to honour these conditions, and falsely assume that they may be doubted or ignored. Neither, of course, do objects and things in reality cease to exist, should we chose – in our daily or scientific practices – to ignore or be ignorant of their properties. But with persons the case is radically different. For, although the cognition of our existence as autonomous persons vis à vis other persons is given *in principle*, i.e. by the ever present logical 'other' of our notion of 'truth', it is also required to be so given and maintained intentionally by others *in practice*. If we had no *desire* to understand others and to make ourselves understood, and did not *commit* ourselves to understand others and to make ourselves understood, there could be no co-operation and communication, no language, nor persons. If so, it must be an inseparable part of the necessary conditions for being and functioning as a person that both psychologically and ethically a person is received and treated by others as a person.

In this contention I may be accused of making the mistake of mixing up fundamental *epistemological principles*, *psychological desires* and *ethical imperatives*. However, I do not deny that such distinctions can be made, nor that they are important. But the point being made is that, when discussing the conditions which are fundamental for being and functioning as a person, it seems of little relevance to ask whether some of these conditions are logically and epistemologically fundamental, as opposed to others which are merely psychological or ethical. If they are equally required and inseparable, interrelated parts of the conditions for persons to be and function as persons, then they may also be said to be equally *logically* required and fundamental. None of these conditions are up for grabs, i.e. they do not depend on whether we prefer to believe they are necessary, nor on whether we feel obliged so to believe.

* * * * * *

It has been the aim of this book to show that psychology is in need of an epistemology which can deal consistently with the wide variety of phenomena that make up our cognition and use of language, and which determine our action and possibilities of action – in everyday as well as in scientific situations. That is, an epistemology which can account consistently for the wide varieties of ways in which we may have objective knowledge of and talk correctly about ourselves and other things existing in a publicly shared world, as well as for our fantasies, thoughts, ideas, emotions, feelings and other internal states, which are personal and only directly observable to ourselves, - but of which, nevertheless, we may have knowledge and talk correctly about in a language we share, and thus may make accessible to others. An epistemology, moreover, which can account for the fact that we can indeed distinguish between what is fact and what is fiction, what exists objectively and what is a product of our fantasies, dreams and imagination. Without such an epistemology there is no hope for psychology to begin to account consistently for how the cognition and use of language by persons - together with their other psychological and social dispositions - determine the wide variety of their action and co-action. With the three principles for cognition and language developed in this text - i.e. the principle of the general correctness of cognition and language, the principle of identity, and the principle of the inter-subjectivity of the notion of 'truth' - it is to be hoped that an outline of such an epistemology may begin to be visible.

What a psychology hoping to be scientific does not need any more of is a naturalist epistemology which attempts to get rid of human cognition, its variety, and dependence on both logical conditions, psychological dispositions and socio-economical circumstances, or which attempts to reduce persons to *objects* indistinguishable from what are describable solely in terms of physics and the dispositions applying to physical things, or to organisms describable solely in biological terms and according to the dispositions applying to organisms. But neither does a scientific psychology need a constructivist epistemology which attempts to get rid of the objectivity of the wide variety of ways in which persons may cognize and talk about themselves, their acts and the world existing independently of this acting, cognition, and talking, but which attempts to reduce both this world, cognition, acting and talking to mere products of our Mind indistinguishable from fantasies, imagination and myths. As argued throughout this text, both these epistemologies are deeply flawed and nonsensical; a psychology which will make do with either epistemology as a foundation for its research and theorizing will necessarily be equally flawed and nonsensical - no more a science than a superstition.

REFERENCES

Adams, F. and Aizawa, K., (1994). Fodorian Semantics. In S. P. Stich and T. A. Warfield, *Mental Representation*. A Reader. Oxford: Blackwell.

Austin, J. L., (1950). Truth. *Proceedings of the Aristotelian Society*, 34.

Austin, J. L., (1962). *How to do things with words*. London: Oxford Univ. Press.

Austin, J. L., (1950). Truth. *Proceedings of the Aristotelian Society*, 34.

Ayer, A. J., (1985). *Ludwig Wittgenstein*. Penguin Books.

Berkeley, G., (1930). *A Treatise Concerning the Principles of Human Knowledge*, Chicago: Open Court Publishing Company

Bohr, N., (1939/1972). The causality problem in atomic physics. In Niels Bohr, *Collected Works*. Amsterdam: North-Holland

Broadbent, d., (1993). *The Simulation of Human Intelligence*. Oxford: Basil Blackwell Ltd.

Brentano, F., (1973). *Psychology from an empirical standpoint*. New York: Humanities. (Originally published, 1874.)

Bruce, V. and Green, P. (1985), *Visual Perception: Physiology, Psychology and Ecology*. London: Lawrence Erlbaum Associates.

Bruner, J., (1977). Early Social Interaction and Language Acquisition. *Studies in mother-infant interaction*, Schaffer, M. (ed.). Academic Press

Bruner, J. (1983). Child's Talk. *Learning to Use Language*. London: W. W. Norton & Company.

Bruner, J., (1986). *Actual Minds, Possible Worlds*. Cambridge Mass: Harvard University Press.

Bruner, J., (1990). *Acts of Meaning*. Cambridge, Mass: Harvard University Press.

Bruner, J., Goodnow, J.J. & Austin, A.G., (1956). *A Study of Thinking*. New York: Wiley.

Burr, V. (1995). *An Introduction to Social Constructionism*. London: Routledge.

Chomsky, N., (1990). On the Nature, Use and Acquisition of Language. *Mind and Cognition. A Reader*. Lycan, W.G. (ed.). Oxford: Blackwell.

Chrisholm, R. M., (1967). Intentionality. In E. Edwards (Ed.), *The encyclopedia of philosophy*, Vol. 4, pp. 201-204). New York: Macmillan.

Churchland, P. M, (1981). Eliminative materialism and propositional attitudes. *Journal of Philosophy*, 78, 67 - 90.
Culler, J., (1986). *Saussure*. London: Fontana Press.
Davidson, D., (1974). Psychology as philosophy. In S.C. Brown (ed.) *Philosophy of Psychology*. London: Macmillan.
Dennet, D., (1978). *Brainstorms*. Montgomery, VT: Bradford Books.
Dennet, D., (1987). *The Intentional Stance*. MA: MIT Press/A Bradford Book.
Dennet, D., (1991). *Consciousness Explained*. Little, Brown & Company.
Descartes, R. (1925), *The Method, Meditations, and Selections from the Principles of Descartes*, Veitch, J. (ed.), Edinburgh: Blackwood.
Descartes, R., (1637/1911). Discourse on the method of rightly conducting the reason and seeking for truth in the sciences. In E. S. Haldane and G. R. T. Ross (Eds), *The philosophical works of Descartes*. Cambridge: Cambridge University Press.
Derrida, J., (1972). *Positions*. Paris: Edition de Minuit.
Donaldson, M., (1992). *Human Minds*. London: Penguin Books Ltd.
Dretske, F., (1981). *Knowledge and the Flow of Information*. Cambridge, Mass.: MIT Press.
Dreyfus, H.L. & Dreyfus, S.E., (1986). *Mind over Machines. The Pover of Human Intuition over Expertise in the Era of the Computer*. London: Basil Blackwell, Ldt.
Duncan, K.D., (1987). Fault Diagnosis Training for Advanced Continuous Process Installations. Rasmussen, J., Duncan, K.D. & Leplat, J. (eds.) *New Technology and Human Error*. Chichester: John Wiley & Sons.
Ebbesen, E.B., Parker, S:, & Konecni, V.J. (1977). Laboratory and field analyses of decisions involving risk. *Journal of Experimental Psychology: Human Perception and Performance*, 3, 576-589.
Ericsson, K., and Simon, H. (1980). Verbal Reports as Data. *Psychological Review*, 87, 3.
Gergen, K. J., (1985). The Social Constructionist Movement in Modern Psychology. *American Psychologist*, Vol. 40, No. 3, p. 266 - 275.
Goodman, N., (1978). *Ways of world making*. The Harvester Press
Goodman, N., (1984). *On minds and other matter*. Cambridge Mass.: Harvard University Press.
Grayling, A. C., (1988). *Wittgenstein*. Oxford: Oxford University Press.
Fodor, J., (1975). *The Language of Thought*. New York: Thomas Y Crowell.
Fodor, J., (1980). Methodological solipsism considered as a research strategy in cognitive psychology. *The Behaviour and Brain Sciences*, 3, 63-109.
Fodor, J., (1981). *Representation*. Cambridge, Mass: MIT Press.
Fodor, J., (1983). *The modularity of mind*. Cambridge MA: MIT Press/Bradford Books.
Fodor, J., (1985). Fodor's guide to Mental Representation. *Mind*, 94, pp. 76 - 100.
Fodor, J., (1986). Banish DisContent. *In: Mind and Cognition. A Reader*. (ed. Lycan, 1990). Oxford: Blackwell.
Fodor, J., (1987). *Psychosemantics: The problem of meaning in the philosophy of mind*. Cambridge, MA: MIT Press.
Fodor, J., (1994). A Theory of Content, II: The Theory. In S. P. Stich and T. A. Warfield, *Mental Representation*. A Reader. Oxford: Blackwell.
Fodor, J., (1998). *Concepts. Where cognitive science went wrong*. New York: Oxford University Press.
Fodor, J. and Pylyshyn, Z. (1981), How Direct Is Visual Perception? *Cognition*, 9, 139-196.
Galileo, G., (1953). *Dialogue concerning the Two Chief World Systems*. Chicago: Chicago University Press.
Gibson, J.J. (1950), *The Perception of the Visual World*. Cambridge, Mass: The Riverside Press.

Gibson, J.J. (1959), Perception as a Function of Stimulus. In: Koch, S. (ed) *Psychology: A Study of Science.* Vol. 1. New York: MacGraw-Hill.

Gibson, J.J. (1960), The Concept of the Stimulus in Psychology. *Amer. Psychol.* 1960, 694-703

Gibson, J.J. (1966) *The Senses considered as Perceptual Systems.* Boston: Houghton Mifflin.

Gibson, J.J. (1979), *The Ecological Approach to Visual Perception.* Boston: Houghton Mifflin.

Gibson, J.J. Purdy, J. & Lawrence, L. (1955) A Method of Controlling Stimulation for the Study of Space Perception: The Optical Tunnel. *J. Exp. Psychol.*, 50, 1-14.

Giorgi, a. (1985). toward the Articulation of Psychology as a Coherent Discipline. In Koch, S. and Leary, E. (eds.), *A Century of Psychology as Science. McGraw-Hill.*

Grayling, A. C. (1988). *Wittgenstein.* Oxford: Oxford Univ. Press.

Hacker, P. M. S. (1990). *Wittgenstein. Meaning and Mind.* Analytical Commentary on The Philosophical Investigations, Vol. 3. Oxford: Blackwell.

Hamlyn, D.W. (1957), *The Psychology of Perception: A Philosophical Examination of Gestalt Theory and Derivative Theories of Perception.* New York: The Humanities Press Inc., 1957

Hamlyn, D. W., (1978). *Experience and the growth of understanding.* International Library of the Philosophy of Education. London: Routledge & Kegan Paul

Harris, R., (1996). *The Language Connection. Philosophy and Linguistics.* Bristol: Thoemmes Press.

Horgan, T., (1994). Computation and Mental Representation. In S. P. Stich and T. A. Warfield, (Eds.), *Mental Representation.* A Reader. Oxford: Blackwell.

Horgan, T. and WD, Woodward J., (1985). Folk Psychology is Here to Stay. *The Philosophical Review, No. 2.*

Hume, D. (1951). Enquiries *Concerning Human Understanding.* L.A. Selby-Bigge (Ed.). Oxford: Oxford Univ. Press

Huxley, J.,(1953). *Evolution in Action.* Chatto & Windus, Ltd.

Koch, S. and Leary, E., (1985), Introduction. In Koch, S. and Leary, E. (eds.), *A Century of Psychology as Science. McGraw-Hill.*

Koffka, K. (1935), *Principles of Gestalt Psychology.* New York: Harcourt Brace.

Kripke, S., (1971). Identity and necessity. In M. K. Minitz (Ed.) *Identity and individuation,* (pp. 135-164). New York: New York University Press.

Kripke, S., (1972). Naming and necessity. In D. Davidson & G Harmon (Eds.), *Semantics of natural language,* (pp. 253-355). Dordrecht: Reidel.

Køppe, S., (1990). *Virkelighedens niveauer.* København: Gyldendal.

Lakoff, G., (1987). *Women, fire, and dangerous things. What categories reveal about the mind.* Chicago: University of Chicago Press.

Lee, D.N., (1976). A theory of visual control of braking based on information about time-to-collision. *Perception,* 5, 437-459.

Lee, D.N., & Lishman, J.R. (1977). Visual control of locomotion. *Scandinavian Journal of Psychology,* 18, 224-230.

Leontjew, A.N., (1977). *Problemer i det psykiskes udvikling.* I - III. København: Rhodos.

Lewis, D., (1971). An argument for the identity theory. In D. M. Rosenthal (ed.) Materialism and the Mind-Body Problem. Englewood Cliff, NJ: Prentice Hall.

Locke, J. (1690/1961). *An Essay Concerning Human Understanding,* 1-2. J. W. Yolton (Ed). London: Everyman's Library.

Marr, D. (1982). *Vision. A computational Investigation into the Human Representation and Processing of Visual Information.* San Fransisco: W. H. Freeman and Company.

Marr, D., and Nishihara, H.K. (1978), Analysis of a cooperative stereo algorithm. *Biol. Cybernetics*, 28, 223 - 229.

Maturana, H. R. AND Varela, F. J. (1980). *Cognition, The Realization of the Living*. Dordrecht: D. Reidel.

Mcginn, C. (1996). *The Problem of Consciousness*. Oxford: Blackwell

Mead, G. H. (1934). *Mind, Self, and Society*. Chicago: Chicago Univ. Press.

Michaels, C. F., & Carello, C. (1981). *Direct Perception*. Englewood Cliffs, New Yersey: Printice-Hall, Inc.

Miller, G.A. (1985). The Constitutive Problem of Psychology. In Koch, S. and Leary, E. (eds.). *A Century of Psychology as Science*. McGraw-Hill.

Miller, D. L. (1973). George Herbert Mead. *Self, language and the world*. Austin: University of Texas Press.

Millikan, R., (1994). Biosemantics. In S. P. Stich and T. A. Warfield, *Mental Representation*. A Reader. Oxford: Blackwell.

Moore, G.E. (1953). *Some Main Problems of Philosophy*. London: George Allen & Unwin, Ltd.

Moore, G. E. (1959). Philosophical Papers. London: George Allen & Unwin, Ltd.

Mortensen, A.T., (1972). *Perception og Sprog*. København: Akademisk Forlag.

Nagel, T., (1986). *The view from Nowhere*. Oxford: Oxford University Press.

Nagel, T., (1994). Consciousness and Objective Reality. In R. Warner and T. Szubka, *The Mind-Body problem*. Oxford: Basil Blackwell Ltd.

Nisbett, R.E. & Wilson, T.D., (1977). Telling more than we can know: verbal reports on mental processes. *Psychological Review*, 84, 231-259.

Oppenheimer, P. & Putnam, H., (1958). Unity of Science as a Working Hypothesis. In H. Feigl, M. Scriven and G. Maxwell (eds), Minnesota Studies in the Philosophy of Science, vol. 2, pp. 3 - 36.

Pais, A., (1991). *Niels Bohr's Times, In Physics, Philosophy, and Polity*. Oxford: Oxford Univ. Press.

Pears, D., (185), *Wittgenstein*. London: Fontana Press.

Piaget, J., (1971). *Biology and Knowledge*. Edinburgh University Press.

Piaget, J., (1971). *Structuralism*. London: Routledge & Kegan Paul.

Polanyi., m. (1968). Lifes irreducible structure. In M. Polanyi: *Knowing and being*, 225-241. London: Routhledge and Kegan Paul.

Praetorius, N. (1978). *Subject and object. An essay on the epistemological foundation for a theory of perception*. Psykologisk Skriftserie, University of Copenhagen. Copenhagen: Dansk Psykologisk Forlag

Praetorius, N. (1982). *Fundamental Principles for a Theory of Consciousness*. Paper read at The Boston Colloquium for the Philosophy of Science, May 1981. Published in *Psyke & Logos*, 1, p. 7 - 26. Copenhagen: Dansk Psykologisk Forlag.

Praetorius, N. & Duncan, K.D., (1988). Verbal Reports: A Problem in Research Design. Goodstein, L.P., Andersen, H.B. & Olsen, S.E. (eds.), *Tasks, Errors and Mental Models*. London: Taylor & Francis.

Praetorius, N. and Duncan, K.D. (1991), Flow representation of plant processes for fault diagnosis. *Behaviour & Information Technology*, 10, 41-52.

Putnam, H., (1975). The meaning of 'meaning'. In H. Putnam, (Ed.), *Mind, language and reality*: Philosophical papers of Hilary Putnam, (Vol. 2, pp. 291-303). Cambridge: Cambridge University Press.

Putnam, H., (1975). Explanation and reference. In H. Putnam, (Ed.), *Mind, language and reality*: Philosophical papers of Hilary Putnam, (Vol. 2, pp. 291-303). Cambridge: Cambridge University Press.

Putnam, H., (1983). *Realism and Reason*, Cambridge: Cambridge University Press.
Putnam, H., (1988). *Representation and Reality*. Cambridge, Mass.: The MIT Press.
Putnam, H., (1990). Meaning and Method. Essays in honour of Hilary Putnam. Edited by G. Boolos. Cambridge: Cambridge University Press.
Pylyshyn, Z., (1980). Cognitive Representation and the Process-Architecture Distinction. *Behavioral and Brain Science*, 3, 1.
Quine, W.V.O., (1960). *Word and object*. Cambridge, MA: MIT Press.
Quine, W.V.O., (1970). On the reasons for the indeterminacy of translation. *Journal of Philosophy*, 67, 178-183.
Reason, J., (1979). Action Not as Planned: The Price of Automatization. In Underwood, G. & Stevens, R. (eds.). *Aspects of Consciousness*. London: Academic Press.
Reed, E. S., (1988). *James J. Gibson and the Psychology of Perception*. New Haven: Yale University Press
Reed, E., & Jones, R. (1982). *Reasons for realism*. Selected Essays of James J. Gibson. London: Lawrence Erlbaum.
Rhees, R. (1968). Wittgentstein's Notes for Lectures on "Private Experience" and "Sense Date". *Philosophical Review*, 77, pp. 275 - 230
Richardson, R. C., (1981). Internal representation: Prolog to a theory of intentionality. *Philosophical Topics*, 121, 171-211.
Rosch, E., Mervis, C.B., Gray, W.D., Johnson, D.M. and Boyes-Graem, P., (1976). Basic objects in natural categories. *Cognitive Psychology*, 8, 382-439.
Russell, B. (1927), *The Analysis of Matter*. International Library of Psychology, Philosophy and Scientific Method. London: Kegan Paul, Trench, Truber & Co.
Russell, B. (1948). *Human Knowledge. Its Scope and Limits*. London: George Allen and Unwin Ltd.
Ryle, G., (1949). *The Concept of Mind*. London: Hutchinson.
Saussure, F. de, (1960). *Cours de Linguistique Generale*. Paris: Payot.
Saussure, F. de, (1983). *Course in General Linguistics*. Translated by Roy Harris. London: Duckworth & Co. Ltd.
Saussure, F. de, (1993). *Troisme Cours de Linguistique General (1910-1911)./Saussure's Third Course of Lectures on General Lingusitics (1910 - 1911)*. Translated and edited by E. Komatsu & Roy Harris. Oxford: Pergamon Press.
Sayre, K.M., (1986). Intentionality and information processing: An alternative model for cognitive science. *Behavioral and Brain Science*, 9, 121-166.
Seager, W., (1991). *Metaphysics of Consciousness*. London: Routledge.
Searle, J. R., (1992). *The Rediscovery of the Mind*. Cambridge, Mass: The MIT Press.
Searle, J. R., (1995). *The Construction of Social Reality*. London: Penguin Books.
Smith Bowen, E., (1964). *Return to Laughter*. Doubleday Anchor. New York: Garden City.
Sperry, R.W.A., (1976). An Emergent Theory of Consciousness. Marx, M.H. & Goodson, F. (eds.), *Theories in Contemporary Psychology*. New York: Macmillan.
Stampe, D., (1977). Towards a Causal Theory of Linguistic Representation. In French, P., Euhling, T., and Wettstein, H. (eds.), *Midwest Studies in Philosophy*, vol. 2. Minneapolis: University of Minnesota Press.
Stern, D. N., (1985). *The Interpersonal world of the infants: A view from psychoanalysis and developmental psychology*. New York: Basic Books.
Stich, S., (1983). *From Folk Psychology to Cognitive Science. The Case Against Belief.* Cambridge, MA: MIT press.
Stich, S. P. and Warfield, T. A., (1994) *Mental Representation. A Reader*. Oxford: Blackwell.
Strawson, P. F., (1950). Truth. *Proceedings of the Aristotelian Society*, 34.

Tooby, J. and Cosmides, L. (1996). Foreword in: S. Baron-Cohen: Mindblindness. *An Essay on Autism and Theory of Mind.* Cambridge, Mass: The MIT Press.

Tye, M. (1992). Naturalism and the mental. *Mind*, 101, 421-41.

Ullman, S., (1980). Against direct perception. *The Behavioural and Brain Sciences*, 3, 373-415.

Warner, R., and Szubka, T., (1994). The Mind-Body Problem. A Guide to the Current Depate. Oxford: Basil Blackwell Ltd.

Whorf, B.L., (1956). Science and linguistics. Caroll, J. (ed.), *Language, Thought and Reality.* New York: Wiley.

Wittgenstein, L., (1922/1961). *Tractatus Logico-Philosophicus.* London: Routledge & Kegan Paul.

Wittgenstein, L., (1936/1968). Notes for Lectures on "Private Experience" and "Sense Data". *Philosophical Review*, vol. 77, pp. 271-320

Wittgenstein, L., (1975). *Philosophical Remarks.* R. Rhees, (Ed.). Oxford: Blackwell.

Wittgenstein, L., (1958a). *The Blue and the Brown Book.* Oxford: Blackwell.

Wittgenstein, L., (1958b). *Philosophical Investigations.* Ed. G.E.M. Anscombe and R. Rees. Oxford: Blackwell.

Wittgenstein, L., (1969). *On Consciousness.* Oxford: Blackwell.

Zinkernagel, P., (1962). *Conditions for Description.* London: Routledge and Kegan Paul.

Zinkernagel. P. (1988). *Virkelighed.* København: Munksgaard.

Index

GPSR Compliance

The European Union's (EU) General Product Safety Regulation (GPSR) is a set of rules that requires consumer products to be safe and our obligations to ensure this.

If you have any concerns about our products, you can contact us on ProductSafety@springernature.com

In case Publisher is established outside the EU, the EU authorized representative is:

Springer Nature Customer Service Center GmbH
Europaplatz 3
69115 Heidelberg, Germany

Batch number: 09661837

Printed by Printforce, the Netherlands